# Mobile Computing Techniques in Emerging Markets:

## Systems, Applications and Services

A.V. Senthil Kumar
*Hindusthan College of Arts and Science, India*

Hakikur Rahman
*University of Minho, Portugal*

**Information Science**
**REFERENCE**

| | |
|---|---|
| Managing Director: | Lindsay Johnston |
| Senior Editorial Director: | Heather Probst |
| Book Production Manager: | Sean Woznicki |
| Development Manager: | Joel Gamon |
| Acquisitions Editor: | Erika Gallagher |
| Typesetters: | Milan Vracarich, Jr. |
| Cover Design: | Nick Newcomer, Greg Snader |

Published in the United States of America by
Information Science Reference (an imprint of IGI Global)
701 E. Chocolate Avenue
Hershey PA 17033
Tel: 717-533-8845
Fax: 717-533-8661
E-mail: cust@igi-global.com
Web site: http://www.igi-global.com

Library of Congress Cataloging-in-Publication Data

Mobile computing techniques in emerging markets : systems, applications and services / A.V. Senthil Kumar and Hakikur Rahman, editors.
    p. cm.
  Includes bibliographical references and index.
  Summary: "This book provides the latest research and best practices in the field of mobile computing offering theoretical and pragmatic viewpoints on mobile computing"--Provided by publisher.
  ISBN 978-1-4666-0080-5 (hardcover) -- ISBN 978-1-4666-0081-2 (ebook) -- ISBN 978-1-4666-0082-9 (print & perpetual access) 1. Mobile computing. I. Kumar, A.V. Senthil, 1966- II. Rahman, Hakikur, 1957-
  QA76.59.M6475 2012
  004--dc23
                                              2011043979


British Cataloguing in Publication Data
A Cataloguing in Publication record for this book is available from the British Library.

All work contributed to this book is new, previously-unpublished material. The views expressed in this book are those of the authors, but not necessarily of the publisher.

# Table of Contents

## Section 1
## Concepts and Architectures

## Section 2
## Applications and Cases

**Section 3**
**Protocols and Technologies**

# Detailed Table of Contents

## Section 1
## Concepts and Architectures

**Chapter 1**
*Hakikur Rahman, University of Minho, Portugal*
*Senthil Kumar, Hindusthan College of Arts and Science, India*

Chapter one is based on a survey of concurrent literatures on mobile computing, its applications, and challenges. The review has put forwards three applications of mobile computing: learning, health, and GIS. The chapter also discusses a few constraints and challenges that have emerged in terms of design and application issues.

**Chapter 2**
*Lars Frank, Copenhagen Business School, Denmark*

Chapter two describes an architecture for mobile integrated calendar systems where performance, local autonomy, and availability are optimized by using relaxed Atomicity, Consistency, Isolation, and Durability (ACID) properties and different asynchronous replication methods. The chapter argues that by using relaxed ACID properties across different database locations, it is possible for the users to trust the data they use even if the distributed database temporarily is inconsistent.

**Chapter 3**
*Stefan Zander, University of Vienna, Austria*
*Bernhard Schandl, University of Vienna, Austria*

Chapter three provides detailed insights into the field of Semantic Web-based context-aware computing for mobile systems. The chapter introduces requirements, enabling technologies, and future directions of such systems. It presents a Semantic Web-based context-sensitive infrastructure that resembles concepts from graph theory and distributed transaction management.

## Section 2
## Applications and Cases

### Chapter 4

Chapter four illustrates the applications of Graph Theory algorithms to study, analyze, and simulate the behavior of routing protocols for Mobile Ad hoc Networks. Specifically, the chapter focuses on the applications of Graph Theory algorithms to determine paths, trees, and connected dominating sets for simulating and analyzing respectively unicast (single-path and multi-path), multicast, and broadcast communication in mobile ad hoc networks (MANETs).

### Chapter 5

Chapter five presents an approach to extend a real world mobile tourist guide running on personal digital assistants (PDAs) with collaborative filtering. The study builds a model of item similarities based on explicit and implicit ratings. The model is then utilized to generate recommendations in several ways. The approach integrates the current user location as the context and experiences gained in two field studies are reported in the chapter.

### Chapter 6

Chapter six provides lessons from case studies of two successful and large scale implementations of mobile health (mHealth) solutions and discusses about their choices that were made in the design and implementation of those solutions. The chapter uses Information Infrastructure Theory as a theoretical lens to discuss reasons why these projects have been able to successfully scale.

### Chapter 7

Chapter seven discusses about various nowcasting approaches in making weather forecasting an application of mobile computing. The study aims to integrate various sensors along with a mobile to provide the capability of data measurement to the vast population which use the mobile and create regional grid networks. The study also uses the mobile for updating weather parameters as well be the focal point of communication of the weather nowcasting information.

**Section 3**
**Protocols and Technologies**

## Chapter 8

*Nerea Toledo, University of the Basque Country, Spain*
*Marivi Higuero, University of the Basque Country, Spain*

Chapter eight presents an approach for addressing vehicle-to-infrastructure communications by means of NEtwork MObility (NEMO) management, including some solutions like, NEMO Basic Support (NEMO BS). The study has described the demanded key features to the NEMO protocols to be applied in the Intelligent Transportation System (ITS) context, including an analysis of the fulfillment of these key features by the NEMO protocols.

## Chapter 9

*Raja Al-Jaljouli, Deakin University, Australia*
*Jemal Abawajy, Deakin University, Australia*

Chapter nine presents a comprehensive taxonomy of various security threats to Mobile Agent System (MAS) and the existing implemented security mechanisms. The chapter discusses different security mechanisms and highlights the related security deficiencies. It describes various security properties of the agent and the agent platform. The chapter also introduces the properties, advantages, and roles of mobile agents in various applications.

## Chapter 10

*Raja Al-Jaljouli, Deakin University, Australia*
*Jemal Abawajy, Deakin University, Australia*

Chapter ten discusses the security protocols presented in the literature that aim to secure the data mobile agents gather while searching the Internet, and identifies the security flaws revealed in the protocols. The protocols are analyzed in the study with respect to the security properties, and the security flaws are identified. The chapter also introduces common notations used in describing security protocols and describes the security properties of the data that mobile agents gather.

Chapter eleven discusses providing an innovative system for tracking and monitoring objects using radio frequency (RF) transmitters and receivers, and querying about these objects using mobile phones. The protocol for the system is presented in the chapter, and a simulation has been applied to check the feasibility of the project under this study before applying it to a real world scenario. A wide range of application in emergency services, landmarks locating, and warehouse management are also explained.

# Foreword

Mobile Computing is a technology that allows transmission of data, via a computer, without having to be connected to a fixed physical link. Mobile voice communication is widely established throughout the world and has had a very rapid increase in the number of subscribers to the various cellular networks over the last decade. An extension of this technology is the ability to send and receive data across these cellular networks. Mobile data communication has become an essential and rapidly evolving technology as it allows users to transmit data from remote locations to other remote or fixed locations. This proves to be the solution to the biggest problem of business people on the move.

Years of research efforts in mobile computing have brought us tremendous knowledge and experience in this broad domain, as well as many hardware/software platforms, frameworks, libraries and toolkits. It is believed that an ultimate goal of mobile computing is to build useful applications and services that meet people's needs. The speed and the quality of expanding and creating a vast variety of multimedia services like voice, email, short messages, Internet access, m-commerce, mobile video conferencing, streaming video and audio has brought true mobile multimedia experiences to mobile customers. However, due to constant changing environments, limited battery life and diverse data types, mobile multimedia implies considerable challenges to operators, infrastructure builders in terms of ensuring fast, reliable services and accommodating the quick growing global customer needs.

Mobile Computing is expected to encompass heterogeneous access technologies and the Internet backbone for providing multimedia services to both mobile and stationary users. It poses significant technical challenges to enable broadband wireless access with seamless and ubiquitous coverage and quality-of-service provisioning. The tremendous advances in wireless communications and mobile computing, combined with the rapid evolution in smart appliances and devices, generates new challenges and problems requiring interactions between different network layers and applications in order to offer advanced mobile services.

However, the advances in wireless communication technologies and the proliferation of mobile devices have enabled the realization of intelligent environments for people to communicate with each other, interact with information-processing devices, and receive a wide range of mobile wireless services through various types of networks and systems everywhere, anytime. A key enabler of this pervasive and ubiquitous connectivity environment is the advancement of software technology in various communication sectors, ranging from communication middleware and operating systems to networking protocols and applications.

The objective of this book is to provide the latest research and the best practices in the field of Mobile Computing. Theoretical and pragmatic viewpoints on mobile computing will provide guidance to the professionals who will use this book to enrich their knowledge and utilize in their practices. A solid base of mobile computing and an expansive vision of this practice will combine to promote the understanding and the successful implementation of mobile computing techniques in emerging markets.

This book will be very useful to individuals, researchers, scientists, academics, students, libraries, journalists and development practitioners working in the field of mobile computing. This book will generate tremendous impetus in terms of mobile computing based research initiations which would be useful for the society.

*Ang Ban Siong*
*Human Rights Ambassador for NGO to United Nations*

**Ang Ban Siong** *is an Honorary Ambassador in the field of Human Rights for NGO to United Nations and also serves as key personnel for many local and international companies located in Malaysia, China, Singapore, Taiwan, Philippines, Indonesia and Thailand. He earned his Bachelor of Information Technology (USQ, Australia), B.Sc. (Hon) in Applied Forensic Accountancy and Fraud Examinations, MBA and DBA degree qualifications from Ansted University as well as Honorary Doctorate degree from University in India. His achievements have been recognized by the Governments of Malaysia and China leading him to be bestowed as a knighted order by Royal of State of Pahang in Malaysia. Also Ambassador Dato' Dr. Ang is a Certified Forensic Accountancy and Fraud Examinations (QFFE) as well as license holder of QFFE. He is associating with many international professional institutions from the USA, UK, Canada, Africa, Europe and Asian countries as Fellow and advisory council member.*

# Preface

Mobile computing can be seen as a computing environment of physical mobility (Talukdar, 2010). Mobile computing allows transmission of data, via a computer, without having to be connected to a fixed physical link (Koudounas & Iqbal, 1996). The concept of mobile computing entails the use of the Internet and Intranets for communicating and computing while on the move. Typically, mobile hosts are condensed versions of multipurpose computers, with small memory, relatively slower processors, and low-power batteries, and communicate over low-bandwidth wireless communication links (Zeng & Agrawal, 2002; Agrawal, Rao & Sanders, 2003).

The communication involves data and voice communication. Mobile data communication is an evolving technology that allows users to transmit data from remote locations to other remote or fixed locations (Koudounas & Iqbal, 1996). This has been proved to be an essential technology for all, especially to the business community who are always on the move. Mobile voice communication has also evolved throughout the world and varying from geographical locations and local demands, notwithstanding the technology or cost, the number of subscribers is rising fast.

The user of a mobile computing environment can access data, information, or other logical objects from any device in any network while being on the move. The devices are laptops, Personal Digital Assistants (PDAs), and mobile phones. A mobile computing system allows a user to perform a task from anywhere using a computing device in the public (using the Web), education (search engines like Google or Yahoo, content search, indexed contents from libraries, etc.), government (basic information, or levies or taxes), corporate (for obtaining business information), personal spaces (for accessing medical record or address book), social networks (using Facebook, LinkedIn, etc.), or infotainment (Youtube, Dailymotion, etc.), or many more. Furthermore, there are many additional things that a mobile computing system can perform that a stationary computing system cannot. These added functionalities are the reason for separately characterizing mobile computing systems (Amjad, 2004; B´far, 2005; Talukdar, 2010).

Many factors have contributed to the emergence and continued growth of mobile computing, including recent advances in hardware and communications technologies. However, this new paradigm shift does not only provide numerous usage advantages, but also created new challenges in computer operating systems development. In terms of mobile computing, these challenges include relatively unusual issues like, security, privacy, inadequate bandwidth, frequent network disconnections, resource restrictions, high cost of technology, quality of technology, lack of reliable standards, power limitations, and foremost, any possibility of health hazard (Welch, 1995; Zeng & Agrawal, 2002; Amjad, 2004).

This book, focusing mobile computing techniques in the emerging market, has tried to provide the latest research and the best practices in the field of mobile computing. In this context, inclusion of theoretical aspects and real life cases on mobile computing would provide guidance to the professionals and researchers who could use this book to be informed about current scenario and applications. With a collection of manuscripts surrounding applications of mobile computing in innovative usage, this book will promote the understanding of the concepts and successful implementation of mobile computing technologies in emerging markets.

The audience of this book includes researchers, scientists, academics, students, librarians, journalists, development practitioners, and individuals. This book will generate remarkable impetus in terms of mobile computing based research initiations, and thus, it will have highly acceptable scholarly value and at the same time potentially contribute to this emerging sector of research.

## ORGANIZATION OF CHAPTERS

The book has been divided into three sections; concepts and architectures, applications and cases, and protocols and technologies. Altogether there are eleven manuscripts covering wider range of concepts, designs, and applications.

Chapter one is based on a survey of concurrent literatures on mobile computing, its applications, and challenges. The review has put forwards three applications of mobile computing: learning, health, and GIS. The chapter also discusses a few constraints and challenges that have emerged in terms of design and application issues.

Chapter two describes an architecture for mobile integrated calendar systems where performance, local autonomy, and availability are optimized by using relaxed Atomicity, Consistency, Isolation, and Durability (ACID) properties and different asynchronous replication methods. The chapter argues that by using relaxed ACID properties across different database locations, it is possible for the users to trust the data they use even if the distributed database temporarily is inconsistent.

Chapter three is providing detailed insights into the field of Semantic Web-based context-aware computing for mobile systems. The chapter introduces requirements, enabling technologies, and future directions of such systems. It presents a Semantic Web-based context-sensitive infrastructure that resembles concepts from graph theory and distributed transaction management.

Chapter four illustrates the applications of Graph Theory algorithms to study, analyze, and simulate the behavior of routing protocols for Mobile Ad hoc Networks. Specifically, the chapter focuses on the applications of Graph Theory algorithms to determine paths, trees, and connected dominating sets for simulating and analyzing respectively unicast (single-path and multi-path), multicast, and broadcast communication in mobile ad hoc networks (MANETs).

Chapter five presents an approach to extend a real world mobile tourist guide running on personal digital assistants (PDAs) with collaborative filtering. The study builds a model of item similarities based on explicit and implicit ratings. The model is then utilized to generate recommendations in several ways. The approach integrates the current user location as the context and experiences gained in two field studies are reported in the chapter.

Chapter six provides lessons from case studies of two successful and large scale implementations of mobile health (mHealth) solutions and discusses about their choices that were made in the design and implementation of those solutions. The chapter uses Information Infrastructure Theory as a theoretical lens to discuss reasons why these projects have been able to successfully scale.

Chapter seven discusses about various nowcasting approaches in making weather forecasting an application of mobile computing. The study aims to integrate various sensors along with a mobile to provide the capability of data measurement to the vast population which use the mobile and create regional grid networks. The study also uses the mobile for updating weather parameters as well be the focal point of communication of the weather nowcasting information.

Chapter eight presents an approach for addressing vehicle-to-infrastructure communications by means of NEtwork MObility (NEMO) management, including some solutions like, NEMO Basic Support (NEMO BS). The study has described the demanded key features to the NEMO protocols to be applied in the Intelligent Transportation System (ITS) context, including an analysis of the fulfillment of these key features by the NEMO protocols.

Chapter nine presents a comprehensive taxonomy of various security threats to Mobile Agent System (MAS) and the existing implemented security mechanisms. The chapter discusses different security mechanisms and highlights the related security deficiencies. It describes various security properties of the agent and the agent platform. The chapter also introduces the properties, advantages, and roles of mobile agents in various applications.

Chapter ten discusses the security protocols presented in the literature that aim to secure the data mobile agents gather while searching the Internet, and identifies the security flaws revealed in the protocols. The protocols are analyzed in the study with respect to the security properties, and the security flaws are identified. The chapter also introduces common notations used in describing security protocols and describes the security properties of the data that mobile agents gather.

Chapter eleven discusses providing an innovative system for tracking and monitoring objects using radio frequency (RF) transmitters and receivers, and querying about these objects using mobile phones. The protocol for the system is presented in the chapter, and a simulation has been applied to check the feasibility of the project under this study before applying it to a real world scenario. A wide range of application in emergency services, landmarks locating, and warehouse management are also explained.

## CONCLUSION

Globally, the number of mobile phones surpassed the number of fixed or wired phones in 2003. This is true in many individual nations among middle-income and low-income countries. Telecom carriers are increasingly using mobile and wireless technologies to address the last mile problems, especially in developing countries (Amjad, 2004). Mobile communications will strive to continue to greatly influence the future life of common people and accommodating these issues, entrepreneurs and researchers are working relentlessly to incorporate innovative usage for providing daily life requirements of the each individual in the society. Future mobile computing will see the ubiquitous nature of service in terms of access and technology.

*A.V. Senthil Kumar*
*Hindusthan College of Arts and Science, India*

*Hakikur Rahman*
*University of Minho, Portugal*

# REFERENCES

Agrawal, M., Rao, H. R., & Sanders, G. L. (2003). Impact of mobile computing terminals in police works. *Journal of Organizational Computing and Electronic Commerce, 13*(2), 73–89. doi:10.1207/S15327744JOCE1302_1

Amjad, U. (2004). *Mobile computing and wireless communications*. NGE Solutions, Inc.

B'far, R. (2005). *Mobile computing principles: Designing and developing mobile applications with UML and XML*. Cambridge University Press, 2005.

Koudounas, V., & Iqbal, O. (1996). *Mobile computing: Past, present and future*. Retrieved July 4, 2011, from http://www.doc.ic.ac.uk/~nd/surprise_96/journal/vol4/vk5/report.html

Talukdar, A. K., Ahmed, H., & Yavagal, R. R. (2010). *Mobile computing: Technology, applications and service creation* (2nd ed.). Tata McGraw Hill Education Private Limited.

Welch, G. F. (1995). A survey of power management techniques in mobile computing operating systems. *ACM SIGOPS Operating System Review, 29*(4), 47–56. doi:10.1145/219282.219293

Zeng, Q.-A., & Agrawal, D. P. (2002). Handoff in wireless networks . In Stojmenovic, I. (Ed.), *Handbook of wireless networks and mobile computing*. John Wiley & Sons, Inc.doi:10.1002/0471224561.ch1

# Acknowledgment

We are indebted to the Editorial Board of the IGI Global for offering us the opportunity to edit this book on mobile computing in such a time of transition of technologies. Particularly to Kristin M. Klinger – Director of Editorial Content, Erika L. Carter – Acquisition Editor, and Joel Gamon –Development Manager for their continuous suggestions, supports and feedbacks. We are also grateful to Editorial Advisory Board Members and all contributing authors of this book for their assistance, contribution, and collaboration. Our indebtedness extends to all reviewers for their extended support during the tiring review process.

We would not have completed this work in time unless continuous support came from our family members.

Finally, our regards go to the colleagues at our university departments for their supports and encouragements, without which, we would not have been able to complete this research publication.

*A.V. Senthil Kumar*
*Hindusthan College of Arts and Science, India*

*Hakikur Rahman*
*University of Minho, Portugal*

# Section 1
# Concepts and Architectures

# Chapter 1
# Mobile Computing:
## An Emerging Issue in the Digitized World

**Hakikur Rahman**
*University of Minho, Portugal*

**Senthil Kumar**
*Hindusthan College of Arts and Science, India*

## ABSTRACT

*With the advent of complex but user friendly mobile communications technologies and transformation of mobile devices being handy for usage, the applications and utilities of mobile devices have come into the palm of almost each and every human being of this modern world. Furthermore, with the unprecedented growth of the Internet and its outreach, the demand and requirement of users are growing fast, ranging from basic livelihood support, to infotainment, to social networking. Applications of mobile devices nowadays do not include only the facilities for calling another cell phone and text messaging, but also connecting to social networks, service providers networks, and servers of various organizations, like academic or business or health sector, thus providing appropriate services to users, meeting daily demands including emergencies. However, all these are dependent on technologies, social, cultural, and economic issues, which this study has explored. This chapter is based on a survey of concurrent literatures on mobile computing, its applications, and challenges. This study has put forwards three applications of mobile computing: learning, health, and GIS. In this aspect, by exploring the background on mobile computing, the chapter discusses a few constraints and challenges that have emerged in terms of design and application issues. Thereafter, before the conclusion, the chapter puts forward a few future research hints.*

DOI: 10.4018/978-1-4666-0080-5.ch001

## INTRODUCTION

Technology provision has evolved to support these days modern society with demanding expectations that has resulted in rapidly growing information systems. The creation of such ever-expanding and innovative systems is based upon a type of symbiotic relationship where technology facilitates new means, and opens up new horizons. Furthermore, in the context of adoptive technology, the expectations, requirements and demands of the society further fuel the development. As such, the fundamental building blocks of information systems are constantly being enhanced or new ones are identified to enable this development. As the world is passing through the new millennium, technology support is such that the society is on the brink of entering a true information age, where technology integration has became an essential element of social, economic, cultural and political systems (Hameed, 2003). Similar to any other human aspects or technological issues as such, a single technology cannot survive in the global competition. Rather, a combination of similar technologies or an integration of similar technologies adopted to face the demand and challenge has evolved throughout the years. Concept of mobile computing is such a technology, which is evolving and at the same time, it incorporates integrated technologies from diversified aspects of human nature satisfying the growing demand. Recent advances in hardware technologies, such as portable computers and wireless communication networks have led to the emergence of mobile computing systems (Dunham & Helal, 1995).

*Mobile Computing is a technology that allows transmission of data, via a computer (or similar device) without having to be connected to a fixed physical link (Koudounas & Iqbal, 1996).*

Furthermore, the availability of lightweight, portable computers and wireless technologies has created a new class of applications called mobile applications. These applications often run on scarce resource platforms, such as Personal Digital Assistants (PDAs), notebooks, and mobile phones, each of which have limited CPU power, memory, and battery life. They are usually connected to wireless links, which are also being characterized by lower bandwidths, higher error rates, and more frequent disconnections (Gaddah & Kunz, 2003). Evidently, mobile computing is a concept which is revolving around demand and capacity factors between human and computers satisfying aspects of requirements and competencies.

Mobile computing is a type of interaction between human and computer where the computing device, such as the computer, even in a normal situation can be thought of being not in static condition. This concept has brought out three aspects of computing; the communication network, the computing hardware, and the software. The mobile communication incorporates issues like communicating in ad-hoc environments among infrastructure networks as well as communication properties, protocols, data formats and physical technologies; the hardware focuses on mobile devices or device components; the software aspect deals with the necessities, uniqueness and sophistications of mobile applications.

Hence, as discussed above, the concept of mobile computing can mean that computing devices, for example, notebook-PCs, PDAs and wearable computers are carried by users rather than contained within a confined environment. In recent years, portable computing devices have become very small and powerful, giving their users access to a variety of applications in personalized form, regardless of the user locations. Each of these devices is intended to stay with a particular user so that the user's profile can be maintained in the portable device and can easily evolve over time, without having to be transferred from place to place in an external environment. Therefore, the mobile computing approach needs to provide both personalization and privacy. However, its users are forced to carry devices, such as PCs (in various forms), PDAs, and smart-phones (or androids), which may not be light and may only have small screens and clamped keyboards. Moreover, this approach is not suitable for context-dependent services because it is difficult for a portable device to sense its environment (Satoh, 2002; 2003).

Apart from these, mobile computing provides great promise to organizations that adopt mobility to gain a competitive advantage, as well as users that demand anytime, anywhere computing capabilities (Gold, 2005). This can be achieved by enhancing the utility of the portable computing device to improve customer service, speed up decision making and maintain a high-quality workforce (Caldwell & Koch, 1998; Nokia, 2006; Thomas, 2007).

Along these contexts, mobile computing is the discipline for creating an information management platform, which is free from spatial and temporal constraints. The freedom from these constraints allows its users to access and process desired information from anywhere in the space. The state of the user, either static or mobile, does not affect the information management capability of the mobile platform. A user can continue to access and manipulate desired data while traveling on plane, on ship, in car or bus or train. Thus, the discipline creates an impression that the desired data and sufficient processing power are available on the spot, where as in reality they may be located far away in a remote place (Satyanarayanan, 1996).

The discipline of mobile computing has its origin in Personal Communications Services (PCS). PCS refers to a wide variety of wireless access and personal mobility services provided through a small terminal (for example, cell phone), with the goal of enabling communications at any time, at any place, and in any form. PCS are connected to Public Switched Telephone Network (PSTN) to provide access to wired telephones. PCS include high-tier digital cellular systems for widespread vehicular and pedestrian services (dynamic locations) and low-tier telecommunication system standards for residential, business, and public cordless access applications (static locations) (Satyanarayanan, 1996).

High-tier digital cellular systems include:

- Global System for Mobile communications (GSM)
- IS-136 TDMA based Digital Advanced Mobile Phone Services (DAMPS)
- Personal Digital Cellular (PDC), and
- IS-95 CDMA-based CDMAOne System.

Low-tier telecommunication systems include:

- Cordless Telephone 2 (CT2)
- Digital Enhanced Cordless Telephone (DECT)
- Personal Access Communication Systems (PACS), and
- Personal Handy phone Systems (PHS) (Lin & Chlamtac, 2001).

In recent years, several wideband wireless systems and special data systems have been developed to accommodate Internet and multimedia services. This chapter will not include these specialized systems in the study. The chapter discusses in detail all essential aspects of mobile computing and uses this platform to elaborate on fundamental concept of mobile computing. However, before going to the background section, this section discusses on infrastructural aspects of mobile computing.

The last few years have seen a true revolution in the telecommunications sector. In addition to the three generations of wireless cellular systems, ubiquitous computing has been possible due to the advances in wireless communication technology and availability of many light-weight, compact, portable computing devices, like laptops, PDAs, digital tablets, cellular phones, and electronic organizers. Various mobile computing paradigms are developed, and some of them are already in daily use for personal applications as well as for business works. Wireless Personal Area Networks (WPANs), covering smaller areas (from a couple

*Figure 1. WLAN architectures (adopted from Agrawal, Deng, Poosarla, & Sanyal, 2003)*

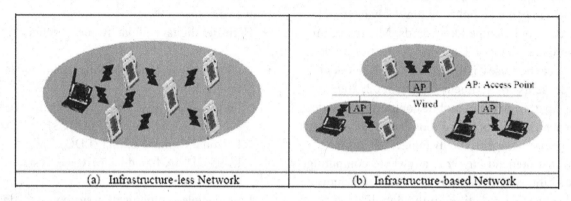

of centimeters to few meters) with low power transmission, can be used to exchange information between devices within the reach of a person. A WPAN can be easily formed by replacing cables between computers and their peripherals, helping people do their everyday chores or establish location aware services. One noteworthy technique of WPANs is a Bluetooth based network. However, WPANs are constrained by short communication range and cannot scale very well for a longer distance. Similar technologies include RFID (Radio Frequency Identification) that uses communication through the use of radio waves to exchange data between a reader and an electronic tag attached to an object, for the purpose of identification and tracking.

Wireless Local Area Networks (WLANs) have gained enhanced usefulness and acceptability by providing a wider coverage range and an increased transfer rates. The most well-known representatives of WLANs are based on the standards IEEE 802.11 (IEEE, 1999), HiperLAN and their variants. IEEE 802.11 has been the predominant standard for WLANs for many years, which support two types of WLAN architectures by offering two modes of operation, ad-hoc mode and client-server mode. In ad-hoc (also known as peer-to-peer) mode (see Figure 1(a)), connections between two or more devices are established in an instantaneous manner without the support of a central controller. The client-server mode (see

Figure 1(b)) is chosen in architectures where individual network devices connect to the wired network via a dedicated infrastructure (known as access point), which serves as a bridge between the mobile devices and the wired network. This type of connection is comparable to a centralized LAN architecture with servers offering dedicated services and clients accessing them. A larger area can be covered by installing several access points similar to the cellular structure with overlapped access areas (Agrawal, Deng, Poosarla, & Sanyal, 2003).

The corresponding two architectures shown in Figure 1 are commonly referred to as infrastructure-less and infrastructure-based networks. Ad hoc network is a collection of wireless mobile hosts forming a momentary network without the aid of any centralized administration or standard support services (Agrawal & Zeng, 2002). Due to its inherent infrastructure-less and self-organizing properties, an ad hoc network provides an extremely flexible method for establishing communications in situations where geographical or terrestrial constraints demand totally distributed network system, such as military tracking, natural disaster, emergency situation, reconnaissance surveillance and instant conference. However, one has to realize that in spite of the flexibility and advantage brought by mobile computing, ultimate prices are there; which are security, vulnerability and instability (Agrawal, Deng,

Poosarla, & Sanyal, 2003). Researchers and researches are intense along these perspectives with the advent of diversified mobile technologies and maturation of mobile computing. However, this study will not exemplify infrastructure aspects, rather keeps focused to a few applications related to data management, education and health, including a few challenges related to technologies, infrastructure and usage.

Now this chapter goes on discussing about the background concept of mobile computing. Thereafter, it put forwards various technologies and applications of mobile computing along with constraints and challenges. Before the conclusion, the chapter sets forth a discussion platform for future issues on mobile computing. As methodology, this study accepts vertical review of literature and case studies.

## BACKGROUND

This study has found many definitions for the term mobile computing. Rebolj et al. (2001) describes mobile computing as the combination of mobile computing devices, mobile networks, and mobile services as illustrated in Figure 2.

Kristoffersent and Ljungberg (1998) identified environment, modality and applications as the three main components of Mobile IT (Information Technology) use. According to them, *Environment* is the physical and social surroundings, *Modality* is the fundamental patterns of motion, and *Application* is the combination of technology, program and data in use, which are being illustrated in Figure 3.

However, Rebolj, Magdič and Čuš Babič, (2001) argue that, although some studies have already tried to survey this fast growing area of information technology the term mobile computing is quite new and has no clear definition. He further argues that, mobile computing does not only involve mobile computing devices (laptops, notebooks, PDAs and wearable computers), which

*Figure 2. Mobile computing (adapted from Rebolj et al., 2001; Irizarry, 2008)*

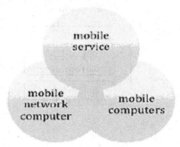

are designed to be carried around, but also the mobile networks to which these computers are connected. Mobile services are the third component (and this study supports), rounding out this definition of mobile computing (Rebolj, Magdič, & Čuš Babič, 2001; 2004). Hence, the definition made by Rebolj et al. prevails.

Although the number of research papers addressing mobile computing is modest, there is no doubt that a great deal of research is going on, perhaps even too fast for papers to be published. As one technology overtakes or integrates with another, and technical solutions are undoubtedly becoming more consistent and reliable, it is more reasonable to concentrate on general concepts and problems. One such problem is the adaptation of existing information systems suitable for efficient integration with mobile computing (Rebolj, Magdič, & Čuš Babič, 2004). But, first of all one has to indentify effective solutions. In this regard, this study agrees with Vizard that, "until then, mobile computing will just remain a troublesome niche application for those, who can afford to pay for it" (Vizard, 2000: 83).

The very concept of mobile computing encircled on science fiction when Coda was first conceived in early 1987. Coda is best known as a system that enables mobile file access through its support for disconnected and weakly connected operation. Today, mobile information access is not only feasible but is beginning to be regarded

*Figure 3. Basic reference model of mobile informatics (adopted from Kristoffersent & Ljungberg, 1998)*

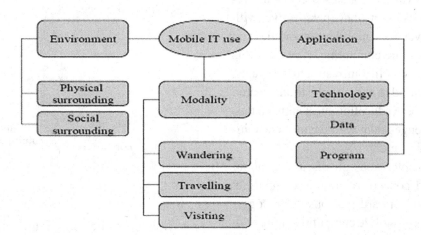

as indispensable. Key elements of Coda's design are being brought into mainstream commercial practice through their adoption in the IntelliMirror component of Windows 2000 (Satyanarayanan, 2002). Satyanarayanan (2002) also mentions that, while much still remains to be done, both Coda and mobile computing have clearly come a long way. In this context, much of the mobile computing depends on the mobility of the device and the user.

Mobility is the characteristic of an object which is mobile. In the field of computing technology the mobile object can be in both computations and communications, according to which two new paradigms are incurred as mobile computa-

tions and mobile communications by extending the features of the objects in these two areas with mobility. The two paradigms then act as the basic components to construct the new research field, mobile computing. This extension is illustrated in Figure 4. It should be mentioned that computation and communication are always interdependent on each other. Hence, mobile computation must base on the support of wireless or wired networks, and at the same time forms the basic techniques for mobile communications. Thus mobile computing revolves around computational and communication aspects in the arena of mobility (Sun & Sauvola, 2002). Figure 4 illustrates this concept of mobile computing based on mobility, while

*Figure 4. Mobility and mobile computing (adopted from Sun & Sauvola, 2002)*

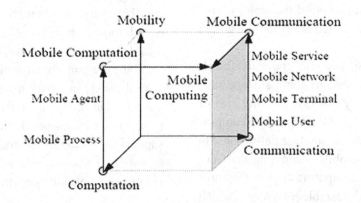

*Figure 5. A nearly fully connected information space (adopted from Kumar, n.d.)*

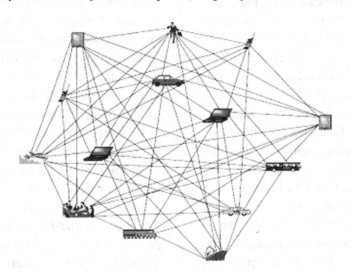

Figure 5 shows a possibility of interconnectivity among various mobile agents in a fully connected environment.

The next section looks into some technological aspects of mobile computing and in doing so; put forwards a few applications of those technologies for learning, health and data management.

## TECHNOLOGIES AND APPLICATIONS

The concept of mobile computing involves the use of the Internet and intranets for communicating and computing while the user is on the move. Typically, mobile hosts are reduced versions of multipurpose computers, with less memory, relatively slower processors, and low-power batteries, and communicate over low-bandwidth wireless communication links. Developments in mobile computing, such as the rapid growth of mobile computing devices and expansion of mobile networks, are enabling the creation of a numerous mobile-based applications (Agrawal, Rao, & Sanders, 2003). The Wireless Application Protocol (WAP)[1] plays an important role in especially GSM-based (Global System for Mo-

bile Communications, originally Groupe Spécial Mobile) networks as it bridges the gap between the mobile world and the Internet world (TCP/IP networks) by optimizing standards for the unique constraints of the wireless environment. It also offers complicated but secured mechanisms and application platforms for various mobile applications (Dahlbom & Ljungberg, 1998).

The availability of lightweight, portable computers and wireless communications has made mobile computing applications more practical. An ever increasing mobile workforce, home working, and the computerization of inherently mobile activities are driving a need for powerful and complex mobile computer systems and applications integrated with fixed systems. Mobile cellular telephony is widely available throughout the world and computers are being integrated with these telephones to form mobile computing devices. Many businesses (not only commercial, but also edutainment, learning, tourism, health service and financial) are dependent on distributed, networked computing systems and are beginning to rely on high-speed communications for multimedia interactions and web-based services. Users are now requiring access to these services while travelling. In addition, new multimedia applications are emerging for web enabled telephones and mobile computers with integrated communications (Agrawal, Rao, & Sanders, 2003).

However, multimedia applications require more sophisticated management of those system components, which affect the Quality of Service (QoS) delivered to the user, than for simpler voice or data-only systems. In this aspect, the underlying concepts of bandwidth, throughput, timeliness (including Jitter), reliability, perceived quality and cost are the foundations of what is known as QoS. Furthermore, portable computers introduce particular problems of highly variable communication quality, management of data location for efficient access, restrictions of battery life, smaller screen size, and cost of connection, which all impact the ability to manage and deliver the required QoS in a mobile environment (Chalmers & Sloman, 1999).

With rapid growth of sophistications in mobile devices such as PDAs and smart phones, the techniques of mobile computing are becoming important and popular day by day. The applications of mobile computing are dominating the research issues, thus leading to pervasive computing and ubiquitous computing (Chien, He, Tsai, & Hsueh, 2010).

In the platform of mobile computing information between processing units flows through wireless channels. The processing units (referred as client in client/server paradigm) are free from temporal and spatial constraints. That is, a processing unit (client) is free to move about in the space while not being disconnected from the server. This temporal and spatial freedom provides a powerful facility allowing users to reach the data site (place where the desired data is stored) and the processing site (the geographical location where a processing must be performed instantly) from anywhere. This capability allows organizations to set their offices at any location (Satyanarayanan, 1996).

Furthermore, there are many other potential applications of mobile computing which will become important in near future, as the power

of portable computing devices increases and the cost of wireless communications decreases. Portable computing devices are commonly used for accessing electronic mail, sending faxes, accessing the web or remote databases, and using cellular telephones or local networks when users are travelling. Palmtop computers and PDAs are being integrated with cellular telephones as part of the increasing convergence between telecommunications and computing. This type of very lightweight device could be used as an electronic "newspaper" capable of delivering selective news that is personalized according to individual user preferences and is more up-to-date than a newspaper (updated each and every moment) (Chalmers & Sloman, 1999).

Web-based news delivery is already available, but due to the heavy content size and cost of communication, users are mainly dependent on desktop workstations in accessing web-based news content. However, with considerable decrease in wireless communication costs are turning towards web-based content. Web-based access to multimedia entertainment, videos, music, and games are becoming popular for mobile computing. Web-enabled cellular phones and in-car communications are beginning to emerge and are considered as a future growth area as context-aware applications.

There are a variety of location-aware applications in which the computing device is able to determine the users´ physical location, for example, Global Positioning Systems (GPS) is used to display current position on maps, receive traffic and weather information, and act as a car-navigation aid, including supply of information like nearest hospitals, hotels, restaurants, places of interests, gas stations for travelers. Similar location-based applications on hand-held devices are being used as guides within museums, art-galleries or tourist spots or towns. These determine current position and provide information on exhibits or buildings near the user, including access to additional information such as a painter or architect's biogra-

*Figure 6. The landscape of mobile computing and commerce (adopted from Turban, 2008)*

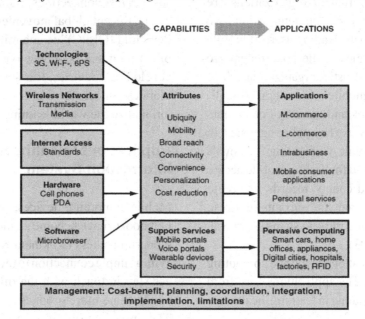

phy or history of the town (Shekar & Lin, 1994; Leonhardt, 1998; Chalmers & Sloman, 1999).

Utility services are becoming potential users of mobile computing services. Emergency services like fire and ambulance can access plans of buildings or details of hazardous chemicals from remote databases. The use of mobile computers by engineers working on power distribution systems or construction sites are having rapid progress. There are many other applications where field workers benefits from use of mobile computing devices for access to detailed 3-D drawings or plans, for example aircraft maintenance engineers, architects, or construction workers. Access to educational material from digital libraries extends the concept of lifelong learning to education and training while on the move in trains or buses. However, the common factor in these applications is likely to be based on multimedia interactions and not just textual data or voice (Katz, 1994; Chalmers & Sloman, 1999). Eventually, a few mobile applications lead to making financial transactions, falling into the category of e-commerce. Figure

6 shows various applications of mobile computing incorporating different technology platforms.

The impact of mobile computing applications has reached almost all sorts of industries and entrepreneurships. For example, mobile inventory management applications that track the location of goods and services are expected to help companies improve delivery times. As early as in 2001, United Parcel Service invested approximately $100 million to upgrade their wireless systems to manage the movement of goods in warehouses. The utility of these applications may be gauged from the fact that the company expected that the $100 million investment to pay for itself within 16 months. Mobile distance learning is one of the promising applications of mobile computing, and universities have begun implementing mobile infrastructures to assist students get access to academic databases from anywhere at any time. On a more advanced scale, product-locating applications are being developed to facilitate consumers locate nearby vendors of specific products and even enable these vendors to compete on a real-time basis for a consumer's business. Many government organizations are also beginning to use mobile computing to improve their service. Among the leading users of mobile computing technologies in the government are the police and criminal justice organizations, since many of them need mobile information to facilitate code enforcement, and the use of computers has been increasingly used by police agencies. In many countries, Mobile Computing Terminals (MCTs) are being used for getting mobile access to federal, state, and county records related to auto registrations, summons, warrants of arrest, and on-line offense-reporting systems (Agrawal, Rao, & Sanders, 2003).

Next, a few applications of mobile computing are being discussed. The applications vary from using of mobile computing in learning, healthcare and data management.

## Applications: Laptops in Classrooms

Use of laptop with wireless connectivity has gained momentum in education and learning. Grace-Martin and Gay (2001) made a classical study on the impact of "laptops in the classroom" and found the following benefits through an extensive literature review: increased student motivation, better developed professional/job skills, increased collaboration among students, better school attendance, better problem-solving, better and/or *more sustained* academic achievement, better writing skills, and *extension* of the school day (students tend to keep working on school-related assignments on their laptops after the regular school day is over, for example, in the evenings at home). However, a few studies have acted to generate some skepticism about these benefits. Indeed, some education technologists have questioned the ostensible hegemony of the optimism regarding Internet technology and laptops in the classroom. It should be noted that a majority of these studies involved K-12 (versus higher education) students in classrooms in which students were provided with *hardwired* connections to the Internet (ver-

sus *wireless* connections) (Grace-Martin & Gay, 2001). Recent global movement on "One Laptop Per Child (OLPC)" is one milestone in the field of learning and skill development, especially in developing countries. It is speculated that, using mobile connectivity OLPC movement will revolutionalize the world learning system.

## Applications: Healthcare Information System

Mobile computing devices, such as laptops and notebooks, have become an indispensable part of modern business (Kakihara & Sørensen, 2006). With a simple connection to the network, employees stay in touch with the office or authorities, and become more productive outside the office. The introduction of smaller, wireless-based handheld computers, PDAs and smart phones has further fuelled the interest in mobility (Fontelo & Chismas, 2005). Similar to other industries, the healthcare industry is also taking advantage of this phenomenon. The drive comes from the desire to provide better healthcare (Wales, 2003).

As mobile computing thrives at a rapid pace, it is likely that this technology will be a key channel in the delivery of that information. The application of mobile computing to healthcare has typically not been as extensive that of other technologies, such as medical imaging. However, developments in mobile computing and communication have enabled this technology to be applied in ways previously unseen. This has enabled platform independent distribution of medical applications and information, particularly in the areas of paramedical and other frontline support. Developments in wireless communications technologies and the move to hand-held mobile devices are also forcing a re-evaluation of existing technology infrastructures within healthcare. Moreover, as society becomes increasingly mobile in almost all aspects of life, the expectation and requirement for a supporting healthcare service, no doubt, are increasing in parallel (Hameed, 2003).

The healthcare industry, initially considered lagging in ICT (Information and Communication Technology) adoption, has witnessed a tremendous growth in the adoption of emerging mobile computing devices (Havenstein, 2005). Mobile computing provides great benefits to the healthcare industry. These include tele-medicine support, improved data collection, timely access to the latest information to interested parties. Since this is now possible to collect data from within and outside an institution, it could lead to productivity improvement and cost savings. However, the portable nature of these mobile computing devices that make them so attractive is not available without some risks. As healthcare providers increasingly moving toward mobile computing, demands for scalability, reliable access and security are becoming prominent (Satyanarayanan, 1996; Herrera, 2006).

In these days, mobile devices are coming with wireless-ready and built-in support for Wireless Local Area Networks (WLANs), Bluetooth and Wireless Wide Area Networks (WWAN) (Karygiannis & Owens, 2002). These access technologies are providing easy access to medical information anywhere and anytime. However, with the ease of access comes great responsibility in terms of utilization of information (Crounse, 2006) and confidentiality. Crumbley (2003) stated that as valuable as mobile devices are to caregivers, they function in an extremely vulnerable environment, and pose new threats to the privacy and security of health information. While protection is desirable (for privacy and economic reasons) to reinforce customer confidence and trust, regulatory implications ought to be mandatory (Grove, 2003). Thomas (2007) argued that, prompted by these economic and regulatory pressures, hospitals and healthcare providers desire new solutions that can address core business needs and manage the huge volumes of time and security sensitive data that are involved in this sector (Portale, 2002).

## Applications: Electronic Data Collection Using GIS

Multiple studies substantiate productivity gains of between 15 and 25% when the mobile workforce is equipped with mobile devices and wireless access. These productivity savings came about in three forms: through time savings, provision of flexibility, and increased quality of work. This can be achieved by moving existing business processes, automated or manual, beyond the organization's premises to wherever and whenever those tasks can be carried out efficiently. One of the key enablers to this move is through the use of GIS (Geographical Information System). In the developed world, as GIS is maturing, it is increasingly being applied to a variety of new tasks within utilities, including planned maintenance, the real-time location of faults on a network, and the dispatch of maintenance teams to rectify those faults.

Figure 7 shows the electronic data collection process and its components. In this case, the active data sourcing device is a PDA which has been used by the field workers and connected via wireless communication to the regional depot (or even to the central office where the corporate GIS database may reside). Here, the "sync-and-go" methodology has been used to download the latest sourced network element specifications to the PDA, and to upload the information sourced by the PDA into the corporate GIS database. This has carried out using the mobile computer's standard "cradle", connected via the USB port to the office computer (Van Olst & Dwolatzky, 2004).

Varying from usage in basic livelihood improvement to advanced science applications, mobile devices are becoming handy and popular day by day. However, due to the nature of operations, characteristics of the devices, preferences of the users, and foremost, the operating environments, mobile devices are facing many constraints and challenges. A few of them are being discussed next.

*Figure 7. The electronic data sourcing procedure (adopted from Van Olst & Dwolatzky, 2004)*

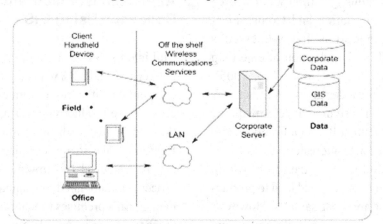

## CONSTRAINTS AND CHALLENGES

Despite immense usage of mobile computing, the environments in which they work are characterized by severe resources constraints and frequent changes in operating conditions. Gupta (2008) mentioned about several articles which have identified the fundamental challenges in mobile computing. Mobile systems are resource poor, less secured, have poor connectivity to the wired infrastructure and have less energy since they are powered by battery.

However, earlier research in this area indicated the limitations of mobile computing in addressing the information management needs. These include problems with bandwidth limitations, screen area for applications, battery capacity, and the inefficiencies of pen-based input (Buyuk-kokten, Garcia-Molina, & Paepcke, 2000). Other problems include client/server/agent adaptation, which is important on slow networks (Papastavrou, Samaras, & Pitoura, 2000), data dissemination on limited bandwidth networks, location-dependent data search, and interface design for mobile devices. The speed at which research on mobile computing is moving may be too fast to capture all the advances being made in the field, especially as it is being applied to various utility industries. For this reason the study limits the discussion to the latest application development platforms and devices surrounding the usage of mobile computing for education, health and data management (Irizarry, 2008).

This section details out a few constraints and challenges of mobile computing. Firstly, it relates to the environments of mobile computing as depicted in Figure 1. As it portrays, mobile computing environment has three main components which affect their performance. These are related to mobile devices, networks and services. However, to complete the picture, this study likes to relate these three parameters with another parameter that has been incorporated in Figure 2, which is mobile user, i.e., human as the fourth parameter. Furthermore, above all, the security issue comes as a major challenge, which incorporates all four parameters mentioned here. Hence, these five major challenges are being discussed here:

### Challenges Related to Devices

For a given cost and level of technology in consideration of weight, power, size and ergonomics impact a shortfall in computational power of mobile devices. Even if mobile elements may improve in absolute ability, they would always be resource-poor relative to their static counterparts (Satyanarayanan, 1993; 1996; 2002).

Due to their limited weight and size, mobile devices have limited resources such as processor speed, memory size, display size, screen resolution and disk capacity. Hence, designers of mobile applications are faced with challenges on how to meet these constraints while satisfying users demand of fast responsive mobile computing environment. Some mobile devices may have different input device such as microphone or a pen point screen or touch screen to interact with users. These limited input devices need to be addressed carefully when designing a mobile user interface (Gupta, 2008).

Power limitations of mobile devices are also to be addressed in terms of their usage, which should recognize a very limited power to use. While battery technology will certainly improve over time, but the sensitivity to power consumption will not reduce. Concerns for power consumption will extent in many levels of hardware and software to be fully effective (Satyanarayanan, 1996; Gupta, 2008).

## Challenges Related to Networks

Wired links normally provide one-to-one communication without interference, whereas wireless links use one-to-many communication that has a considerable noise and interference level and suffer from bandwidth limitations. Simultaneous wireless communications require channel separation, where channel may refer to time, frequency, or code. The channel capacity typically available in wireless systems is much lower than what is available in wired networks. The regulated frequency spectrum further limits the number of users that can be served concurrently (Turban, 2008).

Wireless networks usually have limited bandwidth, high latency and frequent disconnection due to power limitations, spectrum availability and continuous mobility. In this aspect, the most critical parameter is the bandwidth. As bandwidth increases, power consumption increases which shorten the battery life of mobile devices. Hence,

energy restrictions of mobile devices limit the effectiveness of data throughput to and from the device even if wireless networks connections provides stable high bandwidth. Therefore, data access technique need to be designed to overcome these limitations of bandwidth, latency, disconnected operation to satisfy user's expectations (Gupta, 2008).

Mobile devices use battery power, and limited power resources pose further design challenges. High noise levels cause larger bit-error rates. To reduce the error rate, forward error correction algorithms or error detection schemes (such as cyclic redundancy control) followed by buffering and selective retransmission need to be used. Furthermore, one-to-many free space communication is also insecure, since a third party may easily receive the same messages. In this aspect, to improve the security, encryption and decryption procedures require significant power and bandwidth resources (Stojmenovic´, 2002).

In contrast to most stationary computers, which stay connected to a single network, mobile computers encounter more heterogeneous network connections in several ways. Firstly, as they leave the range of one network transceiver and switch to another, they may need to change transmission speeds and protocols. Secondly, in some situations a mobile computer may have access to several network connections at once, for example, where adjacent cells may overlap or where it can be plugged in for concurrent wired access. Thirdly, mobile computers may need to switch interfaces, for example, when going between indoors and outdoors (Gupta, 2008). While inside a premise may offer reliable, high-bandwidth wireless connectivity while others may only offer low-bandwidth connectivity, while on roaming, a mobile client may have to rely on a low-bandwidth wireless network with unpredictable gaps in coverage (Satyanarayanan, 1996; Davies, Friday, Blair & Cheverst, 1996; Davies, Blair, Cheverst & Friday, 1996).

## Challenges Related to Mobility of the Device

Mobility is about moving around or changing places while using the devices. This movement affects three aspects in terms of the device and the user.

Firstly, it affects the physical space due to changes of physical location of the user. This in turn affects the network connection (resulting in poor connectivity or disconnection). To face this challenge, network connection has to adapt to this behavior of user movement by re-connecting user with respect to new location. Hence, data and file operations have also to adapt by supporting disconnection file operations as in the Coda file system (Gupta, 2008).

Secondly, due to the mobility, it affects the information space that consists of a large number of applications and data files scatter in the Internet. Mobility in information space affects the selection of application or data file. To face this challenge, applications and their data have to adapt to users´ operating environment to perform their task efficiently. One way to accomplish this is by caching of data on the local machine as being done in the HTTP/1.1 protocol to reduce redundant web information traffic and thereby improve network utilization.

Thirdly, it is the connection space which is affected by the mobility. It is the huge network of links that connect between various computer platforms. Movement in this space corresponds to selection of route between links and selection of a specific type of platform. To face this challenge, adaptation can be achieved by avoiding congested links and selecting shortest path between two points when delivering information (Gupta, 2008).

## Challenges Related to Security

As discussed above, in ad hoc networks mobile hosts are not bound to any centralized control like base stations or access points. They roam independently and capable to move freely with an arbitrary speed and direction. Thus, the topology of the network may change randomly, frequently and unpredictably. In such networks, information transfer is implemented in a multi-hop fashion, for example, each node acts not only as a host, but also as a router, thus forwarding packets for those nodes that are not in direct transmission range with each other. By nature, ad hoc networks are highly dynamic, self-organizing, but with scarce channels. Besides these risks, ad hoc networks are prone to security threats due to their difference from conventional infrastructure-based wireless networks (Agrawal, Deng, Poosarla, & Sanyal, 2003).

Precisely because getting connected to a wireless link is so easy, the security of wireless communication can be compromised much more easily than that of wired communication, especially if transmission covers a wider area. This increases pressure on mobile computing software designers to include rigorous security measures. Security is further complicated if users are allowed to cross security domains. For example, a hospital may allow patients with mobile computers to use nearby printers but prohibit access to distant printers and resources designated for hospital personnel only (Gupta, 2008).

Secure communication over insecure channels is accomplished by encryption. Security depends on a secret encryption key known only to the authorized parties. Managing these keys securely is difficult, but it can be automated by software (Gupta, 2008). As mentioned here, a fundamental tool to achieve network security objectives is cryptography. However, the challenge of using cryptography in mobile ad hoc networks is the management of cryptographic keys. Since nodes are mobile in wireless networks, their interactions are spontaneous and unpredictable, which makes public key cryptography more appropriate in this setting than conventional cryptography. In this aspect, the most widely accepted solution for the public key management problem is based on public key certificates that are issued by (online) certification authorities and distributed via (online) key distribution servers (Stojmenovic, 2002).

## Challenges Related to Human Factors

Due to their mobile nature, mobile devices are inherently hazardous. A Wall Street stockbroker is more likely to be mugged on the streets of Manhattan and have his laptop stolen than to have his workstation in a locked office be physically subverted. In addition to security concerns, portable computers are more vulnerable to loss or damage (Satyanarayanan, 1996). The author has the bitter experience of mugging his Laptop at the main bus station of an European tourist city with over ten years of works, which were not backed up for about two years. This has resulted in delay of several books and publications in pipeline, including immense inexpressible agony. Furthermore, it has been observed that many people are involved in this sort of activities and has even created physical market or online web links, where often stolen or mugged mobile devices are included.

Deducting from these constraints and challenges the study finds a few issues of future research in the aspect of mobile computing and its usage.

## FUTURE ISSUES

The interaction between man and machine is set to alter dramatically in the near future. With computational power becoming more and more accessible and easy to embed in almost any shape and material, the presence of intelligent devices will grow exponentially by making daily life easier and humanizing future contacts with objects and machines. This appearance of distributed intelligence in and around human lives could be termed as 'Ambient Intelligence' (Beigl, Gray, & Salber, 2001).

A key requirement of mobile computing systems is the ability to access critical data regardless of the location. Data from shared file systems and databases must be made available to programs running on mobile devices. For example, an engineer servicing a jet engine on a parked aircraft needs access to engineering details of that model of engine as well as past repair records of that specific engine. Similarly, a businessman who is continuing his work on the train home needs access to his business records. Yet another example involves emergency medical response to a case of poisoning: the responding personnel will need rapid access to medical databases describing poison symptoms and antidotes, as well as access to the specific patient's medical records to determine the drug sensitivity (Satyanarayanan, 1993). Hence, the research for mobile devices should be focused on delivering a mobile client in a broadband world. The aim should be to allow users to have quick access and optimal interactivity with e-services and web-based applications when they are mobile. This is termed as "High Availability". The future of mobile computing is about streamlined synchronization between a mobile device and a main system via mobile networking (Bijsmans, 2000). The ultimate aim is to maintain an uninterrupted connection between the main system and the mobile device.

Research and applications in various domains have proved the high prospective of mobile computing. The importance of mobile computing is not merely in bringing information to the external terminals of universal information systems, or in having this information and the computing power available anywhere and anytime, but in some important new conditions like the permanent availability of the key project actors in the virtual space. However, these circumstances should not be improperly used to the human detriment. Instead, the computer can take the role of a sophisticated assistant and automatically process as much information as possible (Rebolj, Magdič & Čuš Babič, 2004).

Mobile communications will continue to greatly transform the way of people's life. The tremendous demands from social networks are pushing the booming development of mobile communications faster than ever before, leading to plenty of emerging advanced techniques. Various mobile devices with wider transmission bandwidth, manifold wireless and wired networks, and more powerful processing capability, together with advances in computing technology have brought more and more miscellaneous services to be delivered with more excellent quality (Sun & Sauvola, 2002).

However, the next generation mobile systems need the support of all the advances on new theories, concepts, algorithms, architectures, standards, and protocols. In the near future, more and more Internet based web service can be smoothly accessed with various mobile devices through the wide deployed wireless networks. The 3G (third-generation) mobile communications systems are being deployed for multimedia data applications, while the fourth-generation (4G) mobile communications are paving the way for the future. Eventually, the mobile personal telecommunications and wireless computer networks are converging in the coming new generation of mobile communications. Future mobile communications systems will evolve with the trend of global connectivity through the internetworking and interoperability of heterogeneous wireless networks. However, roaming in such network architectures will be very complex and may cause many new problems. Furthermore, the requirement of smooth and adaptive delivery of real time and multimedia applications will make the design of mobility management scheme as a future challenge (Sun & Sauvola, 2002).

## CONCLUSION

As explored and found during this study, mobile computing is an already emerging technology that enables access to digital resources at any time, from any location. From a narrow viewpoint, mobile computing represents a convenient addition to wire-based geographically dispersed local area distributed systems. However, in broader sense, mobile computing represents the elimination of time-and-place restrictions imposed by desktop computers and wired networks. In forecasting the impact of mobile technology, one could do well to observe recent trends in the use of the wired infrastructure, particularly, the Internet. In recent years, the advent of convenient mechanisms for browsing Internet resources has engendered an explosive growth in the use of those resources. The ability to access them at all times through mobile computing will allow their use to be integrated into all aspects of human life and will pave the door to accelerate the demand for ubiquitous network services. This study has already found a few of these ubiquitous services and has included in the applications section. Along the route, there are constraints and challenges, and researchers and designers are working relentlessly to nullify them. However, the challenge for mobile computing designers will remain as how to adapt the system structures that have worked well for traditional computing so that mobile computing can be integrated there as a best fit (Gupta, 2008). These will require research and study in the aspect of technology (adaption), culture (adoption) and economy (ability).

# REFERENCES

Agrawal, D. P., Deng, H., Poosarla, R., & Sanyal, S. (2003). Secure mobile computing. *Invited Talk, Proceedings of the 4th International Workshop on Distributed Computing (IWDC 2003),* December 28-31, 2003, Kolkata, India.

Agrawal, D. P., & Zeng, Q.-A. (2002). *Introduction to wireless and mobile systems*. Brooks/Cole Publisher.

Agrawal, M., Rao, H. R., & Sanders, G. L. (2003). Impact of mobile computing terminals in police work. *Journal of Organizational Computing and Electronic Commerce, 13*(2), 73–89. doi:10.1207/S15327744JOCE1302_1

Beigl, M., Gray, P., & Salber, D. (2001). Location modeling for ubiquitous computing. *Workshop Proceedings, Ubicomp 2001*. Atlanta.

Bijsmans, J. (2000). *Future mobile computing. Research Report 7.4.2000, Hewlett Packard Laboratories Bristol*. HP Invent, Hewlett-Packard Company.

Buyukkokten, O., Garcia-Molina, H., & Paepcke. A., (2000). Focused Web searching with PDAs. *Computer Networks – The International Journal of Computer Telecommunications Networking, 33*(1-6), 213-230.

Caldwell, D., & Koch, J. L. (1998). *Mobile computing and its impact on the changing nature of work and organization*. Leavey School of Business and Administration. Working Paper. Santa Clara University, Santa Clara, CA.

Chalmers, D., & Sloman, M. (1999). A survey of quality of service in mobile computing environments. *IEEE Communications Surveys. Second Quarter, 1999*, 1–10.

Chien, B.-C., He, S.-Y., Tsai, H.-C., & Hsueh, Y.-K. (2010). An extendible context-aware service system for mobile computing. *Journal of Mobile Multimedia, 6*(1), 49–62.

Crounse, B. (2006). *Mobile devices usher in new era in healthcare delivery: House calls for health professionals*. Microsoft Corporation.

Crumbley, J. (2003). *Trust us*. Access Control and Security systems, Government Security. Retrieved March 23, 2011 from http://securitysolutions.com/mag/security_trust_us/index.html

Dahlbom, B., & Ljungberg, F. (1998). Mobile informatics. *Scandinavian Journal of Information Systems, 10*(1&2), 227–234.

Davies, N., Blair, G. S., Cheverst, K., & Friday, A. (1996). Supporting collaborative applications in a heterogeneous mobile environment. *Computer Communications Special Issue on Mobile Computing, 19*, 346–358.

Davies, N., Friday, A., Blair, G. S., & Cheverst, K. (1996). Distributed systems support for adaptive mobile applications. *Mobile Networks and Applications, 1*(4), 399–408.

Dunham, M. H., & Helal, A. (1995). Mobile computing and databases: Anything new? *SIGMOD Record, 24*(4), 5–9. doi:10.1145/219713.219727

Fontelo, P. A., & Chismas, W. G. (2005). PDAs, handheld devices and wireless healthcare environments: Minitrack introduction. *Proceedings of the 38th Hawaii International Conference on System Sciences*.

Gaddah, A., & Thomas Kunz, T. (2003). *A survey of middleware paradigms for mobile computing*. Carleton University Systems and Computing Engineering Technical Report SCE-03-16, July 2003

Gold, J. (2005). *Managing mobility in the enterprise*. A J. Gold Associates White Paper, July 2005. Retrieved from www.jgoldassociates.com

Grace-Martin, M., & Gay, G. (2001). Web browsing, mobile computing and academic performance. *Journal of Educational Technology & Society, 4*(3), 95–107.

Grove, T. (2003). *Summary analysis: The final HIPAA security rule.* HIPAA Advisory, White Paper, Phoenix Health Systems, February 2003.

Gupta, A. K. (2008). Challenges of mobile computing. *Proceedings of 2nd National Conference on Challenges & Opportunities in Information Technology (COIT-2008), RIMT-IET* (pp. 86-90). Mandi Gobindgarh, March 29, 2008.

Hameed, K. (2003). The application of mobile computing and technology to health care services. *Telematics and Informatics, 20,* 99–106. doi:10.1016/S0736-5853(02)00018-7

Havenstein, H. (2005). *Health care: Doctors and PDAs proved a good match, helping give the industry an early lead with wireless.* Cerner Bridge Medical, Computerworld News.

Herrera, T. (2006). *Solutions for Health Insurance Portability and Accountability Act (HIPAA) compliance.* White Paper, Juniper Networks, Inc.

IEEE. (1999). *LAN standards of the IEEE Computer Society.* Wireless LAN Medium Access Control (MAC) and PHysical Layer (PHY) specification. IEEE Standard 802.11, 1999 Edition.

Irizarry, J. (2008). Potential applications of emerging portable computing platform for information sharing in construction projects. *Proceedings International Conference on Construction and Real Estate Management, (CD-ROM), ICCREM,* Toronto, Canada.

Kakihara, M., & Sorensen, C. (2006). Practicing mobile professional work: Tales of locational, operational, and interactional mobility. *Emerald, 6*(3), 180–187.

Karygiannis, T., & Owens, L. (2002). *Wireless network security 802.11: Bluetooth and handheld devices.* NIST Special Publication 800-848.

Katz, R. H. (1994). Adaptation and mobility in wireless Information Systems. *IEEE Personal Communication, 1*(1), 6-17.

Koudounas, V., & Iqbal, O. (1996). *Mobile computing: Past, present and future.* Retrieved April 16, 2011, from http://www.doc.ic.ac.uk/~nd/ surprise_96/journal/vol4/vk5/report.html

Kristoffersent, S., & Ljungberg, F. (1998). Representing modalities in mobile computing: A model of IT use in mobile settings. In Urban, B., Kirste, T., & Ide, R. (Eds.), *Proceedings of Interactive Applications in Mobile Computing.* Germany: Fraunhofer Institute for Computer Graphics.

Kumar, V. (n.d.). *Mobile computing: A brief history of personal communication system.* University of Missouri-Kansas City, USA. Retrieved April 27, 2011 from http://k.web.umkc.edu/kumarv/cs572/ PCS-history.pdf

Leonhardt, U. (1998). *Supporting location-awareness in open distributed systems.* Ph.D. Thesis, Department of Computing, Imperial College, London, 1998.

Lin, Y.-B., & Chlamtac, I. (2001). *Wireless and mobile network architectures.* John Wiley & Sons.

Nokia. (2006). *The route to true competitive advantage: Today's evolution of workforce mobility.* White Paper. Nokia.

Papastavrou, S., Samaras, G., & Pitoura, E. (2000). Mobile agents for World Wide Web distributed database access. *IEEE Transactions on Knowledge and Data Engineering, 12*(5), 802–820. doi:10.1109/69.877509

Portale, O. (2002). *Healthcare: the mobile opportunity.* Mobile Enterprise, Sun Microsystems, Feature story, November 08, 2002.

Rebolj, D., Magdič, A., & Čuš Babič, N. (2001). Mobile computing in construction. In *Advances in Concurrent Engineering: Proceedings of the 8th ISPE International Conference on Concurrent Engineering: Research and Applications.* West Coast Anaheim Hotel, California, USA, July 28 - August 1, 2001.

Rebolj, D., Magdič, A., & Čuš Babič, N. (2004). Mobile computing – The missing link to effective construction IT. *Proceedings of the International Conference on Construction Information Technology, Langkawi,* February 17-21, 2004, (pp.327-334).

Satoh, I. (2002). Physical mobility and logical mobility in ubiquitous computing environments. *Proceedings of International Conference on Mobile Agents (MA'2002), LNCS.* Springer, October, 2002.

Satoh, I. (2003). Spatial agents: Integrating user mobility and program mobility in ubiquitous computing environments. *Wireless Communications and Mobile Computing, 3,* 411–423. doi:10.1002/wcm.126

Satyanarayanan, M. (1993). *Mobile computing.* In Hot Topics, IEEE Computer, September 1993.

Satyanarayanan, M. (1996). Fundamental challenges in mobile computing. In *Proceedings of the Fifteenth Annual ACM Symposium on Principles of Distributed Computing* (Philadelphia, Pennsylvania, United States, May 23-26, 1996) *PODC '96* (pp. 1-7). New York, NY: ACM Press.

Satyanarayanan, M. (2002). The evolution of CODA. *ACM Transactions on Computer Systems - TOCS, 20*(2), 85-124.

Shekar, S., & Lin, D. (1994). Genesis and advanced traveller information systems (ATIS): Killer applications for mobile computing. *MOBIDATA: An Interactive Journal of Mobile Computing, 1*(1).

Stojmenovic, I. (Ed.). (2002). *Handbook of wireless networks and mobile computing.* John Wiley & Sons, Inc.doi:10.1002/0471224561

Sun, J. Z., & Sauvola, J. (2002). On fundamental concept of mobility for mobile communications. *Proceedings of the 13th IEEE International Symposium on Personal, Indoor and Mobile Radio Communications, 2,* (pp. 799– 803). Lisbon, Portugal, 2002.

Thomas, G. D. A. (2007). *A framework for secure mobile computing in healthcare.* Master's thesis. Nelson Mandela Metropolitan University, South Africa.

Turban, E. (2008). *Mobile, wireless, and pervasive computing. Information Technology for management: Transforming organizations in the digital economy* (pp. 167–206). New York, NY: John Wiley & Sons.

Van Olst, R., & Dwolatzky, B. (2004). Electronic data collection using mobile computing technologies. *GIS Technical, a Proceeding of SATNAC Conference.* Stellenbosch, 2004.

Vizard, M. (2000). The way things work is the fundamental problem with mobile computing. *InfoWorld, 22*(40), 83.

Wales, J. (2003). *The pocket PC's prescription for health care.* Smart Phone & Pocket PC Magazine, May 2003.

## ADDITIONAL READING

Abowd, G. D., & Mynatt, E. D. (2000). Charting Past, Present, and Future Research in Ubiquitous Computing. *ACM Transactions on Computer-Human Interaction, 7*(1), 29–58. doi:10.1145/344949.344988

Ala-Laurila, J., Haverinen, H., Mikkonen, J., & Rinnemaa, J. (2002). Wireless LAN architecture for mobile operators. In Bing, B. (Ed.), *Wireless Local Area Networks: the New Wireless Revolution* (pp. 159–176). New York, NY: John Wiley & Sons.

Bacon, J., Moody, K., Bates, J., Hayton, R., Ma, C., McNeil, A., Seidel, O., & Spiteri, M. (2000). Generic Support for Distributed Applications. *IEEE Computer,* 68-77, March 2000.

Baldauf, M., Dustdar, S., & Rosenberg, F. (2004). A survey on context-aware systems. *International Journal of Ad Hoc and Ubiquitous Computing, 2004, Vol. 2, No. 4,* 2007, 263-277.

Banerjee, N., Wu, W., Das, S. K., Dawkins, S., & Pathak, J. (2003). Mobility Support in Wireless Internet. *IEEE Wireless Communications, 10*(5), 54–61. doi:10.1109/MWC.2003.1241101

Becker, S. A., Sugumaran, R., & Pannu, K. (2004). The use of mobile technology for proactive healthcare in tribal communities. In *Proceedings of the 2004 Annual National Conference on Digital Government Research* (Seattle, WA, May 24 - 26, 2004). *ACM International Conference Proceeding Series.*

Bowden, S., Dorr, A., Thorpe, A., Anumba, C. J., & Gooding, P. (2005). Making the Case for Mobile IT in Construction. *Proceedings of the ASCE International Conference on Computing in Civil Engineering.* Lucio Soibelman, Feniosky Peña-Mora - Editors, July 12–15, 2005,Cancun, Mexico.

Carzaniga, A., & Wolf, A. L. (2001). Content-Based Networking: A New Communication Infrastructure. In *NSF Workshop on an Infrastructure for Mobile and Wireless Systems.* Scottsdale, AZ, October 2001.

Cassidy, T. (2005). *PDAs* Clarify Dictation. Advance for Health Information Professionals. *Advance News magazine, 15(17),* 32.

Ciampa, M. (2005). *Security+ Guide to Network Security Fundamentals* (2nd ed.). Canada: Thomson Learning Inc.

Endsley, M. R., Bolte, B., & Jones, D. G. (2003). *Designing for Situation Awareness: An Approach to User-Centered Design.* Boca Raton, Fl.: Taylor and Francis.

Flinn, J., & Satyanarayanan, M. (1999). Energy-aware Adaptation for Mobile Applications.

Forman, G. H., & Zahorjan, J. (1994). The Challenges of Mobile Computing. *IEEE Computer Society, 27*(4), 38–47. doi:10.1109/2.274999

Ganz, A., Istepanian, S. H., & Tonguz, O. K. (2006). Advanced Mobile Technologies for Health Care Applications. *Journal of Mobile Multimedia, 1*(4), 271–272.

Gorlenko, L., & Merrick, R. (2003). No wires attached: Usability challenges in the connected mobile world. *IBM Systems Journal, 4,* 640.

Imielinski, T., & Badrinath, B. R. (1994, October). Mobile wireless computing. *Communications of the ACM, 37*(10), 19–28. doi:10.1145/194313.194317

In *Proceedings of the 17th ACM Symposium on Operating Systems and Principles.* Kiawah Island, SC, December, 1999.

Irvien, J. M., Tafazolli, R., & Groves, I. S. (2000, Dec.). Future mobile networks. *Electronics & Communication Engineering Journal, 12*(6), 262–270. doi:10.1049/ecej:20000604

Jiang, C., & Steenkiste, P. (2002). A Hybrid Location Model with a Computable Location Identifier for Ubiquitious Computing. In Borriello, G. & Homquist. L.E. (ed.), *UbiComp 2002: Ubiquitous Computing, 4th International Conference.* *Göteborg, Sweden.*

Kondratova, I. (2004). Voice and Multimodal Technology for the Mobile Worker. *ITcon, Special Issue on Mobile Computing in Construction., (9):* 345-353.

Kong, P., Zerfos, H., Luo, S. L., & Zhang, L. L. (2001). Providing Robust and Ubiquitous Security Support for Mobile Ad-Hoc Networks. *Proceedings of the IEEE 9th International Conference on Network Protocols (ICNP '01),* 2001.

Lebeck, A. R., Fan, X., Zheng, H., & Ellis, C. S. (2000). Power Aware Page Allocation. In *Proceedings of the Ninth International Conference on Architectural Support for Programming Languages and Operating Systems*. November, 2000.

Meissner, A., Baxevanaki, L., Mathes, I., Branki, C., Bozios, T., Schoenfeld, W., et al. (2002). Integrated Mobile Operations Support for the Construction Industry: The COSMOS Solution. In *Proceedings of the 6ᵗʰ World Multiconference on Systemics, Cybernetics and Informatics (SCI 2002)*. Orlando, FL, United States, July 14-18, 2002.

Miettinen, M., & Halonen, P. (2006). Host-Based Intrusion Detection for Advanced Mobile Devices. *IEEE, 2*, 72- 76

Mukherjee, A., & Siewiorek, D. P. (1994). *Mobility: A Medium for Computation, Communication and Control*, School of Computer Science, Carnegie Mellon University, IEEE Workshop on Mobile Computing Systems and Applications, Santa Cruz, CA, US, December 1994

Papadimitratos, P., & Haas, Z. (2002). Secure Routing for Mobile Ad Hoc Networks. *Proceedings of the SCS Communication Networks and Distributed Systems Modeling and Simulation Conference*, January 2002.

Pitoura, E., & Bhargava, B. (1995). Maintaining Consistency of Data in Mobile Distributed Environments. *Purdue University, 15ᵗʰ International Conference on Distributed Computing Systems, May 1995* (pp. 404-413).

Rajgopalan, S., & Badrinath, B.R. (1995). Adaptive Location Management for Mobile-IP. *First ACM Mobicom 95,* November 1995.

Satoh, I. (2002). A Framework for Building Reusable Mobile Agents for Network Management. *Proceedings of Network Operations and Managements Symposium (NOMS'2002,* (pp.51-64). IEEE Communication Society, April, 2002.

Satyanarayanan, M., Noble, B., Kumar, P., & Price, M. (1994). Application-Aware adaptation for Mobile Computing. *Carneigie Mellon University, Sixth ACM SIGOPS European Workshop*, July 1994.

Spreitzer, M., & Theimer, M. (1993). Scalable, Secure, Mobile Computing with Location Information. *Communications of the ACM, 36*(7), 1993. doi:10.1145/159544.159558

Treek, D. (2003). An integral framework for information systems security management. *Computers & Security, 22*(4), 337–360. doi:10.1016/S0167-4048(03)00413-9

Williams, T. P. (2003). Applying Handheld Computers in the Construction Industry. *Practice Periodical on Structural Design and Construction, 8*(Issue 4), 226–231. doi:10.1061/(ASCE)1084-0680(2003)8:4(226)

## KEY TERMS AND DEFINITIONS

**Digital Tablet:** A digital tablet or graphics tablet (or digitizer, digitizing tablet, graphics pad, drawing tablet) is a computer input device that allows the user to hand-draw images and graphics, similar to the way one draws images with a pencil and paper. These tablets may also be used to capture data or handwritten signatures or graphics. It can also be used to trace an image from a piece of paper which is taped or otherwise secured to its surface. Capturing data in this way, either by tracing or entering the corners of linear poly-lines or shapes is known as digitizing.

**Mobile Agent:** A mobile agent is a composition of computer software and data that is able to migrate (move) from one computer to another autonomously and continue its execution on the destination computer. A mobile agent is a type of software agent, with the feature of *autonomy, social ability, learning*, and most importantly, the *mobility*. More specifically, a *mobile agent* is a process that can transport its state from one environment or entity to another, with its data being intact, and be capable of performing appropriately and satisfactorily in the new environment.

**Mobile Communication:** A mobile phone or simply, mobile (also called cell phone or hand phone) is an electronic device used for mobile telecommunications (mobile telephony, text messaging or data transmission) over a cellular network of specialized base stations known as cell sites. Mobile communications let one operate without the need for a fixed phone line providing greater flexibility in business operations, faster customer responsiveness and savings in staff time.

**PDA:** PDA stands for Personal Digital Assistant and it is a lightweight, usually pen-based consumer electronic device that looks like a hand-held computer but instead performs specific tasks; can serve as a diary or a personal organizer or a personal database or a telephone or an alarm clock etc.

**WLAN:** A Wide Area Network (WAN) is a computer network that covers a broad area (for example, any network whose communications links cross metropolitan, regional, or national boundaries). This is in contrast with personal area networks (PANs), local area networks (LANs), campus area networks (CANs), or metropolitan area networks (MANs) which are usually limited to a room, a building, a campus or a specific metropolitan area (such as, a city or town or metropolis) respectively.

## ENDNOTE

[1]    http://www.wapforum.com/

# Chapter 2
# Architecture for Integrated Mobile Calendar Systems

**Lars Frank**
*Copenhagen Business School, Denmark*

## ABSTRACT

*In central databases the consistency of data is normally implemented by using the ACID (Atomicity, Consistency, Isolation, and Durability) properties of a DBMS (Data Base Management System). This is not possible if distributed and/or mobile databases are involved, and the availability of data also has to be optimized. The objective of this chapter is to describe an architecture for mobile integrated calendar systems where performance, local autonomy, and availability are optimized by using relaxed ACID properties and different asynchronous replication methods. By using relaxed ACID properties across different database locations it is possible for the users to trust the data they use even if the distributed database temporarily is inconsistent. It is also important that disconnected locations can operate in a meaningful way in so-called disconnected mode.*

## INTRODUCTION

When using DBMS to manage the database an important aspect involves transactions which are any logical operation on data and per definition database transactions must be atomic, consistent, isolated, and durable in order for the transaction to be reliable and coherent. The ACID properties

DOI: 10.4018/978-1-4666-0080-5.ch002

of a database are delivered by a DBMS to make database recovery easier and make it possible in a multi user environment to give concurrent transactions a consistent chronological view of the data in the database. The ACID properties are consequently important for users that need a consistent view of the data in a database. However, the implementation of ACID properties may influence performance and slow down the availability of a system in order to guarantee that all users have a consistent view of data even in case of failures. In several situations, the availability and the response time will be unacceptable if the ACID properties of a DBMS are used without reflection. This is especially the case in distributed and/or mobile databases where a failure in the connections of a system should not prevent the system from operating in a meaningful way in disconnected mode.

Information systems that operate in different locations can be integrated by using more or less common data and/or by exchanging information between the systems involved. In both situations, the union of the databases of the different systems may be implemented as a database with so called relaxed ACID properties where temporary inconsistencies may occur in a controlled manner. However, when implementing relaxed ACID properties it is important that from a user's point of view it must still seem as if traditional ACID properties were implemented, which therefore will keep the local databases trustworthy for decision making. In the following part of the introduction, author gives an overview of how relaxed ACID properties may be implemented and used in central and mobile databases integrated by using relaxed ACID properties.

The Atomicity property of a DBMS guarantees that either all the updates of a transaction are committed/executed or no updates are committed/executed. This property makes it possible to re-execute a transaction that has failed after execution of some of its updates. In distributed databases, this property is especially important if data are replicated as inconsistency will occur if only a subset of data is replicated. The Atomicity property of a DBMS is implemented by using a DBMS log file with all the database changes made by the transactions. The global Atomicity property of databases with relaxed ACID properties is implemented by using compensatable, pivot and retriable subtransactions sequentially. By applying these subtransactions it is allowed to commit/execute only part of the transaction

and still consider the transaction to be atomic as the data converge towards a consistent state. The global Consistency property is not defined in databases with relaxed ACID properties because normally such databases are inconsistent and this inconsistency may be managed in the same way as the relaxed Isolation property.

The Isolation property of a DBMS guarantees that the updates of a transaction cannot be seen by other concurrent transactions until the transaction is committed/executed. That is the inconsistencies caused by a transaction that has not executed all its updates cannot be seen by other transactions. The Isolation property of a DBMS may be implemented by locking all records used by a transaction. That is the locked records cannot be used by other transactions before the locks are released when the transaction is committed. The global Isolation property of databases with relaxed ACID properties is implemented by using countermeasures against the inconsistencies/ anomalies that may occur. The Durability property of a DBMS guarantees that the updates of a transaction cannot be lost if the transaction is committed. The Durability property of a DBMS is implemented by using a DBMS log file with all the database changes made by the transactions. By restoring the updates of the committed transactions it is possible to recover a database even in case it is destroyed. The global Durability property of databases with relaxed ACID properties is implemented by using the local Durability property of the local databases involved.

Data replication is normally used to decrease local response time and increase local performance by substituting remote data accesses with local data accesses (Frank, 2005). At the same time, the availability of data will normally also be increased as data may be stored in all the locations where they are vital for disconnected operation. These properties are especially important in mobile applications where the mobile user often may be disconnected from data that are vital for the normal operation of the application. The major disadvantages of data replication are the additional costs of updating replicated data and the problems related to managing the consistency of the replicated data.

In general, replication methods involve n copies of some data where n must be greater than 1. The basic replication designs storing n different copies of some data are defined to be n-safe, 2-safe, 1-safe or 0-safe, respectively, when n, 2, 1 or 0 of the n copies are consistent and up-to-date at normal operation. In mobile computing, using only the 1-safe and 0-safe replication designs is recommended, as these use asynchronous updates which allow local system to operate in disconnected mode.

As described earlier the objective of this chapter is to describe an architecture for mobile integrated calendar systems where performance, local autonomy and availability are optimized by using relaxed ACID (Atomicity, Consistency, Isolation and Durability) properties and different asynchronous replication methods implemented. The architecture also supports integration with stationary systems that use information from a user's calendar to submit the user's needs and receive offers that support these needs. Therefore, it should be possible to receive committed information about the appointments made by the user's interaction with the stationary supply systems with offers. Such stationary systems could be meeting arrangement systems, reservation systems like travel and hotel reservation systems, students teaching schemas, web services, etc. It is important that the design allows both calendar users and stationary users to make changes to already committed appointments, as these are needed in the appointments of the user's real life.

The objective of this chapter is also to describe the necessary tools/methods that are used for integrating the mobile databases like integrated calendar systems. In order to make it easier to implement countermeasures against isolation anomalies, at the end of the chapter the author describes the design rules he had used for selecting the replication designs in the mobile calendar system.

## RELATED WORK

The transaction model used in this paper is *the countermeasure transaction model* (Frank & Zahle, 1998). This model uses the relaxed Atomicity property (Garcia-Molina & Salem,1987; Mehrotra, Rastogi, Korth, & Silberschatz, 1992; Weikum & Schek, 1992; Zhang,1994). How to use countermeasures against consistency anomalies are described by Frank (2011c). The model also uses countermeasures against the isolation anomalies. These anomalies are described by Berenson et al. (1995) and Breibart et al. (1992). In the countermeasure transaction model the problem of implementing the Durability property is solved as describe by Breibart et al. (1992). The replication designs used in this chapter are described in more details by Frank (2005 ; 2010a). The design rules used for choosing among the different replication designs from a costs point of view are described by Frank (2008).

The Pull Update Propagation method was first described by Frank (1988 ;1985). Pacitti and Simon (2000) have evaluated different versions of the Push Propagation. The *immediate-immediate* strategy propagates an update each time an update is executed. The *immediate-wait* strategy propagates the updates of a transaction when the transaction is committed. The *deferred* strategy supported by most commercial DBMSs propagates the updates of several transactions in a single message. The evaluation of the different UP (Update Propogation) methods described in this chapter is taken from Frank and Zahle (1998). A great deal of research has tried to solve the special problems of mobile transactions.

Special middleware for mobile computing have been described and evaluated by Capra, Emmerich and Mascolo (2002). However, Capra et al. (2002), reject message-orientated middleware as such middleware requires too much memory for persistent queues of messages received, but not processed. The author do not agree and in the countermeasure transaction model, the queuing problem is solved by switching to pull communication when pushed messages are not acknowledged after a few attempts. Anyway, this solution does not solve the problem of overloading a mobile host, but it enables the mobile host to process the lost messages later. Serrano-Alvardo, Roncancio and Adiba (2004) have given an overview and compared some of the best known transaction models designed for mobile computing.

How to use relaxed ACID properties in integrating E-governance applications are described in general by Frank (2011a) and a more specific description of integrating heterogeneous EHR (Electronic Health Records) is described by Frank and Pape-Haugaard (2011). Frank (2010b) has also described architecture for integrated mobile logistics management and control as part of a distributed ERP system. How to use countermeasures against consistency anomalies are described by Frank (2011c).

The rest of the paper is organized as follows: First, the author describes how relaxed ACID properties are implemented in the transaction model used in this chapter. Next, the author explains the details about how to implement the relaxed atomicity property followed by the description of the most important replication designs used in mobile databases. Illustration of how the different replication techniques may be used to make integrated mobile calendar systems is given next. The next section elaborates upon the rules used for choosing among the different asynchronous replication designs. Concluding remarks are presented finally.

## THE TRANSACTION MODEL

A *multidatabase* is a union of local autonomous databases. *Global transactions* (Gray & Reuter,

1993) access data located in more than one local database. In recent years, many transaction models have been designed to integrate local databases without using a distributed DBMS. The countermeasure transaction model (Frank & Zahle, 1998) has selected properties from these transaction models and integrated the use of countermeasure in order to reduce the problems caused by the missing isolation property in a distributed database that is not managed by a distributed DBMS. In the countermeasure transaction model, a global transaction involves a *root transaction* (client transaction) and several single site *subtransactions* (server transactions). Subtransactions may be nested transactions, i.e. a subtransaction may be a *parent transaction* for other subtransactions. All communication with the user is managed from the root transaction, and all data is accessed through subtransactions. The following subsections will give a broad outline of how relaxed ACID properties are implemented.

### The Atomicity Property

An updating transaction has the *atomicity property* and is called *atomic* if either all or none of its updates are executed. In the countermeasure transaction model, the global transaction is partitioned into the following types of subtransactions executed in different locations:

- The *pivot* subtransaction that manages the atomicity of the global transaction. The global transaction is committed when the pivot subtransaction is committed locally. If the pivot subtransaction aborts, all the updates of the other subtransactions must be compensated.
- The *compensatable* subtransactions that all may be compensated. Compensatable subtransactions must always be executed before the pivot subtransaction is executed to make it possible to compensate them if the pivot subtransaction cannot be commit-

ted. A compensatable subtransaction may be compensated by executing a *compensating* subtransaction.

- The *retriable* subtransactions that are designed in such a way that the execution is guaranteed to commit locally (sooner or later) if the pivot subtransaction has been committed.

The global atomicity property is implemented by executing the compensatable, pivot and retriable subtransactions of a global transaction sequentially. For example, if the global transaction fails before the pivot has been committed, it is possible to remove the updates of the global transaction by compensation. If the global transaction fails after the pivot has been committed, the remaining retriable subtransactions will be (re)executed automatically until all the updates of the global transaction have been committed.

## The Consistency Property

A database is *consistent* if its data complies with the consistency rules of the database. If the database is consistent both when a transaction starts and when it has been completed and committed, the execution has the *consistency property*. Transaction *consistency rules* may be implemented as a control program that rejects the commitment of transactions, which do not comply with the consistency rules.

The above definition of the consistency property is not useful in distributed databases with relaxed ACID properties because such a database is almost always inconsistent. However, a distributed database with relaxed ACID properties should have *asymptotic consistency*, i.e. the database should converge towards a consistent state when all active transactions have been committed/compensated. Therefore, the following property is essential in distributed databases with relaxed ACID properties:

If the database is asymptotically consistent when a transaction starts and also when it has been committed, the execution has the *relaxed consistency property*.

## The Isolation Property

The isolation property is normally implemented by using *long duration locks*, which are locks that are held until the global transaction has been committed (Frank & Zahle, 1998). In the countermeasure transaction model, long duration locks cannot instigate isolated global execution as retriable subtransactions may be executed after the global transaction has been committed in the pivot location. Therefore, *short duration locks* are used, i.e. locks that are released immediately after a subtransaction has been committed/aborted locally. To ensure high availability in locked data, short duration locks should also be used in compensatable subtransactions, just as locks should be released before interaction with a user. This is not a problem in the countermeasure transaction model as the traditional isolation property in retriable subtransactions is lost anyway. If only short duration locks are used, it is impossible to block data. (Data is *blocked* if it is locked by a subtransaction that loses the connection to the "coordinator" (the pivot subtransaction) managing the global commit/abort decision). When transactions are executed without isolation, the so-called *isolation anomalies* may occur. In the countermeasure transaction model, relaxed isolation can be implemented by using countermeasures against the isolation anomalies. If there is no isolation and the atomicity property is implemented, the following isolation anomalies may occur (Berenson et al., 1995 ; Breibart et al., 1992).

- *The lost update anomaly* is by definition a situation where a first transaction reads a record for update without using locks. Subsequently, the record is updated by another transaction. Later, the update is overwritten by the first transaction. In extended transaction models, the lost update anomaly may be prevented, if the first transaction reads and updates the record in the same subtransaction using local ACID properties. Unfortunately, the read and the update are sometimes executed in different subtransactions belonging to the same parent transaction. In such a situation, a second transaction may update the record between the read and the update of the first transaction.

- *The dirty read anomaly* is by definition a situation where a first transaction updates a record without committing the update. Subsequently, a second transaction reads the record. Later, the first update is aborted (or committed), i.e. the second transaction may have read a non-existing version of the record. In extended transaction models, this may happen when the first transaction updates a record by using a compensatable subtransaction and later aborts the update by using a compensating subtransaction. If a second transaction reads the record before it has been compensated, the data read will be "dirty".
- *The non-repeatable read anomaly* or *fuzzy read* is by definition a situation where a first transaction reads a record without using locks. Later, the record is updated and committed by a second transaction before the first transaction has been committed. In other words, it is not possible to rely on the data that have been read. In extended transaction models, this may happen when the first transaction reads a record that later is updated by a second transaction, which commits the update locally before the first transaction commits globally.
- *The phantom anomaly* is a situation where a first transaction reads some records by using a search condition. Subsequently, a second transaction updates the database

in such a way that the result of the search condition is changed. In other words, the first transaction cannot repeat the search without changing the result. Using a data warehouse may often solve the problems of this anomaly (Frank, 2003).

The countermeasure transaction model (Frank & Zahle, 1998) describes countermeasures that eliminate or reduce the problems of the isolation anomalies. In this chapter, it is assumed that the following countermeasures are used:

The *Reread Countermeasure* is primarily used to prevent the lost update anomaly. Transactions that use this countermeasure read a record twice using short duration locks for each read. If a second transaction has changed the record between the two readings, the transaction aborts itself after the second read. In the replicated databases used in this chapter the *Reread Countermeasure will only function if there is a primary copy location*.

The *Commutative Updates Countermeasure* can prevent lost updates merely by using commutative operators. Adding and subtracting an amount from an account are examples of commutative updates. If a subtransaction only has commutative updates, it may be designed as commutable with other subtransactions that only have commutative updates as well. In the replicated databases used in this chapter the *Commutative Updates Countermeasure is recommended when there is no primary copy location*.

The *Pessimistic View Countermeasure* reduces or eliminates the dirty read anomaly and/or the non-repeatable read anomaly by giving the users a pessimistic view of the situation. In other words, the user cannot misuse the information. The purpose is to eliminate the risk involved in using data where long duration locks should have been used. The pessimistic view countermeasure may be implemented by using:

- The pivot or compensatable subtransactions for updates that "limit" the users' op-

tions. That is concurrent transactions cannot use resources that are reserved for the
compensatable/pivot subtransaction.

- The pivot or retriable subtransactions for
updates that "increase" the users' options.
That is, concurrent transactions can only
use increased resources after the increase
has been committed.

## The Durability Property

Updates of transactions are said to be *durable* if
they are stored in a stable manner and secured by
a log recovery system. In case a global transaction
has the atomicity property (or relaxed atomicity),
the global durability property (or relaxed durability
property) will automatically be implemented, as it
is ensured by the log-system of the local DBMS
systems (Breibart et al., 1992).

## RELAXED ATOMICITY IMPLEMENTATION BY USING UPDATE PROPAGATION

*Update Propagation* (UP) is used for updating
remote data with distributed atomicity and durability properties synchronized with the updates of
the parent transaction. Frank (2010a) describes
the implementation of UPs in the following way:

The parent transaction makes the UP "call"
by storing a so-called *transaction record* in persistent storage at the parent location. The parent
transaction id, the id of the subtransaction and the
parameters of the subtransaction are stored in the
transaction record. If the parent transaction fails,
the transaction record will be rolled back and
the subtransaction is not executed. If the parent
transaction is committed, the transaction record
is secured in persistent storage, and it is said that
the UP has been *initiated*. After the initiation of
the UP, the transaction record will be sent by the
UP tool to the location of the corresponding subtransaction. The data transfer may be implemented

by using push and/or pull technology. UPs have
the following properties, which are in contrast to
RPCs (Remote Procedure Calls):

- If a parent transaction initiates several
UPs, the corresponding, stored programs
may be executed in parallel.
- A stored program initiated from a UP has
atomicity synchronized with the parent
transaction, i.e. either both or none are
executed.
- The remote stored program does not return
control to the parent transaction. That is
UPs may be used for executing retriable
subtransaction after the pivot subtransaction is committed.

Most DBMS products have a UP tool. However,
they can only be used to propagate updates between
the vendors' own DBMS products. Therefore, it is
important to know how to implement a UP tool.
In the implementations, only short duration locks
are used, i.e. the implementations must usually
be integrated with countermeasures against isolation anomalies. In general, the problem is that a
parent transaction at location A wants to transfer
the subtransaction id and parameters to a stored
program at location B. The subtransaction id, the
parameters, and the id of the parent transaction are
stored in the transaction record. The main steps in
transferring the transaction record and executing
the corresponding stored program are as follows:

1. The parent transaction writes the transaction record in a DBMS file at location A to
ensure that the transaction record has the
same atomicity and durability as the other
local DBMS updates of the parent transaction. Subsequently, all location A updates
are committed. Therefore, the subtransaction
related to the UP will not be executed if the
parent transaction fails before commitment
of the location A updates. When the transaction records have been committed, there

are two main methods for transferring the transaction records in the transaction file at location A.

2a.  At location B, one may periodically select transaction records from the location A transaction file. After a transaction record has arrived at location B, the corresponding subtransaction may be executed, which is referred as a *Pull Update Propagation.*

2b.  At location A, one may insert the transaction record from the location A transaction file in a location B transaction file. After the transaction record has been stored in the location B transaction file, the corresponding subtransaction may be executed, which is referred as a *Push Update Propagation.*

In the following subsections, the Pull and Push Update Propagation methods will be described in more details and the methods will be compared. Next, the special problems of implementing UPs in mobile computing are described. Finally, the author makes conclusions about UP implementation.

## Pull Update Propagation

The transaction records stored at location A have a sequence number as a primary key. If the parent transaction at location A fails, the transaction record at the location A transaction file will be backed out, and the sequence number may be reused. This is very straight forward since each site has the local ACID properties. In the location B database, there is a counter record containing the sequence number of the last transaction record successfully transferred and executed at location B. At location B, there is a general transfer program, which is restarted periodically. When the program is restarted, it will begin to read the counter record containing the last successfully executed subtransaction. Then, the program will search for the next transaction record in the transaction file at location A. (A RPC may be used, since it will be restarted periodically as described above, in case it should

fail). When the next transaction record arrives at location B, the counter record will be updated and the corresponding subtransaction will be executed. Following this operation, the location B updates the sub transaction and the update of the counter record will be committed.

If the transfer program or the subtransaction fails before this commitment, the counter record is backed out together with the updates of the subtransaction. Because the counter record is backed out, the same transaction record will be transferred to location B the next time the transfer program is started. In other words, if a transaction record in the location A transaction file is committed, the corresponding subtransaction will sooner or later be executed at location B.

The sequence number of the next transaction record may be stored in a "sequence number record" in the location A database. This sequence number record may constitute a bottleneck, as it will be updated for each transaction record sent to location B. The problem may be reduced by locking the sequence number record as the last update of the parent transaction. (The problem may be further reduced by using a time-stamp instead of a sequence number, as the time-stamp does not have to be updated by the DBMS. If a time-stamp is used, it is still necessary to use an exclusive lock while getting the time-stamp, because the transaction file must be updated in an ascending sequence of the primary key, which is in this case the time-stamp).

## Push Update Propagation

This method does not use a sequence number or time-stamp to control the transfer and execution of subtransactions. Instead, location B has a transaction file to ensure that a subtransaction will only be executed once at location B. The Push Update Propagation includes the following steps:

1.  The location A updates, including the transaction record, are committed.

2. The parent transaction at location A initiates step 3, which sends the transaction record to location B. The transfer will only be delayed if step 2 fails, as step 3 will be restarted automatically.

3. This step is a general transfer program, which is restarted periodically with "short" time intervals. This step may also be started from step 2. The program will start sending the transaction records in the location A transaction file to location B. For each transaction record in the location A transaction file, the following steps are executed:

   i. When the transaction record arrives at location B, the program is checks whether the transaction record already has been inserted into the location B transaction file. If the transaction record has already been stored, the location B updates have already been executed and committed locally. If the transaction record has not been inserted, the insert and the location B updates will be executed and committed locally. After the local commitment of the location B updates, a RPC will be executed and start the following stored procedure at location A.

   ii. At location A, the transaction record corresponding to the location B subtransaction will be deleted from the location A transaction file, i.e. the submission of the transaction record in step 3 cannot be repeated any more. When the delete operation has been committed, the Push Update Propagation of a single transaction record is finished.

## Pull versus Push Update Propagation

The Pull and Push Update Propagations have different properties. Therefore, it is often necessary to implement both methods in a distributed system. The most important differences are listed below:

- Pull Update Propagations are batch oriented as transaction records are only transferred periodically. Push Update Propagations are not batch oriented as a transaction record normally is sent immediately (periodic transfers are only used in cases of error).

- Pull Update Propagations have a smaller execution cost than Push Update Propagations, because a Pull Update Propagation does not involve deleting the transaction record in the location A transaction file.

- The programming cost of Pull Update Propagations is smaller than the programming cost of Push Update Propagations as the programming of a Pull Update Propagation does not involve deleting the transaction record in the location A transaction file.

- The Pull Update Propagation of transaction records is single threaded, i.e. the transaction records are transferred and the corresponding subtransactions are executed in sequence order. The Pull Update Propagation may be designed to be multithreaded if there is a counter record for each thread. The Push Update Propagation may be designed as multithreaded if step 2 in the transfer algorithm always initiates transfer of the newest transaction record while the periodic restart of step 3 always transfers the oldest transaction record in the location A transaction file. If the transaction transfer is single threaded there might be a potential bottleneck. On the other hand, if the transaction transfer is multithreaded, the subtransactions related to the transaction records are not always executed in sequential order, which may increase the isolation anomalies.

*Figure 1. Push update propagation after disconnection*

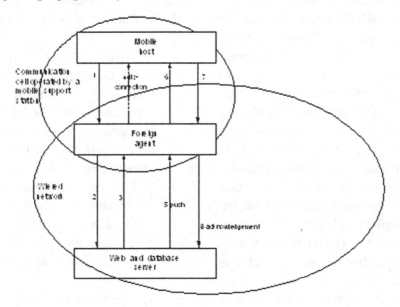

- The updates of the Pull UP have two potential bottlenecks since "the counter record" and "the sequence number record" must be updated for all transaction transfers.

## Update Propagation in Mobile Computing

A *mobile transaction* is a transaction where at least one mobile host is involved. In this situation, *disconnection* and *location dependency* (*the movement problem*) will normally be problems even if relaxed atomicity is used. The objective of this section is to describe how these problems are solved in the countermeasure transaction model.

First, the author describes how update propagation can be used to resubmit subtransactions/ messages until they have been committed in the receiving location. This solves the disconnection problem. The update propagation tool can also reduce the location dependency problem as messages are not lost and, therefore, can be picked up from a new location. Next, the author describes how the integration of update propagation and mobile IPs can be used to solve the location dependency

problem in general. When both the disconnection problem and the location dependency problem have been solved, the availability is high compared with the traditional mobile transactions that do not use countermeasures to relax the isolation property.

## The Disconnection Problem

The communication properties of the update propagation are interesting to mobile computers as messages managed by the UP are tolerant to disconnection failures and manual interruptions by the mobile user. This will be illustrated in the following examples. The first example will show how a Push update propagation resubmits a message until a commit/ acknowledge message has been returned. The last two examples will illustrate how a client can use a Pull update propagation when inquiring about stored messages.

### Example

Figure 1 illustrates how update propagation can push a message to a mobile host even if the communication line or mobile host has failed for a period of time. Before the communication starts,

*Figure 2. Pull update propagation after disconnection*

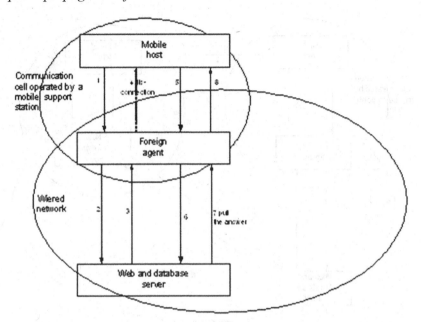

the mobile host has received an IP address from a foreign subnet managed by a *mobile support station* directly connected to the wired network. This IP address can be used as long as the mobile host stays within the communication cell covered by the mobile support station. The propagation is carried out as follows:

1. The mobile host submits a query to a web server via a foreign agent connected to the mobile support station.
2. The foreign agent routes the query to the web server.
3. The web server answers the mobile host and initiates an update propagation in the database server that later resubmits the answer until the mobile host sends a commit/ acknowledge message.
4. The mobile host does not receive the answer as it is disconnected.
5. The push update propagation tool can resubmit the answer periodically until the mobile host returns a commit/ acknowledge message. However, the author recommend

that the database server only resubmits the answer a few times because an unanswered push update propagation can be overtaken by a pull update propagation at any time as illustrated in the next example.

6. The mobile host receives the answer.
7. The mobile host returns a commit/acknowledge message.
8. The database server receives the commit/ acknowledge message and stops resubmitting the answer.

## Example

Figure 2 illustrates how an update propagation can pull a message to a mobile host even if the communication line or mobile host has failed for a period of time. In this example, the first 4 steps are the same as in the previous example as the steps do not activate the update propagation tool.

1. The mobile host submits a query to a web server via a foreign agent connected to the mobile support station.

*Figure 3. Pull update propagation to a new cell*

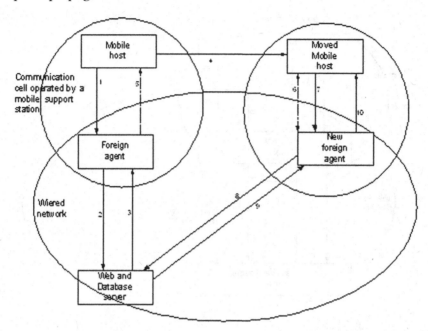

2. The foreign agent routes the query to the web server.

3. The web server answers the mobile host and initiates an update propagation that later resubmits the answer until the mobile host sends a commit/ acknowledge message.

4. The mobile host does not receive the answer as it is disconnected.

5. The pull update propagation tool is activated and asks for stored messages to the mobile host.

6. The database server receives the message from the mobile host and finds the answer initiated in step 3.

7. The answer is resubmitted from the database server to the mobile host.

8. The mobile host receives the answer.

## Example

The Push update propagation cannot resubmit a message to a mobile host that has left the cell from which the query was submitted. However, using a

Pull update propagation enables the mobile host to receive the answer as illustrated in Figure 3. In this example, the first 3 steps are the same as in the previous example.

1. The mobile host submits a query to a web server via a foreign agent connected to the mobile support station.

2. The foreign agent routes the query to the web server.

3. The web server answers the mobile host and initiates an update propagation that later resubmits the answer until the mobile host sends a commit/ acknowledge message.

4. Let us suppose that the mobile host moves to another cell managed by a new mobile support station.

5. The mobile host does not receive the answer as it is has moved to another cell.

6. Now, the mobile host receives its new IP address in the new cell.

*Figure 4. Mobile IP with location independency*

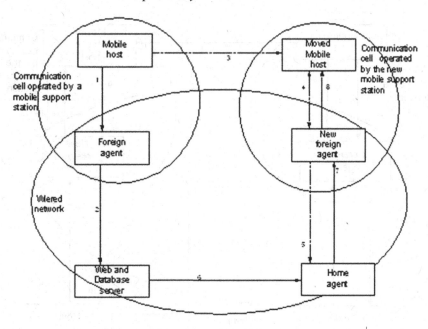

7. The pull update propagation tool is activated and asks for stored messages to the mobile host.
8. The database server receives the message from the mobile host and finds the answer initiated in step 3.
9. The answer is resubmitted from the database server to the mobile host.
10. The mobile host receives the answer.

## The Location Dependency Problem

If a mobile host uses a *mobile IP address*, all messages to the mobile host will be sent to the mobile host's fixed home address, from where the messages will be submitted to the last IP address used by the mobile host in its communication with its home agent. If the mobile host has left this address, all messages will be lost until the home agent receives the mobile host's new IP address. The next example and Figure 4 illustrate how the mobile IP technique can solve the location dependency problem.

*Example*

1. The mobile host submits a query to a web server via a foreign agent connected to the mobile support station.
2. The foreign agent routes the query to the web server.
3. Let us suppose that the mobile host moves to another cell managed by a new mobile support station.
4. Now the mobile host receives its new IP address in the new cell and sends it to its home agent.
5. The home agent receives the new IP address.
6. The database server answers the mobile host by submitting the answer to the home agent.
7. The answer is routed to the new foreign agent.
8. The mobile host receives the answer.

A mobile IP address solves the problem of location dependency, but not the problem of disconnection. Therefore, UP and mobile IP should be integrated. However, the mobile IP and the UP

*Figure 5. Mobile IP integrated with Push update propagation*

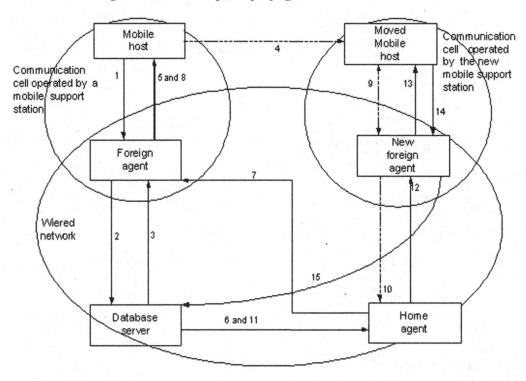

tool are expensive transfer methods for a message, and they are only necessary in case of failure. Therefore, an important message to a mobile host should first be sent directly to the mobile host and only stored as a UP message. If the mobile host answers within a time limit, the UP message to the home agent is deleted. Otherwise, the message will automatically be delivered indirectly by the UP message to the home agent. This is illustrated in the following example and in Figure 5.

*Example*

1. The mobile host submits a query to a web server via a foreign agent connected to the mobile support station.
2. The foreign agent routes the query to the web server.
3. The web server answers the mobile host and initiates a queued retriable message to the home agent, i.e. the answer is only sent to

the home agent if the update propagation tool has to resubmit the answer.

4. Let us suppose that the mobile host moves to another cell managed by a new mobile support station.
5. The answer cannot be delivered to the mobile host as it has moved in step 4.
6. Sooner or later the UP tool resubmits the answer as it has not received an acknowledgement from the answer sent in step 3.
7. The home agent routes the answer to the foreign agent.
8. The answer still cannot be delivered to the mobile host as the host moved in step 4.
9. Now the mobile host receives its new IP address in the new cell and sends it to its home agent.
10. The home agent receives the new IP address.
11. Sooner or later the UP tool resubmits the answer as it has not received an acknowl-

edgement from the answers sent in steps 3 and 6.

12. The answer can now be routed to the new foreign agent.
13. The mobile host receives the answer.
14. The mobile host returns an acknowledgement in steps 14-15 to stop the UP tool resubmitting the answer.

## Conclusions about Relaxed Atomicity Implementation

Normally, distributed and/or long duration transactions do not have the traditional ACID properties because this will reduce availability and performance of accessing data. In this situation, author have recommended using the relaxed atomicity property. The distributed relaxed atomicity property manages the workflow of a distributed transaction in such a way that either all the updates of the transaction are executed (sooner or later) or all the updates of the transaction are removed/compensated. In mobile computing, author's transaction model solves the disconnection problem by using update propagations that resubmit subtransactions/messages until they have been committed in the receiving location. This solution may cause a overflow in persistent queues of messages received, but not processed. The queuing problem is solved by switching from push to pull communication. By integrating the update propagation tool with mobile IPs (Perkins, 1997), it is also possible to solve the location dependency problem. Relaxed atomicity may also be implemented by using different types of SOA services corresponding to the different types of subtransactions, Frank (2011b). This is described in the next section where data are replicated by using the transaction model.

## IMPORTANT REPLICATION METHODS FOR MOBILE COMPUTING

The n-safe and quorum safe replication designs are not suited for mobile systems as it should be possible to operate in disconnected mode. Therefore, the author describe only the asynchronous replication methods in the following.

### Basic 1-Safe Design

In the *Basic 1-Safe Design* (Grey & Reuter, 1993), the primary transaction manager goes through the standard commit logic and declares completion when the commit record has been written to the local log. In the basic 1-safe design, the log records are asynchronously spooled to the locations of the secondary copies as illustrated in Figure 6. In case of a primary site failure in the basic 1-safe design, production may continue by selecting one of the secondary copies but this is not recommended in real time systems as it may result in *lost transactions*.

### 0-Safe Design with Local Commit

The *0-Safe Design with Local Commit* is defined as n table copies in different locations where each transaction will first go to the nearest database location, where it is executed and committed locally (see Figure 7).

If the transaction is an update transaction, the transaction propagates asynchronously to the other database locations, where the transaction is re-executed without user dialog and committed locally at each location. This means that all the table copies normally are inconsistent and not up to date under normal operation. The inconsistency must be managed by using countermeasures against the isolation anomalies. For example, to prevent lost updates in the 0-safe design, all update transactions must be designed to be commutative (Frank & Zahle, 1998).

*Figure 6. Basic 1-Safe database design*

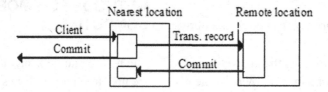

## 1-Safe Design with Commutative Updates

The *1-Safe Design* can transfer updates to the secondary copies in the same way as the 0-safe design. In such a design, lost transactions cannot occur because the transaction transfer of the 0-safe design makes the updates commutative. The properties of this mixed design will come from either the basic 1-safe or the 0-safe design. The *1-Safe Design with Commutative Updates* does not have the high update performance and capacity of the 0-safe design. On the other hand, in this design the isolation property may be implemented automatically as long as the primary copy does not fail. Normally, this makes it much cheaper to implement countermeasures against the isolation anomalies, because it is only necessary to secure that "the lost transactions" are not lost in case of a primary copy failure.

## 0-Safe Design with Primary Copy Commit

The *0-Safe Design with Primary Copy Commit* is a replication method where a global transaction first must update the local copy closest to the user by using a compensatable subtransaction. Later, the local update may be committed globally by using a primary copy location. If this is not possible, the first update must be compensated. The primary copy may have a version number that also are replicated. This may be used to control that an updating user operates on the latest version of the primary copy before an update is committed globally in the location of the primary copy. If an update cannot be committed globally in the primary copy location, the latest version of the primary copy should be send back to the updating user in order to repeat the update by using the latest version of the primary copy.

## Implementation of Internet Replication Services with Relaxed ACID Properties

In order to implement integration flexibility between the different calendar and ERP (Enterprise Resource Planning) modules it is not acceptable that the different modules communicate directly with each other. All communication between different modules must be executed by applications offered as for example SOA services. In order also to implement relaxed ACID properties between the

*Figure 7. 0-Safe database design with local commit*

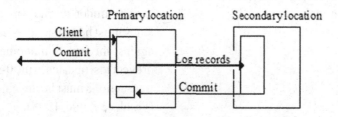

modules each module should offer the following types of services:

- Read only services that are used when a health unit wants to read data managed by another health unit.
- Compensatable update services that are used when a health unit wants to make compensatable updates in tables managed by another health unit.
- Retriable update services that are used when a health unit wants to make retriable updates in tables managed by another health unit.

The SOA services may be used to implement the asynchronous replication designs in the following way:

- All types of the 1-safe designs can be implemented in the following way: Either the primary copy location uses the retriable services of the secondary copy locations directly to transfer the replicated records, or the primary copy location can use the retriable service of a distribution center to transfer the replicated records indirectly. The first solution with direct replication transfer is fast but in this case each module must know how to contact the different secondary locations. In the solution with indirect transfers it is only the distribution center that should know how to contact the different secondary locations.
- The 0-safe design with local commit can be implemented in the same ways as the 1-safe designs except that all locations should function as a primary copy location when they have records to replicate.
- The 0-safe design with deferred commit can be implemented in the following way: First a compensatable update is executed locally in the initiating location. Next, a retriable service call to a primary copy

location is executed, and if the update is committed the primary copy location can transfer the replicated records in the same ways as the 1-safe designs. However, if the update is rejected in the primary copy location, a retriable service call is executed in order to remove the original update from the initiating location.

## DESIGN OF INTEGRATED MOBILE CALENDAR SYSTEMS

Figure 8 illustrates the most important tables in an integrated mobile calendar system. A stationary database server belonging to an enterprise may function as the backup and recovery system for the tables of the mobile calendar users. The stationary database server is also used to store all retriable updates used by the different replication designs. All the tables illustrated in Figure 8 are fragmented as different users only need to know the appointments and the related information that the participant is involved in. Another reason for fragmentation is that different fragments of the tables should be replicated by using different replication methods as later illustrated.

In this chapter, author assume that the Appointment table has the following attributes:

- Appointment ID.
- Appointment-name with a short description of the appointment.
- Time for the appointment.
- Date for the appointment.
- State of the appointment.
- Duration of the appointment.
- Location ID used as foreign key to the location of the appointment.
- Father-appointment ID used for hierarchical appointments.

Most of these attributes are not mandatory as the calendar may also be used as a personal

*Figure 8. ER-diagram of the most important tables in an Integrated Calendar System*

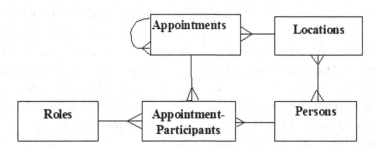

notebook for private matters that the user has to remember. Such non-distributed personal "appointments" need not even be related to a person, as a non-distributed appointment always is owned by the owner of the physical calendar device where it is stored. Such personal "appointments" should, at most, be replicated to the stationary database server for back up reasons.

In contrast, meetings are real appointments where most of the attributes of the calendar database may be used. Meetings arranged by an employer of some calendar users can theoretically be replicated using the basic 1-safe design with the primary copy in the stationary database server belonging to the enterprise and secondary copies in all the locations of the mobile participants. However, such a solution is very inflexible as none of the users can update the calendar if they are disconnected from the central database server. It is also inflexible if the primary copy is stored in the mobile location of the user that first created the meeting appointment, as none of the other users ought to have the possibility to change the meeting appointment if the creating user is isolated. It may be useful, for example, to have more updating users if the head of a meeting creates the meeting appointment while a secretary may be responsible for making all practical arrangements around the meeting. A third solution is to replicate by using the 0-safe design with local commit. This solution is the most flexible but the author do not recommend using the 0-safe design with local commit in this situation as there is a risk of the "lost update

anomaly" (Berenson et al., 1995), where the first of two concurrent updates is overwritten by the second. Therefore, the author recommend that meeting appointments are replicated by using the 0-safe design with primary copy commit in the stationary database server and secondary copies in all the locations of the mobile participants. With this replication design it is possible for all persons with a proper authorization role to update the meeting appointment. As explained in earlier, a version number in the appointment table may be used to control that an updating user operates on the latest version of the primary copy before an update is committed globally. This makes it possible to prevent the lost update anomaly.

By using the Roles table, it is possible to define different types of participants in a meeting. This is important as the different types of participants in a meeting may have different authorities to make changes in the meeting arrangement depending on their role.

Invited participants without update rights should only have a copy of the meeting appointment and its related entities. However, all participants should have the right to confirm whether they are attending the meeting or not. Therefore, the Appointment-Participants table should have a State attribute where it is possible to write whether the participant intends to participate or not. The author recommend that this State attribute should be replicated by using the basic 1-safe design with the primary copy in the location of the participant because the participant knows best the state of his/

hers participation. On the other hand, the author also recommend that creation and deletion of records in the Appointment-Participants table should have the same replication design as the table fragment with the corresponding meeting appointments because inviting participants is part of the work of arranging a meeting. For the same reasons, fragments of the tables with Locations, Persons, and Roles should also have the 0-safe design with primary copy commit in the stationary database server and secondary copies in all the locations of the mobile participants. However, it should only be possible for a user to delete Locations, Persons, and Roles that the user has created himself. Therefore, all these tables should have an Identification attribute with the ID of the creator and a Privacy attribute to indicate whether the record is for public or private use.

The author recommend that some fragments of the tables with Locations, Persons, and Roles have the basic 1-safe design with the primary copy in the central location, as every employee in an enterprise should have a list of the other employees and their location in the enterprise. In the same way, it may be useful to standardize the name of the roles used for meetings and the locations of the meeting rooms of the enterprise. As such information is known in advance, it does not matter whether the updates are delayed and therefore, the affordable basic 1-safe design can be recommended.

As an example of a hierarchical appointment one may assume a parent appointment with the Appointment-name "Travel from A to B". This appointment has two children called "Departure from A" and "Arrival at B". These children should have Location relationships to location A and location B. However, if the travel has a return ticket, it is very important that the entire travel arrangement can be viewed as a single structured appointment in order to make it possible to make travel orders by describing the entire travel arrangement as a single object that can be sent to, for example, an Internet travel agent in order to

receive different travel possibilities from which the user can select a specific travel arrangement that can be stored in the appointment database of the calendar system with a confirmed State value. Therefore, author recommend storing any travel arrangements as a hierarchical appointment where the parent appointment is the travel arrangement itself. The second level is the single tickets of the travel arrangement. The third level of the hierarchical travel appointment is the single location appointments from which the journey begins and ends. By describing a travel arrangement once and for all in a travel object, it is possible to use the same travel object when searching for different offers at different travel home pages. That is, the travel object should first be non-replicated as it is the user that makes the first overall description of the travel arrangement. Later when the user has received non-confirmed offers that support the specifications of the user, one of these offers can be accepted and the others deleted. When a specific travel arrangement is selected and committed in the location of the travel agent, it should have the State value "committed" and be replicated as 0-safe with the primary copy in the location of the travel supplier, as the supplier is the only one that can commit changes in a confirmed travel arrangement. However, for some types of tickets it is still possible for a user to change a travel arrangement. Therefore, as in the previous meeting example the 0-safe design with the primary copy commit for all travel appointments is recommended. However, in case of travel appointments using the stationary database server belonging to the user's enterprise as the primary copy location cannot be recommended because it is only the travel supplier that can commit travel changes globally. In any case, the stationary database server belonging to the user's enterprise may be used as a redistribution center that ensures that retriable updates are replicated to all the participants of an appointment and not only to the user(s) filed in the Supplier's database. Appointments with hospitals, physicians, dentists, advocates, hotel reservations etc., should be designed in the same way as travel objects as such appointments may also be hierarchical and consist of several sub-appointments that only can be committed globally in the location of the service supplier. That is, the service-appointment objects should at first be non-replicated and later on when an appointment is confirmed by the service supplier, the State of the appointment should then be shifted from non-committed to committed and it should be replicated by using the 0-safe design with the primary copy commit in the location of the service supplier.

For Airline companies and hotels there are very detailed standards for the formats of data and how to exchange service offers and acceptances. For most other lines of business there are also some standards. For service offers and appointments with detailed standards it is relatively easy to make subtypes to the Appointments that describe the details of the appointments in a calendar system. However, there may be lines of businesses and individual service suppliers that do not follow any standards. By using research results within the area the Semantic Web it should be possible automatically to transform such offers to the formats used in a calendar database. Another way to deal with non-standardized offers is to let the user mark the attributes of accepted appointments one by one in order to store the attributes of an appointment correct in the calendar database. In practice, the author recommend a combination of the two methods where a user has to acknowledge or correct a semantic interpreted appointment created on the basis of the non-standardized appointment.

Web services are often used in mobile and ubiquitous computing to find services that the user needs from a supplier close to the current physical location of the user. Such service registrations may be integrated into the calendar system, and the user may be alarmed when a hit is found. If the service provider needs time slots for making a deal, the situation is the same as described above where the service supplier has limited time slots

and therefore, should manage the primary copy commit. However, often the user only needs the address of the supplier in order to go there and in such situations the offer of the supplier does not need to be confirmed or updated in the location of the user. That is, the user can keep the original non replicated service request and the corresponding answers without confirming any of them. Therefore, the answers of such requests are only stored as snapshot copies that never need updates or confirmation. As such appointments at most are copied to the central database server for back up reasons, they may be deleted by the user even if the user is disconnected.

Students that follow a course or study can also use a hierarchical appointment structure to describe the lectures and examinations they have to follow. At the top of such a study structure author recommend to have the study itself. The second level is the course of study. The third level of a study description is normally a schema with the weekly lessons of the courses. However, such a structure does not fit into a calendar system and therefore, the schema should automatically be changed to the single lessons of the courses that fit perfectly into the structure of a calendar system. As the hierarchical structure is managed from the institution of the study, it should be feasible to download the entire study structure from the home page of the study. Later it should be possible to receive changes to the planned lessons. Therefore, the author recommend that the hierarchical appointment structure of the study should be basic 1-safe with the primary copy in the location of the study and secondary copies in the locations of the students. If a student does not want to participate in a lecture, it is possible for the user to store this information in the State attribute of the Appointment-Participants table for the lecture(s) involved. It is possible to replicate the state of the participants to the location of the study. However, teaching institutions like universities do not care and therefore, the replication of the participants' states should be optional.

## RULES FOR CHOOSING REPLICATION DESIGNS IN MOBILE SYSTEMS

Frank (2008) has described design rules for selecting the different replication designs from a development cost point of view. However, in mobile computing the availability is normally more important and therefore, the following changes are made in the design rules of Frank (2008) in order to optimize the availability of mobile users:

1.  The first recommendation is to replicate data only when it is necessary for availability reasons or convenient for economical reasons. Therefore, replication is only recommended if at least one of following conditions is fulfilled:
    a.  It is not possible to operate in disconnected mode without using replicated data.
    b.  The costs of data replication are less than the benefits of replication.
2.  The 0-safe design with local commit is recommended in situations where it is important to operate in disconnected mode and where it is possible to implement sufficient local countermeasures against the isolation anomalies and the consistency property. This is normally the case when facts like supplier deliveries are recorded.
3.  The 0-safe design with primary copy commit is recommended in situations where it is important to operate in disconnected mode and where it is not possible to implement sufficient local countermeasures against the isolation anomalies and the consistency property. This is for example in the case when the consistency property says that the stocks of products must not go negative. If the primary copy of stocks are stored in stationary locations it is possible from a mobile location to execute compensatable services in the primary copy locations, and

if all these are committed it is possible for the mobile location to commit the global transaction.

4.  The 1-safe designs are recommended in situations where it does not give any meaning to operate in disconnected mode. Therefore, e.g. mobile updates have to wait until the primary copy location has committed the updates. This is for example in the case when the consistency property says that the balance of a customer must not exceed the credit limit of the customer. If the updates are commutative as it is the case of money transactions, author will recommend using the 1-safe design with commutative updates as this gives higher availability. If the updates are not commutative as it is the case in replacement updates of for example customer addresses author will recommend using the basic 1-safe design.

## FUTURE ENHANCEMENT

Frank (2008) has described design rules for selecting the different replication designs from a development cost point of view. In this chapter, the author have changed the design rules in order to optimize the availability of mobile users. The author believe that many more types of mobile applications will be designed and implemented by using the replication designs used in this paper. In theory, all centralized multiuser systems can be redesigned as mobile applications integrated with stationary applications. Therefore, it is important to analyze how it is possible to make standardized replication implementations or middleware that can implement the use of the different replication designs by using Meta data descriptions. In the future, an analization of how it is possible to standardize such development by using both stationary and mobile (heterogeneous) ERP modules as examples will be undergone.

## CONCLUSION

The architecture for the mobile calendar systems described in this paper uses relaxed ACID properties and different types of asynchronous replication designs to integrate the mobile functionalities with the functionalities of the different stationary supply systems for the mobile calendar systems. By using the data replication design rules described in this chapter it has been possible to optimize the availability of the calendar system in such a way that many of the functionalities of the calendar system can also operate in disconnected mode.

## ACKNOWLEDGMENT

This research is partly funded by the 'Danish Foundation of Advanced Technology Research' as part of the 'Third Generation Enterprise Resource Planning Systems'(3gERP) project, a collaborative project between Department of Information at Copenhagen Business School, Department of Computer Science at Copenhagen University and Microsoft Business Systems.

## REFERENCES

Berenson, H., Bernstein, P., Gray, J., Melton, J., O'Neil, E., & O'Neil, P. (1995). A critique of ANSI SQL isolation levels. In *Proceedings of ACM SIGMOD Conference* (pp. 1-10).

Bernstein, P., Hsu, M., & Mann, B. (1990). Implementing recoverable requests using queues. *SIGMOD Record*, •••, 112–122. doi:10.1145/93605.98721

Breibart, Y., Garcia-Molina, H., & Silberschatz, A. (1992). Overview of multidatabase transaction management. *The VLDB Journal*, 2, 181–239. doi:10.1007/BF01231700

Capra, L., Emmerich, W., & Mascolo, C. (2002), Middleware for mobile computing. In *Tutorial Proceedings of International Conference of Networking 2002*, Springer (pp 1-39).

Frank, L. (1985). *Databaser*. Copenhagen: Samfundslitteratur.

Frank, L. (1988). *Database theory and practice*. Addison-Wesley.

Frank, L. (2003). *Patent application*. Copenhagen Business School.

Frank, L. (2005). Replication methods and their properties. In Rivero, L. C., Doorn, J. H., & Ferraggine, V. E. (Eds.), *Encyclopedia of database technologies and applications*. Hershey, PA: Idea Group Inc.doi:10.4018/978-1-59140-560-3.ch092

Frank, L. (2008). Architecture for mobile ERP systems. In *Proceedings of 7th International Conference on Applications and Principles of Information Science (APIS2008)* (pp.412-415).

Frank, L. (2010a). *Design of distributed integrated heterogeneous or mobile databases* (pp. 1–157). Germany: LAP LAMBERT Academic Publishing AG & Co. KG.

Frank, L. (2010b). Architecture for integrated mobile logistics management and control. In *Proceedings of 2nd International Conference on Information Technology Convergence and Services (ITCS 2010)*. IEEE Computer Society.

Frank, L. (2011a). Architecture for ERP system integration with heterogeneous e-government modules. In Chhabra, S., & Kumar, M. (Eds.), *Strategic enterprise resource planning models for e-government: Applications & methodologies*. Hershey, PA: IGI Global. doi:10.4018/978-1-60960-863-7.ch007

Frank, L. (2011b). Architecture for integrating heterogeneous distributed databases using supplier integrated e-commerce systems as an example. In *Proceedings of the International Conference on Computer and Management (CAMAN 2011)*. Wuhan, China.

Frank, L. (2011c). Countermeasures against consistency anomalies in distributed integrated databases with relaxed ACID properties. In *Proceedings of Innovations in Information Technology (Innovations 2011)*. Abu Dhabi, UAE.

Frank, L., & Andersen, S. K. (2010). Evaluation of different database designs for integration of heterogeneous distributed electronic health records. In *Proceedings of the International Conference on Complex Medical Engineering (CME2010)*. IEEE Computer Society.

Frank, L., & Pape-Haugaard, L. (2011). Integration of health records by using relaxed ACID properties between hospitals, physicians and mobile units like ambulances and doctors. *International Journal of Handheld Computing Research, 2(4)*.

Frank, L., & Zahle, T. (1998). Semantic ACID properties in multidatabases using remote procedure calls and update propagations. *Software, Practice & Experience, 28*, 77–98. doi:10.1002/(SICI)1097-024X(199801)28:1<77::AID-SPE148>3.0.CO;2-R

Garcia-Molina, H., & Salem, K. (1987). Sagas. In *ACM SIGMOD Conference* (pp. 249-259).

Gray, J., & Reuter, A. (1993). *Transaction processing*. Morgan Kaufman.

Mehrotra, S., Rastogi, R., Korth, H., & Silberschatz, A. (1992). A transaction model for multidatabase systems. In *Proceedings of International Conference on Distributed Computing Systems* (pp 56-63).

Pacitti, E., & Simon, E. (2000). Update propagation strategies to improve freshness in lazy master replication databases. *The VLDB Journal, 8*, 305–318. doi:10.1007/s007780050010

Perkins, C. (1997). Mobile IP. *IEEE Communication Magazine*.

Serrano-Alvarado, P., Roncancio, C., & Adiba, M. (2004). A survey of mobile transactions. [DAPD]. *Distributed and Parallel Databases, 16*, 1–38. doi:10.1023/B:DAPD.0000028552.69032.f9

Weikum, G., & Schek, H. (1992). Concepts and applications of multilevel transactions and open nested transactions. In Elmagarmid, A. (Ed.), *Database transaction models for advanced applications* (pp. 515–553). Morgan Kaufmann.

Zhang, A., Nodine, M., Bhargava, B., & Bukhres, O. (1994). Ensuring relaxed atomicity for flexible transactions in multidatabase systems. In *Proceedings of the ACM SIGMOD Conference* (pp.67-78).

## ADDITIONAL READING

Fekete, A., Liarokapis, D., O'Neil, E., O'Neil, P., & Shasha, D. (2005). Making snapshot isolation serializable. *ACM Transactions on Database Systems, 30*(2), 492–528. doi:10.1145/1071610.1071615

Frank, L. (2008). *Databases and Applications with Relaxed ACID Properties*. Doctoral dissertation from Copenhagen Business School, Samfundslitteratur.

Frank, L. (2008). Smooth and Flexible ERP Migration between both Homogeneous and Heterogeneous ERP Systems/ERP Modules. In *Proceedings of the 2nd Workshop on 3rd Generation Enterprise Resource Planning Systems*. Copenhagen business School.

Frank, L. (2008). A Transaction Model for Mobile Atomic Transactions. In *Proceedings of 22nd International Conference on Advanced Networking and Applications, Workshops/Symposia* (pp. 868-873). Gino-wan, Okinawa, Japan.

## KEY TERMS AND DEFINITIONS

**ACID Properties:** The properties imply that a transaction is *Atomic*, transforms the database from one *Consistent* state to another consistent state, is executed as if it were *Isolated* from other concurrent transactions, and is *Durable* after it has been committed.

**Atomic Transaction:** A transaction which either performs all the updates or removes them (the "A" in the ACID properties). The atomicity property makes it easier to do database recovery.

**ERP (Enterprise Resource Planning):** Standard software packages for business applications like sale, e-commerce, procurement, logistics, production management etc.

**Isolation Anomalies:** Inconsistencies that occur when transactions are executed without the isolation property (the "I" in the ACID properties).

**Relaxed ACID Properties:** Application implemented properties that from a user point of view functions as if the traditional ACID properties are implemented.

**SOA (Service Orientated Architecture):** SOA is a special way to call a remote application/service without knowing its internet IP address. This gives flexibility as the IP address of an application/service may change in case of a major site failure.

**System Integration:** The property that data or applications from one system can be used by another system without manual intervention.

**Transaction Abort:** Removal of the updates of a transaction. Transaction aborts are normally executed by a DBMS (Data Base Management System).

**Transaction Compensation:** Removal of the updates of a transaction. Transaction compensation is normally executed by an application program in order to simulate the atomicity property in relaxed ACID properties.

# Chapter 3
# Semantic Web–Enhanced Context–Aware Computing in Mobile Systems:
## Principles and Application

**Stefan Zander**
*University of Vienna, Austria*

**Bernhard Schandl**
*University of Vienna, Austria*

## ABSTRACT

*The goal of this chapter is to provide detailed insights into the field of Semantic Web-based context-aware computing for mobile systems. Readers will learn why context awareness will be a central aspect of future mobile information systems, and about the role semantic technologies can play in creating a context-aware infrastructure and the benefits they offer. The chapter introduces requirements, enabling technologies, and future directions of such systems. It presents a Semantic Web-based context-sensitive infrastructure that resembles concepts from graph theory and distributed transaction management. This infrastructure allows for an efficient acquisition, representation, management, and processing of contextual information while taking into account the peculiarities and operating environments of mobile information systems. Authors demonstrate how context-relevant data acquired from local and remote sensors can be represented using Semantic Web technologies, with the goal to replicate data related to the user's current and future information needs to a mobile device in a proactive and transparent manner. In consequence, the user is equipped with contextually relevant information anytime and anywhere.*

DOI: 10.4018/978-1-4666-0080-5.ch003

## INTRODUCTION

The availability and power of mobile devices has significantly increased data-centric mobile applications. Mobility not only influences the types of information we need, but also how we access it, and which tools and mechanisms to process them are at our disposal. To improve mobile application development and the usability of mobile devices in general it is crucial to understand the information needs of mobile users, as well as the interaction metaphors they apply (Sohn, Li, Griswold, & Hollan, 2008).

It is known that mobile search differs substantially from desktop search in terms of intra-query diversity (Kamvar & Baluja, 2006). The diversity of queries initiated in mobile setting is significantly lower compared to queries issued from the desktop. Additionally, query categorization reveals that context searching is similar to desktop, although query exploration is significantly lower. Further differences can be observed with respect to the effort needed to set up a query, and the total amount of queries initiated in one browsing session. Therefore, plain Internet access is often not sufficient for adequately addressing information needs of mobile users. Their situational contexts and current activities cannot be sufficiently addressed. Despite the benefits offered by mobile internet access, issues remain that hinder users from satisfying their information needs. These include impedimental interaction with the device while browsing for information, as well as the extensive attention needed for interaction and information seeking tasks (Sohn et al., 2008). Further, it was observed that mobile users sometimes do not know how to address a specific information need although they had the required resources and tools at their disposal (Sohn et al., 2008).

It has been shown that 72% of mobile information needs can be attributed or related to context (Sohn et al., 2008). This finding indicates that introducing context awareness into mobile information processing infrastructure can be of significant benefit to end-users. Context allows information processing tasks to focus on the user's information needs, depending on their current situation. In a setting where data from various (internal and external) sources are processed on a mobile device, contextual information can help to determine the importance of data in relation to user tasks and activities.

## Context Awareness in Mobile Information Systems

Context awareness in its simplest form describes a system's capability to conceive aspects of the physical and virtual environment it is operating in and tailor its behavior and responses according to the computational analysis of such aspects. Context awareness can therefore be defined as a system's capability of using contextual data for providing relevant information and services with respect to the current situation of the user (Dey, 2001). A system can be denoted as "context-aware" when it is able to adapt its behavior to ongoing activities, as well as the operational environment where it is used in (Schilit & Theimer, 1994; Abowd, Dey, Brown, Davies, Smith, & Steggles, 1999). Context-aware applications and systems are able to react according those changes in an intelligent and user-related manner (Anagnostopoulos, Tsounis, & Hadjiefthymiades, 2007).

This can be accomplished by sensing and interpreting changing conditions, resources, and processes (Dey, 2000; Fournier, Mokhtar, Georgantas, & Issarny, 2006). Context awareness can also be thought of as the machine-equivalent to the human capability of judging a situation and taking appropriate actions (Kim et al., 2004). From a technical viewpoint, context awareness refers to the accurate extraction, combination, and interpretation of contextual information, gathered from various multi-modal sensors (Biegel & Cahill, 2004), and the objective of context-aware computing lies in the identification of the set of relevant features that describe and represent a given situation with the greatest possible accuracy (Schmidt, Beigl, & Gellersen, 1998).

Research in the domain of mobile computing (cf. Kaenampornpan & Ay, 2004) has attempted to use context awareness for overcoming the technical limitation imposed by current mobile devices in terms of small screen sizes and limited interaction possibilities. Users often are confronted with multiple simultaneous activities and information channels, where context-aware computing promises to improve user interaction by reducing explicit user inputs and attention (Kaenampornpan & Ay, 2004). Context awareness can be therefore considered as a methodology for facilitating human computer interaction by lowering explicit cognitive load and user attention.

Another field where context awareness can increase the quality of information management is proactive information provisioning. One can can rightfully expect that an information system should be capable of providing information relevant to the user's current task. However, a system that is capable of doing this without the need for the user to explicitly issue search and retrieval operations can bring significant benefit, because often users are not capable of explicitly expressing their information needs. This is especially true for mobile environments and their limited interaction possibilities. In the following section, an example scenario where a user of mobile devices is supported by such a proactive information provisioning system is presented.

## Motivating Example

John is a representative of a medium-size software company. His tasks include to regularly contact potential customers in order to create awareness for his company's most recent products, to maintain relations with already existing customers in order to ensure their support and maintenance plans still work for them, and to represent the company at exhibitions, industry conferences, and relevant meetings.

His company maintains a customer relationship management software, a product database, a shared calendar system, an company-internal Wiki system as collaboration platform, and a shared file server to store all kinds of documents. Because of his job, John is often required to travel to abroad places. In consequence, he heavily relies on mobile infrastructure to get his work done. He has a powerful laptop, which he uses as his primary working device, as well as a mobile phone, which is used as his personal information management device.

When he is on travel, it is crucial for him to be equipped with all relevant information for his business meetings and other activities. However, he can never be sure to have online access to his company's network from wherever he goes, since certain limitations are in place: missing network coverage or security restrictions may prevent him from establishing a connection via the cellular network, and even if he manages to setup a connection, it may be slow and unreliable. For this reason, John often relies on local replicas of relevant data, which he stores on his laptop and (to a far lesser extent) on his mobile phone. However, because of the limited storage capacity of these devices and the necessary infrastructure, he cannot synchronize all data from all systems mentioned before, so he has to carefully select subsets of these data, which is a tedious and error-prone task.

This selection needs to be done before each trip, since he needs different information every time: this includes data about the (potential) customers he is going to meet (this includes organizations as well as persons), the locations and venues he is going to (including points of interest to visit in his spare time), latest information about the products he is trying to sell (which requires close cooperation with his company's product managers and development department), and data needed for his trip planning and administration (including timetables and travel accounting information).

A system that would be able to automatically select data for replication from a variety of systems would be highly desirable for John, since it would save him several hours of preparation time before each longer trip. Such a system could make use of a number of data sources, which provide valuable hints about which data could be of importance during his trip. First, John organizes all his upcoming appointments and travel plans in his digital calendar, which contains dates, locations, and participants of meetings. Additional information about people and organizations can be found in John's personal address book, as well as in the company's customer relationship management system. There, references to products that customers will use are mentioned; these refer to entries in the product database.

Additionally, the system could infer potential selling options from the interest topics that are stored for leads and potentials. Further, it can lookup information about locations and points of interests that John will visit from external public data sources, e.g., the Linked Open Data cloud. Further, it can find (via keyword lookups) articles from the company-internal Wiki system and the shared file server and replicate all these data to John's both mobile devices. For this purpose the system could rank each information item according to its assumed relevance, and replicate data according to the mobile devices' capacities.

During his trip, John will update and extend the replicated data with upcoming information (e.g., contact data and interests of new potential customers). Whenever his devices have sufficient network connection to his company network, the system should automatically synchronize his devices, upload changes, and update his local replicas according to possible changes in his context. After he has returned from the trip, all information is synchronized back to their origin systems, ensuring that no data are lost. If the system is able to track John's actual usage of replicated information during the trip, it can utilize this implicit feedback to adjust its relevance ranking algorithms, and therefore improve the selection for his next trip.

This scenario shows the big amount of effort that is carried out by knowledge workers repeatedly, in order to ensure they are equipped with all information they need to fulfill their work tasks. It is known that knowledge workers (which can make up to 35% of all company employees in developed countries) loose between 15% and 25% of their work time in unproductive activities, like searching and finding information and relevant documents (Kearney, 2001), and 7.5% of documents within an enterprise information ecosystem are routinely lost (Bergman, 2005). A system that is able to reduce these percentages by proactively providing the user with potentially useful information will generate a substantial business benefit, especially in terms of work time required for information acquisition and organization.

## BACKGROUND

This section is intended to provide fundamental background knowledge about technologies and concepts that are necessary for building a Semantic Web-based context-management and processing infrastructure for mobile devices. Authors provide an introduction to the concepts of context and context awareness and ascertain the two main streams wherein context is either considered a representational issue reflecting environmental aspects, and as an emergent phenomenon that is continuously re-negotiated between communicating partners and thus cannot be determined beforehand, especially not at the design time of a mobile system. Authors also show, how such dynamically evolving contexts can be represented and processed using technologies and concepts from the Semantic Web, while maintinaing its unpredictable and dynamic characteristics—a requirement neglected by most context-aware computing approaches (Teo, 2008). The section also gives a brief introduction to the general idea of the Semantic Web, its main constituents and involving technologies, as well as its most prominent knowledge representation languages. It concludes with a discussion of possible areas where context processing and management can be substantially enhanced by the deployment of Semantic Web technologies, leading to an approach that authors denote as *Semantic Web-enhanced Context-aware Computing*.

## Context and Context Awareness

The notions of context and context awareness have been subject to controversial discussion and differing perception across communities. Several definitions have been proposed to context, depending on its actual usage as well as on the domain in which it was utilized. The word context is derived from the Latin word *con*, which means 'together' or 'with', and *texere*, which is the present infinitive of *'texö'*, meaning 'to weave' or 'intertwine'. A well-known and widely-used definition of context has been proposed by Abowd and his colleagues (1999) which define context as follows:

*Context is any information that can be used to characterize the situation of an entity. An entity is a person, place, or object that is considered relevant to the interaction between a user and an application, including the user and applications themselves.*

This definition describes context as a set of situations and actions (cf. Anagnostopoulos, Tsounis, & Hadjiefthymiades, 2007) that are subject of dynamic and frequent changes including the states of the involved entities (contextual information fragments). Bolchini, Curino, Quintarelli, Schreiber, and Tanca (2007) define context as an abstract process rather than a profile that determines how humans interweave their experiences with the environment surrounding them to give meaning to something. Other authors (e.g., Dourish, 2004; Euzenat, Pierson, & Ramparany, 2008) highlight that context is never universal in that it encompasses all information constituting a specific situation but always defined relative to a concrete situation. Coutaz, Crowley, Dobson, and Garlan (2005) characterize context as *"not simply a state of a predefined environment with a fixed set of interaction resources"* rather than part of an interaction process with the polymorphic and varied environment that is composed of *"reconfigurable, migratory, distributed, and multi-scale resources"*. Schilit and Theimer (1994) refer to context as an entity that comprises location, identity, objects, as well as changes that apply to those objects. Ryan, Pascoe, and Morse (1998) add the dimension of environment and time whereas Dey, Abowd, Pinkerton, and Wood (1997) define context relative to the user's emotional states, their believes, intentions, and concers, taking into account the people in the closest proximity of the user. Context in general can be defined as *"everything that surrounds a user or device and gives meaning to something"* (Teo, 2008).

However, due to the different proliferations on the notion of context, there has not been a clear consensus established on what context exactly is (Dourish, 2004), but there is a common agreement on what it is about: Context is concerned with an evolving, structured, and shared information space that is designed and utilized to serve a particular purpose (Coutaz et al., 2005).

Contextual information, which is a constituent part of surround contexts, can be considered as any information that may be used for describing the situation the user or a device is currently operating in. Contextual information is gathered through a variety of different technologies and considered as computational abstractions over distinguishable virtual or real-word aspects that have a specific relationship to the current task at hand. Contextual information can be static (e.g., the date of a person's birthday) or dynamic (e.g., a person's location). Dynamic context information is usually captured indirectly using sensors. Additional characteristics for classifying contextual information have been proposed by Henricksen, Indulska and Rakotonirainy (2002).

Contextual information can be distinguished according to the way they are acquired: (1) *Explicitly* acquired contextual information is manually specified by the user and refers to information such as established social relationships or fields of interest. (2) *Implicit* contextual information is acquired via communication with hardware or software sensors, which capture specific aspects of the surrounding context by using sensing technologies or by monitoring user and system behavior. Most context frameworks acquire contextual information implicitly, especially from locally deployed physical sensors or embedded ubiquitous sensors. The challenge thereby is to identify the set of features that describe a given situation with the greatest possible accuracy and relevance (Biegel & Cahill, 2004). Deriving reliable information from multiple heterogeneous sources in uncertain and rapidly changing environments is mandatory for context awareness in the domain of mobile computing (Korpipaa, Mantyjarvi, Kela, Keranen, & Malm, 2003).

Consequently, two forms of context awareness can be found in information systems (Gellersen, Schmidt, & Beigl, 2002): *Direct awareness* shifts the process of context acquisition onto the device itself, usually by embodying sensors that autonomously obtain contextual information; e.g., location ascertainment using the device-internal GPS sensor. *Indirect awareness*, in contrast, captures contextual information by communicating with sensors or services via the surrounding environment or infrastructure. For instance, to capture the social context of a user, a mobile device may request data from social communities or portals; to track the user's location, a remote geocoding service (based on the user's IP address) may be employed.

## Positivist and Epistemological View on Context

The technical or positivist school treats context as a conceptualization of human action and their

interaction with the system. As a consequence, context is considered as a set of features of the environment surrounding the user's tasks at hand, which can be computationally captured, represented, and processed. Especially technical disciplines consider context a representational issue and concentrate on sensorial or static data such as location, time, identity etc. (cf. Schmidt et al., 1998) putting emphasize on its codification and representation (Mihalic & Tscheligi, 2007; Teo, 2008). Following this argumentation, context is instance-independent, separable from user activities, and can be scoped in advance (Dourish, 2004). This form of perceiving context has its root in the information systems discipline since it adheres very well to existing software methodologies (Teo, 2008) but is contradictory to the phenomenological view of context that is grounded on subjective and qualitative analysis. This view considers context and activity inextricably connected and fundamental in giving meaning to something and emphasizes the aspect of relevance since context as such can not be defined or determined in advance as its scope is dynamically defined and changes frequently and unpredictably (Dourish, 2004). It should therefore be considered as a *relational* or *occasioned property* to emphasize its dynamic, fluent, and relative character as context arises in the course of action (Dourish, 2004):

*Context isn't just "there", but is actively produced, maintained and enacted in the course of the activity at hand. [...] Context isn't something that describes a setting; it's something that people do. It is an achievement, rather than an observation; an outcome, rather than a premise.*

The epistemological view considers context as an interactional feature inspired by "sociological investigations of real-world practices" (Dourish, 2004). Context is inextricably linked to the process in which it is conceived, which renders the positivist and phenomenological view on context incompatible. As Widjaja and Balbo (2005) point out:

*Action and context are inseparable and should be analyzed as a whole. This implies that context-awareness should be examined in its relationship and consequences to the supported activity or actions.*

As a consequence, the dynamic aspects of context are entirely neglected by technically oriented disciplines, as context is a constitutional part of interaction processes that emerge opportunistically. Context should be rather considered as an emergent phenomenon or feature of interaction that is inextricably linked with user activities (Widjaja & Balbo, 2005; Teo, 2008) and continuously renegotiated between communicating partners (Dourish, 2004; Coutaz et al., 2005; Mihalic & Tscheligi, 2007) wherefore a pre-determination of the relevance of contextual elements is impossible – especially at design time of a system (Euzenat et al., 2008).

Therefore, the context descriptions exposed by a context processing and management framework must be flexible, extensible, and use open vocabularies in order to facilitate the dynamic and emergent nature of context and should not be restricted to static schemas or single vocabularies. Static context descriptions are not able to deal with unknown context information at run time, but require links between different context vocabularies to be specified at design time (Euzenat et al., 2008). The ability to dynamically handle and integrate new types of context information into existing structures is therefore a fundamental requirement of a context framework where open and well-accepted vocabularies help in describing contextual information to guarantee their evolution and accurateness.

## Context and Context Awareness in Information Systems

The notion of context is mainly used in the information systems discipline for two reasons (cf. Dourish, 2004): (1) contextual information is encoded to increase the accuracy of information retrieval processes and (2) contextual information is utilized for adopting systems to the environments in which they are used. In mobile computing, in contrast, context is used for (1) intelligent service provision, (2) realizing adaptive user interfaces, and (3) increasing the accuracy of information retrieval processes since context-aware information systems in general provide a "more natural and less obtrusive way of interaction" (Mihalic & Tscheligi, 2007). This is achieved by filtering the flow of information (i) from the device to the user to decrease information overload as well as (ii) from the user to the device, where user-generated data is augmented with contextual information predominantly in an automated and transparent manner. A variety of research endeavors elaborated on the nature of context and proposed a multitude of—mostly taxonomical—classification schemes to manage and comprehend such information systematically (Bradley & Dunlop, 2005). A context model in general is used for the definition and storage of contextual data in a machine-processable form. The most prominent approaches were summarized by Strang and Popien (2004), ranging from simple key-value models to complex logic or ontology-based models.

The diversity of contextual information is also reflected in the architectures proposed for context management and processing, which differ in their technical capabilities, acquisition techniques, context representations and reasoning capabilities etc. Despite the wide variety of different context management and processing architectures, a common architecture is identifiable, which is depicted in Figure 1.

*Sensors Layer*: The Sensors Layer is the bottom most layer and comprises the set of sensors that a context framework exploits. Such sensors can be classified into three groups (Indulska & Sutton, 2003): (i) physical or hardware sensors for gathering physical context information, (ii) virtual sensors that capture data from software applications and services by observing user interaction and user behavior, and (iii) logical sensors that are responsible for deriving higher-level context information usually by augmenting physically or virtually acquired contexts with additional information from repositories or other data sources.

*Raw-data Retrieval Layer*: This layer is concerned with the retrieval of raw sensorial data usually by encapsulating driver logic or sensor application programming interfaces (APIs) in dedicated wrapper components (cf. Widgets as proposed by Dey, Abowd, and Salber (2001)). It offers common interfaces that encapsulate specific drivers or sensor APIs thus making them exploitable for upper layer components. This layer exposes query-functionality that abstracts from low-level raw sensorial data and provides higher-level representations and functions.

*Pre-processing Layer*: In case sensorial data are too coarse (Baldauf et al., 2007) or need to be clustered for further processing, this is handled by the pre-processing layer. It also provides appropriate abstractions of contextual data whose representation is too technical (e.g. a bit-wise representation of contextual data).

Interpretation, aggregation, and reasoning tasks are also handled by this layer together with quantization algorithms for aligning raw-sensorial data with the elements of a framework's context model.

*Storage/Management Layer*: This layer serves as a context repository as it organizes contextual information and offers interfaces to client applications. Components in this layer expose different operation models where contextual information can be either requested in a polling-based style

*Figure 1. Generic conceptual architecture of Context frameworks (adapted from Baldauf et al., 2007)*

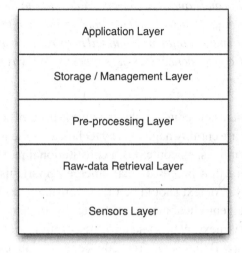

where the client applications issues requests to a context server in regular time (synchronous) or via a subscription mechanism where clients are notified by the context server whenever new contexts are available (asynchronous). Due to the dynamic and unpredictable nature of context (cf. Dourish, 2004; Coutaz et al., 2005; Teo, 2008) the use of an asynchronous communication style (cf. Xia, Yakovlev, Clark, & Shang, 2002) is suggested (Baldauf et al., 2007).

*Application Layer*: The application layer is the uppermost layer and hosts the client applications that request and process contextual information from a context repository or a framework respectively. Contextual information is usually consumed by applications in this layer in order to adopt their actions with respect to user-related tasks. One popular example is increasing the background luminosity level in case the user enters an outdoor area.

Context can be acquired from a multitude of different sources, e.g., by applying sensors or sensor networks, by deriving information from the underlying communication infrastructure or network, or by acquiring status information and user profiles directly from the device the user currently uses (Baldauf et al., 2007). In general, every piece of information or any content can be considered context-relevant if it provides information related to the user's current tasks and activities (Wojciechowski & Xiong, 2006). The way contextual data are acquired significantly influences the architectural style of a context-aware system (Baldauf et al., 2007). Context acquisition architectures can be broadly classified into three different groups (cf. Chen, 2004): (1) *proprietary architectures* that directly access locally deployed sensors where sensor and driver logic is directly implemented in application code limiting context reuse and exchange, (2) *middleware infrastructures* which employ a layered system architecture (e.g. Boehm et al., 2008; Chen & Joshi, 2004; Luther, Fukazawa, Wagner, & Kurakake, 2008) encapsulating sensor specifics in dedicated components and expose uniform interfaces for context utilization, and (3*) context server architectures* (e.g. Hofer, Schwinger, Pichler, Leonhartsberger, Altmann, & Retschitzegger, 2003) that operate similar to database management systems and offer remote access to contextual information hosted within a context repository. The advantages and limitations of each architectural style have been summarized by Chen (2004) and Baldauf et al. (2007).

Different classification schemes for context management architectures have been proposed by Dey et al. (2001), Baldauf et al. (2007) and Euzenat et al. (2008). Those works distinguish between a (1) direct integration of context management into context-aware applications, (2) context management services, and (3) context-aware devices and services augmented with context management functionalities. By directly integrating context management functionality into mobile applications, such applications need to know in advance which sensors are available and how to communicate with them. As a consequence, context data semantics are limited to the applications in which they were interpreted and can, in most cases, not

be shared between applications, which render a dynamic integration of new sensors in running frameworks difficult. Sensors have to be queried by each application separately, which impede the process of context aggregation, exchange, and dissemination. In contrast, if context management functionality is implemented in devices or sensors, their functionality can be shared among components and integrated during run-time (e.g. Biegel & Cahill, 2004; Euzenat et al., 2008).

The usage and utilization of the notion of context in information systems raises some technological as well as human-related challenges. Dey and his colleagues (2001) identified the poor understanding on what context constitutes, how it is to be represented in information systems, and the lack of conceptual models, methods, and tools that would promote the design of context-aware mobile applications as one of the main reasons why context awareness has only insufficiently found its way into the essence of mobile and ubiquitous computing yet. This additionally hampers empirical investigations in human-computer interaction and interaction design. Many works also indicated the non-existent availability of a general model of context and context awareness as one of the main problems of context-sensitive systems. This fact in particular concerns mobile computing where these concepts are used ambiguously across communities and reflect specific application domain peculiarities (cf. Bolchini et al., 2007; Raptis, Tselios, & Avouris, 2005). It also hampers context-aware application development for mobile systems since a widely accepted and well-defined programming model does not exist wherefore sensor-logics are often hard-wired into application code and application developers have to deal with low-level interactions between sensors and context acquisition components (Dey et al., 2001). Therefore, the semantics of contextual information are limited to the applications in which they were acquired and are represented using proprietary formats. Newer approaches (Biegel & Cahill, 2004; Boehm et al., 2008) employ more flexible designs for context processing and representation where sensor-logics or sensor-specific application programming interfaces (APIs) are encapsulated in specific components that can be mutually shared, or employ middleware infrastructures (Henricksen, Indulska, McFadden, & Balasubramaniam, 2005; Huebscher & McCann, 2005) for facilitating communication and interoperability between context processing components while using knowledge representation languages from the Semantic Web such as RDFS (Resource Description Framework Schema) or OWL (Web Ontology Language) for representing contextual information (e.g. Pawar, Halteren, & Sheikh, 2007; Mihalic & Tscheligi, 2007; Euzenat et al., 2008).

## Problems of Context-Aware Computing

A fundamental problem of context-aware computing is that of *context ambiguity* (Dey, 2001) and *context imperfection* (Henricksen et al., 2002), which refers to the implicit assumption shared by many context-aware computing approaches that the computational context is a 1-to-1 reflection of the real-world context. Evidently, this assumption is wrong since the way context is conceived by individuals differs substantially from the way it is acquired and represented electronically (Dey, 2001; Dourish, 2004). A logical consequence of that misperception is that a context framework can only work on a more or less accurate context representation where the degree of accuracy is determined by numerous technical and soft factors. The unpredictable and relative nature of context renders a determination of all contextual aspects that constitute a specific context at a system's design time difficult, if not impossible, since context is always defined relative to the situation in which it is used. An electronic representation of context can therefore never be universal in that a context model contains all information that characterizes a given situation; instead it only represents a relevant subset of the constituting real-world context (Dey,

2001; Henricksen et al., 2002; Dourish, 2004). The problem is that a 1- to -1 relation between a situation and the describing context information does, in most cases, not exist wherefore a situation can be represented by multiple context models with a specific degree of accuracy depending on individual viewpoints. Several methodologies such Bayesian networks, Case-based reasoning, Stochastic models, or Machine learning techniques have been proposed for defining precise transitions between different context descriptions with as little ambiguity and overlapping as possible. However, such approaches contribute towards increasing the accuracy of context acquisition and context representation but do not help in identifying all constituting aspects of a specific situation. Context-aware computing in general is only an approximation to a real-world situation rather than a 1-to-1 reflection of it.

Another problem of context-aware computing is that most architectures are targeted towards a specific application or domain (Euzenat et al., 2008). The difficulties hereby are that certain high-level context interpretations are not absolute characterizations per se, since the concept 'high temperature' for instance depends on the context or situation in which it was acquired. This dependency makes it difficult to share context information in an application and domain-independent manner since implementing new application behaviors based on context characterizations made for one application might not be appropriate for another one (Euzenat et al., 2008). Some authors such as Kaenampornpan and Ay (2004), Cai and Xue (2006) and Teo (2008) therefore focus on describing and representing context in an application independent manner by means of concepts from *activity theory* (Nardi, 1995; Kuutti, 1995) using collaborative plans (Grosz & Kraus, 1996), *task analysis* (Redish & Wixon, 2003; Diaper, 1990), *aspect-oriented context modeling* and *modularization* (Carton, Clarke, Senart, & Cahill, 2007) or *situational reasoning* (Boehm et al., 2008; Luther, Fukazawa, Wagner, & Kurakake,

*Figure 2. Example of RDF graph*

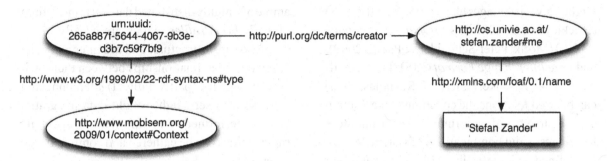

2008) to decouple context from specific application domains and provide abstract contextual concepts (e.g., "business meeting") that adhere to upper-level ontologies. Context and contextual information need a uniform representation to be effectively managed, integrated, and processed by reducing the ambiguities inherently attached (Fournier et al., 2006).

## Semantic Web

The *Semantic Web* (Berners-Lee, Hendler, & Lassila, 2001) is the idea of expressing rich, machine-processable knowledge using the Web infrastructure. Descriptions about *resources* (which are entities of any kind, including digital objects like documents and media, physical objects like humans, cars, or buildings, and abstract concepts like locations, topics, and time periods) are published in a structured format and using vocabularies that follow a well-defined semantics. Applications can consume these descriptions, interpret them, merge them with descriptions from other sources, and infer new knowledge or determine the truth-value of statements. In analogy to the World Wide Web, which nowadays serves as one underlying knowledge-provisioning infrastructure for a wide variety of human knowledge workers, the Semantic Web is envisioned to serve as an underlying information-provisioning infrastructure for information-centric applications,

whereas the information processing is performed semi-automatically by computer programs.

Broadly, the Semantic Web consists of a stack of technologies that build on each other. The core technologies are shared with the World Wide Web: *Uniform Resource Identifiers* (URIs) (Berners-Lee, Fielding, & Masinter, 2005) to identify resources, and *HyperText Transfer Protocol* (HTTP) (Fielding et al., 1999) for the transportation of information. On top of these core technologies, the *Resource description framework* (RDF) (Klyne & Carroll, 2004) is used as the abstract data model in which all information is represented.

RDF is a triple-based graph model, with the *statement* being the atomic unit of information. Each statement consists of three elements (*subject*, *predicate*, and *object*), where the predicate identifies the relationship that is asserted to exist between the subject and the object. URIs can be used for all three elements; whenever the same URI is used in multiple statements these statements are referring to the same resource (or real-world entity). Therefore, a set of RDF statements that shares common URIs can be interpreted as connected graph (cf. Figure 2). RDF itself is an abstract data model; in turn, there exist several serialization formats that can be used to exchange RDF statements between parties, including RDF/XML (Becket, 2004), Turtle (Beckett & Berners-Lee, 2008), and RDF/JSON.

Based on RDF a set of technologies has been developed that aims to make more "knowledge" out of the data represented in RDF graphs. Higher-level languages like *RDF Schema* (RDFS) (Brickley & Guha, 2004), *Web Ontology Language* (OWL) (Motik, Patel-Schneider, & Parsia, 2009), and *Rule Interchange Format* (RIF) (Boley, Hallmark, Kifer, Paschke, Polleres, & Reynolds, 2010) can be used to define the constraints of a domain of discourse based on formal logics; using these languages, valid combinations of statements can be defined axiomatically. This allows implicit knowledge to be discovered based on asserted information, to detect inconsistencies in knowledge bases, or to determine the truth-value of statements, given a set of background knowledge. A query language (SPARQL) (Prud'hommeaux & Seaborne, 2008), which resembles similarity to the SQL language for relational data, can be used to formulate structured information needs, which are evaluated against a set of RDF graphs.

One line of development within the ongoing Semantic Web research field is *Linked Data* (Bizer, Heath, & Berners-Lee, 2009), which denotes the practice of publishing data on the Web according to simple core principles (Bizer et al., 2009). Its core idea is to denote resources (which includes real-world entities as described above) exclusively using HTTP URIs, and to allow data consumers to directly de-reference these URIs (i.e., to fetch their representations using HTTP GET methods). Upon this de-referencing, structured information about the resources is returned, which contains links to other relevant resources. These other resources can then, in turn, be retrieved by the client, allowing it to navigate through a global information network based on the "follow-your-nose" principle.

Since 2007, when the Linked Data W3C community project[1] was established, a significant amount of data has been published according to the Linked Data principles. This includes popular data sets of general interest like *DBpedia* (consisting of structured information extracted from Wikipedia pages), geographic information (like *Geonames*), media-related content like *BBC Programmes*, and

bibliographic information (e.g., *DBLP*). An example of a highly distributed data set is the entirety of all *FOAF Profiles*, which are usually served on private infrastructure, and are interconnected based on social relationships between their owners. Through the (partly indirect) interconnection of these data sets, light-weight data integration can be performed, and information about the same entities can be gathered and combined from heterogeneous, distributed sources.

## Semantic Web-Enhanced Context-Aware Computing

For managing context information systematically, a common structure for representing contextual information needs to be established (Korpipaa et al., 2003). The Resource description framework, discussed in the previous section, has proven to be an appropriate representation framework for representing complex contextual constellations and facilitating the sharing and exchange of context descriptions based on ontological semantics (Zander & Schandl, 2010). It can be used for codifying the semantics of contextual information as well as the relationships among them in a well defined, uniform, and systematic way. On top of RDF, RDF Schema (RDFS) offers a simple set of common language properties that can be used for building context descriptions which can be shared among different context providers and consumers collaboratively (Korpipaa et al., 2003). However, the set of RDFS language elements is not sufficient for expressing rich contextual constellations, wherefore the use of more expressive languages such as OWL is suggested (Mihalic & Tscheligi, 2007; Euzenat et al., 2008). Generally, the use of ontologies as a key component for building a context-aware computing framework is broadly acknowledged (Luther et al., 2008; Preuveneers et al., 2004; Euzenat et al., 2008; Korpipaa, Malm, Salminen, Rantakokko, Kyllonen, & Kansala, 2003), and it has been shown that Semantic Web technologies are sufficiently mature and performant to be deployed on mobile devices (Zander & Schandl, under review).

A context ontology serves as a uniform representation of contextual information and enables a systematic management of context-relevant aspects; it should be separated from application logic (Korpipaa et al., 2005). A number of approaches use single ontologies for context information representation and for the transformation of raw, low-level context data into high-level context descriptions (Ramparany, Euzenat, Broens, Bottaro, & Poortinga, 2006; Korpipaa et al., 2005). Some of these ontologies refer to the analogy of physical objects, i.e., their concepts refer to objects in the real world (the studied context).

Several works have already demonstrated that ontologies are appropriate means for expressing and representing contextual information since they incorporate some characteristics that are essential for mobile and ubiquitous environments (Euzenat et al., 2008; Chen & Joshi, 2004; Heath, Dzbor, & Motta, 2005; Heath, Motta, & Dzbor, 2005). Ontologies are highly expressive and widely adopted knowledge representation techniques, and a multitude of open software tools for their design, creation, management, and storage are available (Pawar et al., 2007). Ontologies offer a well-defined set of concepts and relationships to model the domain of interest, which can be adopted by context management frameworks to integrate and share contextual knowledge from other domains to facilitate context exchange and reuse. Ontologies are based on knowledge representation languages that are open with respect to evolutions of the domain they describe. This allows ontologies to be adapted and extended according to domain-internal changes.

Building a context awareness computing infrastructure on the principles and technologies of the Semantic Web has several implications and advantages: due to the fact that ontologies are based on the open world assumption, context ontology evolution is a central aspect in context management and allows for adapting and modifying a context ontology according to changed conditions. Due to the fact that RDF is a system- and application-independent framework for modeling data and its close relatedness to Web technologies, it is well suited for the data exchange between components (e.g., using HTTP). This facilitates interoperability among context frameworks and services since established and well-known vocabularies together with their inherent semantics can be understood and used across systems.

Since ontologies provide a common structure for representing and describing the relationships and semantics of context-relevant information in a machine-processable way, they can be conceived as a general approach for systematic management of context information. In such a setting, RDF can be used as description syntax to enable the communication and sharing of context information between collaboratively communicating partners, i.e., applications, services, and devices. These descriptions are represented as *labeled multi graphs*, where the contained entities are referred to through HTTP URIs. Its open architecture allows for the integration of different context-relevant vocabularies so that context descriptions can dynamically grow and become more elaborated to better reflect intra-domain evolutions.

Ontologies help in expressing application or service needs, and in aligning them to acquired context information wherefore only relevant information is extracted. This simplifies query processing since a context consumer can limit queries to relevant information, instead of processing the entire context description. In cases of incompatibility of context descriptions, ontology matching algorithms help in reconciling differences in description semantics. Euzenat (2005) therefore suggests the use of *ontology alignment services* to identify correspondences between incompatible context descriptions. Such services perform query transformations and reflect domain and information space evolutions (Euzenat et al., 2008).

Semantic Web technologies allow for mapping low-level sensor data to high-level ontological concepts so that collected context-relevant information is transformed and embedded in a controlled context description. Based on ontological semantics, new facts can be deducted by applying aggregation and reasoning heuristics. In this way, Semantic Web languages such as RDFS and OWL allow for aggregating heterogeneous and autonomously acquired context information both on a syntactic and semantic layer. By transforming sensorial data into RDF statements, context acquisition components are not required to anticipate possible queries beforehand. Instead, the requesting context consumers determine the data that are relevant for them.

In the fields of Semantic Web and Linked Data, a number of vocabularies and ontologies have emerged that are of interest for the representation of contextual and situational information (e.g., time[2] and location[3], technical parameters[4], or social aspects[5]). The elements (terms and concepts) of those vocabularies are well known across communities and expose a well-defined and commonly understood semantics that allows for information integration and exchanges especially in heterogeneous system and network infrastructures. Additionally, such vocabularies are continuously maintained by communities to guarantee their accurateness and evolution. By re-using such vocabularies and (implicitly) connecting context descriptions to external Linked Data sources, we gain two benefits: first, if context descriptions are distributed (either to the public or within a closed environment, e.g., a corporate network) they can be directly combined with already existing data, and existing tools can be directly applied to contextual information without the need to adapt existing software. Second, data from external sources can be imported and used to enrich the context descriptions, leading to a richer semantics, which facilitates more powerful processing and reasoning.

# RELATED WORK

## DBpedia Mobile

*DBpedia Mobile* is a location-centric mobile client application that visualizes data sets from the *DBpedia project* (Auer, Bizer, Kobilarov, Lehmann, & Zachary, 2007; Lehmann et al., 2009) in a Fresnel-based Linked Data browser (Pietriga, Bizer, Karger, & Lee, 2006) based on the user's current position (Becker & Bizer, 2008, 2009a). DBpedia is a community-driven effort for extracting structured information from the Wikipedia project and exposing this information as RDF data on the Web under the GNU Free Documentation License for both machine and human consumption through a set of application programming and query interfaces. An information extraction framework (cf. Auer et al., 2007) converts Wikipedia content to RDF and represents it as a large multi-domain RDF graph, which can be utilized by Semantic Web applications. Furthermore, such data sets can be linked to other data sets exposed on the Web to build a large network of interlinked data sources – the so-called *Web of Data* (Bizer, Heath, & Berners-Lee, 2009).

The DBpedia Mobile client takes the GPS coordinates retrieved from the device's GPS sensor and requests data from the DBpedia project that represent information about objects (POIs) located in the user's immediate vicinity. Those displayed data sets serve as a starting point for exploring related data that are interlinked with the displayed resources. In this respect, DBpedia Mobile serves as a starting point for exploring the *Geospatial Semantic Web* (Egenhofer, 2002). When a resource is selected, the DBpedia Mobile server retrieves, aggregates, and caches related information from interlinked data sources (e.g. reviews about the selected resource extracted from the *Revyu service* (Heath & Motta, 2008)) before they are sent to the mobile device. In case the selected resource exposes RDF links to other resource (cf. Volz, Bizer, Gaedke, & Kobilarov, 2009), the user is able to navigate to and browse related Linked Data source (Becker & Bizer, 2009a).

DBpedia Mobile is realized as a Java Script application and requires a Document Object Model (DOM) Level 1 and 2 capable browser to make use of the *Google Maps API*. RDF data is not processed directly on a device; instead, data such as the currently visible view area and filter settings are sent to the DBpedia Mobile server that features the *Marbles engine* (Becker & Bizer, 2009b) and uses the *Sesame RDF framework* (Broekstra, Kampman, & Harmelen, 2002), which then transforms those data into SPARQL queries (Prud'hommeaux & Seaborne, 2008). Displayed information as well as related resources are retrieved by dereferencing resource URIs (Becker & Bizer, 2009a) and hosted within an instance of the *Virtuoso triple store* (Erling & Mikhailov, 2007) to which such SPARQL queries are issued. To retrieve additional information about interlinked resources, DBpedia Mobile makes use of the *Sindice* (Oren, Delbru, Catasta, Cyganiak, Stenzhorn, & Tummarello, 2008) and *Falcons* (Cheng & Qu, 2009) Semantic Web search engines.

DBpedia Mobile also supports simple and SPARQL-based information filtering as well as publishing location-related information such as pictures or reviews to the Web of Data, which is then interlinked with related DBpedia resources (Becker & Bizer, 2009a). Therefore, each user that registers for the DBpedia Mobile client is provided with *a personal resource URI* that is used for all content contributions of the respective user. Before a view is generated, the server dereferences interlinked resource URIs and retrieves additional data from the Web of Data using the previously mentioned Sindice and Falcons Semantic Web search engines, performs some form of data augmentation, and stores the aggregated data as *NamedGraphs* (Carroll, Bizer, Hayes, & Stickler, 2005) in the Virtuoso triple store (Becker & Bizer, 2009a).

## mSpace Mobile

*mSpace Mobile* is a mobile Semantic Web application developed within the *mSpace-project*[6] that uses a multi-faceted column-based browser for exploring data sets that have a direct or indirect relation to the user's current location. mSpace Mobile offers related information about the user's physical environment such as nearby points of interests, amenities, etc. Considered contexts are *time*, *space*, and *subject*. For instance, users are able to receive additional information about the movies currently playing at nearby cinemas.

mSpace Mobile is built upon the mSpace Software Framework (Schraefel et al., 2005) designed for the management and exploitation of distributed semantically related resources. The basic design principle behind the mSpace framework is to offer users a software tool that allows for exploring a multi-dimensional information space in multiple ways by leveraging protocols and languages from the Semantic Web accounting for its scalability as a distributed data source. Users are able to navigate along associated items in pre-defined contextual dimensions that have a particular relationship to a selected subject.

In contrast to traditional location-based information systems that use single and proprietary data sources, mSpace Mobile exploits data from freely available semantic data sources that publish information by using RDF such as the Open Guide to London (Wilson, Russell, Smith, Owens, & Schraefel, 2005). mSpace Mobile further supports context transitions, i.e., shifting the focus between contextual entities where data retrieval tasks are dynamically adjusted according to the selected (context-relevant) information item. This is a significant difference since most comparable applications merely consider location as the main context. In this respect, the information item the user is interested in becomes the new context.

Although, mSpace Mobile is designed for mobile usage, it employs a distributed client-server-based architecture where the server-side abstracts over multiple triple stores and is responsible for dereferencing and integrating resources (Wilson et al., 2005). mSpace Mobile employs a three-layered architecture: (i) the *mSpace Application layer* hosts functionalities and components for building mobile client applications on top of the mSpace Framework and for generating the queries issued to the mSpace Query Server. The clients themselves are separated from the query generation and translation steps. (ii) The *mSpace Query Server* offers query services that can be utilized by mobile applications in order to query for context-relevant resources. The Query Server uses SOAP, HTTP, and.NET Web Service technologies for its communication with the client applications as well as with the mSpace Knowledge Server. (iii) The *mSpace Knowledge Server* abstracts over a configureable set of RDF repositories and provides facilities for building links between resources residing in different repositories. For this purpose, the Knowledge Server makes use of the RDQL (Seaborne, 2004) query language for querying RDF data sources. However, RDQL is officially superseded by SPARQL as the de facto standard query language for RDF data. The default data repository is the *3store triple store* (Harris & Gibbins, 2003). The Knowledge Server further allows for combining inherently isolated data sources. In this respect, we can observe an analogy to the Linked Data approach (cf. Bizer, 2009; Bizer et al., 2009) whose objective is to establish semantically meaningful links between RDF resources by using well-known Semantic Web concepts such as assigning unique URIs to resources in order to make them identifyable and thus dereferenceable (Lewis, 2007). The Knowledge Server thus exposes a WWW approach for data provisioning in that it consists of a variable amount of data providers, each being controlled separately, that scale with the availability of potential sources.

## IYOUIT

A rather promising project that aims to combine Context-aware functionality and Semantic Web technologies for mobile information systems is IYOUIT, which was developed by DoCoMo Euro-Labs Munich in cooperation with the Telematica Instituut Enschede. IYOUIT describes itself as a *"[...] prototype service to pioneer a context-aware mobile digital lifestyle and its reflection on the Web"* (Boehm et al., 2008). It is built on a distributed infrastructure incorporating semantic technologies and languages to allow for a qualitative interpretation and evaluation of user activities, which are acquired locally through the mobile device, processed on a server, and reflected in a community portal on the Web. Those activities are captured by quantitative sensors and mapped to qualitative data abstractions using formal ontologies. Formalized domain knowledge together with classification and ontology-based reasoning mechanisms are used to support the process of deriving meaningful interpretations of gathered raw sensor data as well as the recognition of behavioral patterns (cf. Luther, Mrohs, Wagner, & Kellerer, 2005; Luther et al., 2008).

The idea of IYOUIT is that users can establish relationships among each other and hence building social networks for sharing context data through the IYOUIT Web portal. Those social networks are represented as OWL ontologies so that Description Logic-based reasoning (cf. Baader, Calvanese, McGuinness, Nardi, & Patel-Schneider, 2003) can be applied for (i) detecting inconsistencies and contradictions in social network data to maintain data accuracy, and (ii) deducing, i.e., revealing implicit relationships between users in a social network.

IYOUIT is designed as a service consisting of a mobile client application and a server infrastructure, the *Context Management Framework* (cf. Boehm et al., 2008). The mobile client application was developed for the Nokia Series-60 devices and automatically captures and collects information about the entities surrounding a user – the user's context – in order to share personal experiences. Such information is, for instance, the places one visited, the people met, or personal information such as recently read books. IYOUIT uses the sensors deployed locally on a mobile phone to collect such information automatically and send it to the Context Management Framework server onto which such data are interpreted, aggregated, further processed, and stored. IYOUIT also employs interface to Web 2.0 applications such as Flickr and Twitter to collect personal information and sharing it online.

The context management framework represents a layered architecture and network of distributed and interconnected components for collecting, managing, and distributing context information proactively (Boehm et al., 2008): It allows for the implementation of flexible services that track, for instance, the position of "friends", identify frequently visited places, collect information about publicly available WLAN hotspots, local weather information, photos, and reflects this information (so-called *context streams*) on the Web portal as well as on mobile clients (Boehm et al., 2008). Its objective is to transform quantitative raw context data (e.g., sensor outputs) into qualitative, human-interpretable statements reflecting the user's current situation by context aggregation, combination, and reasoning (Boehm et al., 2008).

The framework includes a *Privacy Manager* for controlling the distribution of sensitive personal information, an *Identity Manager* for connecting and authentication to 3rd party applications such as Flickr or Twitter, a *Relation Manager* being responsible for reason about the social networks of users, and an *Ontology Manager* for utilizing domain-specific knowledge being formalized in core ontologies. It further employs the concept of *Context Providers* representing components that wrap a specific context source and contain aggregation heuristics for abstracting over low-level quantitative data. An overview of the constituting

components is given in Boehm et al. (2008) and Boehm, Koolwaaij, and Luther (2008).

A set of core context ontologies based on the Web Ontology Language (OWL) (McGuinness & Harmelen, 2004) has been developed for context clustering and context data transformations, which are utilized by context providers. Those context ontologies are exclusively used for high-level context elements due to the fact that ontologies are not well suited for handling large amounts of data (Weithöner, Liebig, Luther, Bohm, Henke, & Noppens, 2007). Reasoning is performed to make implicit knowledge explicit where each relationship has an additional attribute that indicates whether a relation has been explicitly asserted or inferred (Boehm et al., 2008).

## ContextTorrent

A similar system to the one presented in this chapter has been developed within the ContextTorrent-project (Hu, Dong, & Wang, 2009). ContextTorrent is a semantic context management framework that offers access to semantically represented context information for local and remote context-aware applications. It is developed for the Google Android platform and makes use of the mobile Java-based XML parser *NanoXML*. ContextTorrent was inspired by semantic desktop research in which concepts and technologies from the Semantic Web are used to enhance personal information management (cf. Boardman & Sasse, 2004) by providing context-relevant information in an automatic and proactive fashion to support the user's long-term memory (cf. Sauermann, Bernardi, & Dengel, 2005; Franz, Ansgar, & Staab, 2009).

ContextTorrent offers a controlled interface for the exploitation and utilization of semantically represented context data where a mobile device takes the role of a context provider as well as a context consumer. An *overlay peer-to-peer network* allows for connecting mobile devices and mobile applications for largescaled local or remote context query and provision (Hu et al.,

2009). The underlying infrastructure allows for building dynamically established semantic links between related context fragments. An *ontology-based semantic modeler* represents contextual data as RDF resources using an adapted OWL/RDF parser and maintains the links between semantically related context fragments in a dynamic fashion. Those semantic relationships are stored in an *object-oriented database* specifically designed for resource-constrained mobile devices which exposes minimal overheads and better resembles ontological representations and corresponding schema evolutions (Hu et al., 2009).

As opposed to other context management frameworks which distinguish between low-level and high-level or inferred versus aggregated contexts, the management infrastructure employed by ContextTorrent treats all contexts equally regardless of their type or origin, where context-relevant information is considered a semantic resource irrespectively whether the information stem from an external source, a sensor, or the internal context repository. A *N-gram based matching algorithm* (Miller, Shen, Liu, & Nicholas, 2000) is used to rank context information according to its relevance to issued context queries.

ContextTorrent uses the Web Ontology Language (OWL) (McGuinness & Harmelen, 2004) for describing the relationships among context entities, which are represented as Semantic Web resources that have a unique Uniform Resource Identifier (URI) assigned to make them distinguishable and referable (cf. Sauermann & Cyganiak, 2008). ContextTorrent further allows for the dynamic binding of shared context data to applications by making use of the Android's concepts of Intents and Intent Filters in order to enable context sharing between applications at runtime (cf. Meier, 2010). ContextTorrent distinguishes between static and dynamic contexts where the classification is not based on the type of context data but rather on the frequency of context value changes.

For processing RDF and OWL data, the open-source Java-based XML parser NanoXML has been ported to the Android Dalvaik Virtual Machine (DVM). Although NanoXML is a lightweight and fast RDF/XML parser, it lacks sophisticated RDF processing and management capabilities compared to other mobile RDF frameworks such as Androjena, µJena, or MobileRDF (cf. Zander & Schandl, 2010).

## Discussion

Current context-aware applications in the Semantic Web domain treat context as static resources and neglect to consider its inherently dynamic nature, which should not only be reflected through context property value updates but also in what constitutes a specific context model. Therefore, a context model as well as its ontological schemas should be able to dynamically grow and become more elaborated. Although these projects make use of Semantic Web technologies such as RDFS and OWL, they employ client-server architectures for context processing, management, and storage and require a consistent and stable network connection to be fully operational. However, in case of network failures or a loss in network connectivity those applications become practically useless. On the contrary, ContextTorrent as well as the context framework proposed in this chapter allow for processing contextual information directly on a mobile device where no transfers outside the system are necessary for context aggregation, reasoning, and storage. Another distinct aspect is that context acquisition and context representation is not limited to a predefined set of contextual aspects, i.e., the context descriptions created by the framework are dynamic and include as many aspects as could be acquired. This offers context consumers the possibility to query for the data they are interested in leading to a greater flexibility in elaborating on contextual constellations.

*Figure 3. Architecture of the MobiSem context framework*

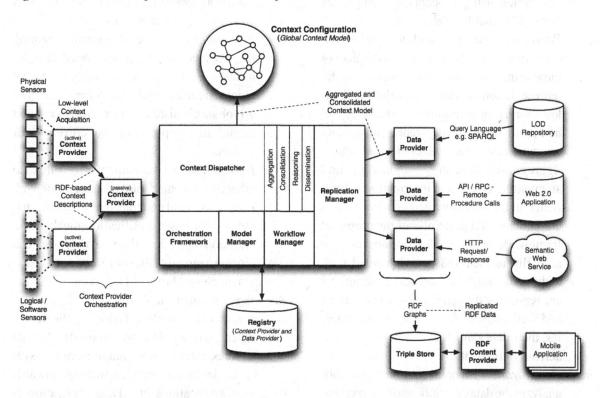

## ARCHITECTURE OF A SEMANTIC WEB-BASED CONTEXT FRAMEWORK FOR MOBILE SYSTEMS

### Overview

In general, a framework-based approach for context management is broadly suggested to facilitate and speed-up context-aware application development and allow for gathering, aggregating, and interpreting contextual data in a structured, well defined, and controlled way (Fournier et al., 2006). The MobiSem framework (cf. Figure 3) extends this idea in that it has been designed specifically to operate on mobile systems, and to use Semantic Web technologies to acquire, interpret, aggregate, store, and reason on contextual information, independent of any application or infrastructure. Semantic Web technologies and

practices, which are designed as an information processing infrastructure for heterogeneous environments, can help to solve some of the issues described before, and are therefore highly relevant for the design and development of ubiquitous and mobile context-aware systems. In the following, an overview of the main concepts and functionalities of the proposed framework is given.

1.  *Acquisition.* The framework allows for acquiring context-relevant data from a wide variety of sources, ranging from locally deployed hardware and software sensors, over sensors located in ubiquitous environments, towards Web 2.0 APIs for retrieving data related to a user's personal or social networks. Raw sensor data are represented in form of an RDF-based context description and transformed into a high-level context

description using concepts and languages from the Semantic Web.

2. *Representation.* The architecture of the context framework has been designed to impose minimal to none restrictions on the representation of contextual data and allows for using individual vocabularies and heuristics for explicitly representing data semantics and reason on contextual data.

3. *Aggregation.* Reusing, exchanging, and augmenting contextual information to built richer and more elaborated context models are fundamental principles of the proposed framework. Context descriptions can be mutually refined and complemented with additional data. Since context descriptions are represented as RDF graphs using open and well-known Semantic Web vocabularies, their data semantics can be understood across components.

4. *Orchestration.* An orchestration framework analyzes the data descriptions of context providing components called *context providers* and use them as a basis for cascading context providers in *directed acyclic network graphs* to dynamically route context descriptions between them. The orchestration framework has been designed to allow for integrating individual orchestration rules and metrics, where the results are stored in an *adjacency matrix*. This matrix is dynamically and transparently re-created whenever a new context provider is integrated into the framework.

5. *Consolidation.* To maintain context consistency, accurateness, and completeness, context descriptions are collected in a central place to guarantee consistency among descriptions and acquisition processes while taking into account technical and operating system peculiarities of mobile platforms. The framework also incorporates compensation strategies to minimize the impact of malfunctioning or unavailable context providers and sustain a proper functionality of acquisition and aggregation processes.

6. *Reasoning.* A lightweight forward-chaining rule reasoner has been developed in order to consolidate context descriptions, detect inconsistencies, and transform them into a global, consistent, and coherent context model that represents the current user context.

7. *Dissemination.* User context is forwarded to other components deployed in the framework in an automated and transparent manner using a push-based notification mechanism. Additionally, external context services can request a model of the user's current context via an elementary HTTP server.

8. *Replication.* Data is replicated in a transparent and automated fashion to the device where no restrictions are imposed to the data sources, which might range from file systems, databases, web repositories, towards web applications etc. Data replication is completely decoupled from context acquisition where mobile applications can access data replicas from a local triple store in a controlled and uniform way.

9. *Storage.* The framework incorporates a persistence layer that allows for storing replicated data sets in a local database from where they can be accessed and utilized. The persistence layer also includes support for named graphs and contains projections for transforming RDF graphs into a relational database scheme and vice versa.

10. *Data Access and Provision.* Access to replicated data stored in the local SQLite database is provided via an Android content provider (cf. Meier, 2010) that has been adapted for RDF data provision and storage. The RDF content provider assigns a unique URI to each replicated data sets and allows other mobile applications to access and utilize data replicas.

*Table 1. Types of context sensors*

|  | Local | Remote |
|---|---|---|
| **Hardware** | Device-internal hardware sensors for capturing physical context (e.g., location, inclination, orientation, etc.) | External ubiquitous sensors deployed in ubiquitous environments that allow for acquiring mostly physical context |
| **Software** | Locally deployed software or logical sensors for monitoring application and user behavior to deduce current and future mobile user information needs | Software sensors that wrap interfaces or APIs of remote data sources mostly for complementing primarily acquired context information (e.g., Web services, APIs, online repositories, etc.) |

To realize these concepts, the authors synthesized concepts from graph theory, distributed transaction management, and the Semantic Web to build an architectural infrastructure for mobile systems that combines context acquisition and data replication processes in order to replicate data related to the user's current and future information needs in a transparent and proactive manner. The framework cascades context providers within so-called *orchestration trees* that resemble concepts of workflow schemes. They allow for a controlled execution of combined context acquisition tasks in a decoupled manner while maintaining data and process consistency. This form of context provider orchestration allows for an efficient acquisition and aggregation of contextual information while taking into account the mobile operating system peculiarities (Forman & Zahorjan, 1994; Krogstie, Lyytinen, Opdahl, Pernici, Siau, & Smolander, 2004).

The approach used exposes two significant advantages compared to existing server-based approaches: contextual information is acquired, processed, and disseminated directly on a mobile device and does not depend on the availability of external systems. Therefore, the transmission of potentially confident data to external systems is avoided, which reduces security and privacy issues since highly private data such as contact information or appointments can be processed locally and do not need to be transferred outside a mobile system. Additionally, the conceptual structure of context and data providers leaves much freedom to the user to implement individual security and

privacy policies, e.g., to incorporate only innocuous data in generated context descriptions. The framework also serves as a technical infrastructure for the deployment of high-level context processing and recognition services to enable situation awareness (cf. Anagnostopoulos, Ntarladimas, & Hadjiefthymiades, 2007; Springer, Wustmann, Braun, Dargie, & Berger, 2008).

## Components

**Context Providers**: As illustrated in Table 1, the context framework is able to acquire contextual information from a variety of context sources and sensors. Those sources are wrapped by context providers, which employ two operation modes. Primary context providers are self-contained components that operate independently and autonomously, and provide contextual information in a proactive manner. They become active whenever a change in the corresponding context source is detected. In contrast, complementary context providers react according to updates received from primary context providers and complement contextual information acquired by primary context providers by taking those context descriptions as input data for initiating their acquisition tasks.

A context provider captures a specific and relevant contextual aspect of the user's current context. A contextual aspect is represented in structured and well-defined ways using semantic technologies (RDF, RDFS, and OWL) to facilitate context information exchange and integration. This allows them to be enriched and complemented with additional and related information in order to enhance the richness and accuracy of the contained information.

**Context Dispatcher**: The context dispatcher is the central component within the context framework. It handles the entire communication between context providers and propagates context models between them. It is notified whenever a new context model has been acquired by a context provider. The context dispatcher collects such updated context models, aggregates them, performs additional processing like inference and consolidation, and creates the so-called *context configuration* (see next page), which is a complete, accurate, and consistent representation of the user's current context, and is subsequently propagated to *data providers*. Currently, reasoning and consolidation is performed on the basis of hard-coded rules; in order to dynamically deploy new reasoning rules to the context dispatcher, a lightweight rule-based reasoner has been developed and integrated into the framework.

**Context Description Queue**: The context description queue is part of the context dispatcher and handles the communication between the context providers and the context dispatcher in an asynchronous manner. It implements specific logic for the management and exchange of context models to enable the recovery of lost descriptions and also serves as a compensation mechanism in case of temporarily unavailable context sources or malfunctioning context providers. Therefore, it not only buffers the most recent context updates, but also stores previously committed context updates. This allows the context framework to revert to previous context models to sustain the proper execution of context acquisition processes.

**Model Manager**: The model manager is used for storing and tracking the context providers' context models that have been acquired in the course of context acquisition processes. For every context provider deployed in the framework, the model manager stores the most recently updated

context model together with status information about the corresponding provider.

**Orchestration Framework**: The orchestration framework dynamically orchestrates compatible context providers in form of a direct acyclic graph based on the type of context information they provide. This graph is called *orchestration tree* and resembles the concepts of a workflow scheme that allows for a controlled execution of combined context acquisition tasks in a decoupled manner while maintaining data and process consistency. By analyzing such data descriptions, the orchestration framework computes compatibility measures so that relationships between context providers can be inferred. Every context provider must therefore provide an RDF description about the type of data it acquires and the vocabularies it uses for representing that data. The orchestration framework operates completely independent from other framework components and is initiated automatically whenever a new context provider is deployed in the framework.

**Workflow Manager**: The workflow manager is responsible for the management and coordination of the context providers' acquisition processes where compatible context providers are orchestrated in an orchestration tree. Those orchestration trees are executed in *context acquisition workflows* that control and manage the acquisition tasks of the involved context providers to sustain a deterministic and consistent behavior. Context updates are propagated through context acquisition workflows and routed between compatible context providers. When all context acquisition tasks have finished, the workflow manager resets the corresponding entries in the model manager and notifies the context dispatcher that a new global context model can be created.

**Registry**: The registry is the central storage component where all context and data provider deployed in the framework must register in order to be integrated in acquisition and replication workflows. It automatically notifies the orchestration framework whenever a new context provider has been registered so that it can be automatically orchestrated with compatible providers.

**Context Configuration**: The context configuration represents an aggregated version of all context providers' context models. It is created by the context dispatcher when all context acquisition workflows completed and the updated context models of their containing context providers are available.

**Replication Manager**: The replication manager controls and orchestrates all data replication tasks. It operates completely decoupled from the other framework components and gets notified by the context dispatcher whenever a new context configuration has been created. The replication manager is responsible for the instantiation of the data provider control threads which control and monitor data replication tasks and propagates the context configuration to each data provider. The replication manager also receives notifications about changed data replicas in order to initiate write-back and synchronization operations.

**Data Providers**: Data providers replicate any kind of RDF data to the mobile device; they can request data from virtually any external data sources or generate data replicas themselves and operate completely decoupled and independent from each other. Each data provider is assigned a unique identifier that is used as part of the addressing scheme to store data in the local triple store. Data providers adjust and initiate their data replication tasks based on the analysis of the context configuration that they receive by the replication manager. For instance, a data provider can analyze data regarding the current location of a device and retrieve information about nearby points of interest.

**Triple Store**: Our triple store implementation is designed to be a lightweight, efficient storage and retrieval mechanism for RDF triples. It abstracts over the concrete storage mechanism that is used by the mobile platform and provides support for named graphs persistence, and RDF serialization and de-serialization. It employs a normalized table

layout (cf. Abadi, Marcus, Madden, & Hollenbach, 2009) where resources, literals, and blank nodes are stored in separate tables and provides support for named graphs. A discussion regarding other database layouts for storing RDF triples including their advantages and limitations can be found in Hertel, Broekstra, and Stuckenschmidt (2008).

**RDF Content Provider**: Replicated data sets are provided to external applications via the RDF content provider. This provider is an individual implementation of an Android content provider (see Rogers, Lombardo, Mednieks, & Meike, 2009) for the system-wide provision of RDF data and contains the projections for transforming RDF graphs into the relational database schema of the local SQLite database and vice versa. It exposes a common interface applications can use for performing query, update, insert, and delete operations on replicated data. It has been extended with named graphs support (cf. Carroll et al., 2005) where each data replica is stored using a unique URI. Content providers expose configurable content URIs that can be used for addressing specific parts of replicated data.

## Orchestration of Context Providers

To facilitate this kind of cooperation between decoupled context providers, the context framework dynamically routes data between context providers based on the type of context information they provide. Therefore, the orchestration framework analyzes the data description of each context provider. Figure 4 depicts an excerpt of an exemplary data description, which consists of sets of mandatory and optional namespaces as well as terms, which can be processed as input data by the respective context provider, as well as namespaces and terms that the context provider uses in its output data. Further details regarding the data description vocabulary elements are given in Zander and Schandl (under review).

The orchestration framework can be configured to either perform a loose orchestration on the

*Figure 4. Exemplary data description for a complementary context provider for extracting contact data from calendar entries*

```
1    <urn:uuid:b772a3a2-46d4-4c43-8f71-7080915ddba7>
2        a    ddesc:ContextProvider ;
3      ddesc:input [
4        ddesc:vocabulary [
5          ddesc:namespace <http://www.semanticdesktop.org/ontologies/ncal#> ;
6          ddesc:concepts [
7            ddesc:mandatory ncal:Attendee, ncal:Calendar, ncal:Event ;
8            ddesc:optional ncal:Organizer, ncal:EventStatus
9          ] ;
10         ddesc:properties [
11           ddesc:mandatory ncal:member, ncal:method ;
12           ddesc:optional ncal:eventStatus
13         ]
14       ]
15     ] ;
16     ddesc:output [
17       ddesc:vocabulary [
18         ddesc:namespace <http://xmlns.com/foaf/0.1/> ;
19         ddesc:concepts [
20           ddesc:mandatory foaf:Organization, foaf:Person
21         ] ;
22         ddesc:properties [
23           ddesc:mandatory foaf:knows, foaf:status, foaf:name
24         ]
25       ]
26     ] .
```

namespace level, or a detailed one by considering concepts and properties given by the context providers' data descriptions. When a new context provider is found in the system, the orchestration framework analyzes its data description and based on its configuration integrates the context provider in the orchestration graph. While running completely decoupled from the context framework, rebalancing the orchestration graph does not affect context acquisition tasks as such. The matching value for each pair of context providers is computed by a *matching algorithm* based on onfigurable scores for correspondences on the namespace, concept, and property levels. The matching algorithms performs an arithmetic matching based on data similarities and is additionally capable of including RDFS semantics such as rdfs:subClassOf relationships. For instance, if one context provider emits foaf:Person

instances and another context provider requires foaf:Agent instances as input data, the matching algorithm detects the compatibility between these differing concepts since foaf:Person is a subclass of foaf:Agent according to the FOAF ontology (Brickley & Miller, 2007).

## Formal Model for Orchestrating Context Providers

The idea of orchestrating context providers is to augment and complement contextual information to build richer, more expressive and elaborated context descriptions while maintaining data and processing consistency, accurateness, and completeness. Therefore, context providers are orchestrated in a *directed tree* called *orchestration tree* based on the type of context information they acquire and provide.

Let $\Pi$ be the set of all primary context providers $P_i$ with $\Pi := \{P_i \,|\, i = 1,..., x\}$ and X the set of all complementary context providers $C_j$ with X:= $\{C_j \,|\, j = 1,..., y\}$. A context source that is wrapped by a context provider is denoted as $S_k$; let $\Sigma$ be the set of all potential context sources $S_k \in \Sigma$ and let A be the set of all context providers $A_k$ where A:= $\{A_k \,|\, k = 1,..., x + y\}$ that are deployed on the context framework and wrap a specific context source $S_k$ irrespectively of their concrete type. Therefore, a mapping can be defined as $A_k \in A \rightarrow S_k \in \Sigma$ between context providers and context sources. Additionally, we can state that the set of context providers $\mathbb{A}$ is the union of the sets of all primary and complementary context providers $(\mathbb{P} \cup \mathbb{C})$ where $\mathbb{P}$ and $\mathbb{A}$ can be equal in case only primary context providers are deployed on the framework, whereas $\mathbb{C}$ is always a real subset of $\mathbb{A}$ since complementary context providers always require a primary context provider to be operational:

$$\mathbb{A} = \mathbb{P} \cup \mathbb{C} \text{ with } \mathbb{P} \subseteq \mathbb{A} \text{ and } \mathbb{C} \subset \mathbb{A} \text{ and } \mathbb{P} \cap \mathbb{C} = \varnothing \quad (1.1)$$

Context providers specify the data they provide in a data description (see Zander & Schandl, under review) that serves as a basis for defining relations between compatible context providers in terms of the contextual data they acquire and require for performing their acquisition tasks. These relations are analyzed by the orchestration framework and recorded in an adjacency matrix (cf. Figure 6). For each primary context provider $P_i \in \mathbb{P}$ an orchestration tree $O^P_i$ can be derived from the adjacency matrix (cf. Figure 7). An orchestration tree $O$ is a directed tree whose root element is always a primary context provider $P_i$ and the adjacent nodes represent complementary context providers $C_j$ that complement the data acquired by the corresponding primary context provider $P_i$. Therefore, there is always a 1:1 relationship between a primary context provider $P_i$ and it corresponding orchestration tree $O^P_i$ where a projection $P_i \mapsto O^P_i$ exists for all $P_i \in \mathbb{P}$ and $O^P_i \in \mathbb{O}$.

Generally, an orchestration tree $O^P_i$ can be represented as a directed acyclic graph $(V, E, \alpha, \omega)$, where the vertices $V$ represent context providers, and the edges $E$ relations between them in the form of exchanged context models, where the context model of a predecessor node is consumed by its successors. $\alpha$ and $\omega$ are projections and

*Figure 5. Example of an adjacency tableau for eight context providers*

|       | $\alpha$ | $\omega$ |
|-------|----------|----------|
| $r_1$ | $P_1$    | $C_1$    |
| $r_2$ | $P_1$    | $C_2$    |
| $r_3$ | $P_2$    | $C_5$    |
| $r_4$ | $C_2$    | $C_3$    |
| $r_5$ | $C_2$    | $C_4$    |

*Figure 6. Example of an adjacency tableau and corresponding adjacency matrix*

|       | $P_1$ | $P_2$ | $P_3$ | $C_1$ | $C_2$ | $C_3$ | $C_4$ | $C_5$ |
|-------|-------|-------|-------|-------|-------|-------|-------|-------|
| $P_1$ | 0     | 0     | 0     | 1     | 1     | 0     | 0     | 0     |
| $P_2$ | 0     | 0     | 0     | 0     | 0     | 0     | 0     | 1     |
| $P_3$ | 0     | 0     | 0     | 0     | 0     | 0     | 0     | 0     |
| $C_1$ | 0     | 0     | 0     | 0     | 0     | 0     | 0     | 0     |
| $C_2$ | 0     | 0     | 0     | 0     | 0     | 1     | 1     | 0     |
| $C_3$ | 0     | 0     | 0     | 0     | 0     | 0     | 0     | 0     |
| $C_4$ | 0     | 0     | 0     | 0     | 0     | 0     | 0     | 0     |
| $C_5$ | 0     | 0     | 0     | 0     | 0     | 0     | 0     | 0     |

$$= \begin{pmatrix} 0 & 0 & 0 & 1 & 1 & 0 & 0 & 0 \\ 0 & 0 & 0 & 0 & 0 & 0 & 0 & 1 \\ 0 & 0 & 0 & 0 & 0 & 0 & 0 & 0 \\ 0 & 0 & 0 & 0 & 0 & 0 & 0 & 0 \\ 0 & 0 & 0 & 0 & 0 & 1 & 1 & 0 \\ 0 & 0 & 0 & 0 & 0 & 0 & 0 & 0 \\ 0 & 0 & 0 & 0 & 0 & 0 & 0 & 0 \\ 0 & 0 & 0 & 0 & 0 & 0 & 0 & 0 \end{pmatrix}$$

*Figure 7. Orchestration graphs derived from the adjacency matrix*

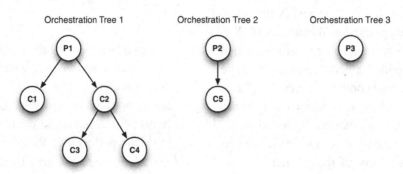

specify concrete relations between predecessor and successor context providers.

More specifically, an orchestration tree $O = (V, R, \alpha, \omega)$ can be described as follows:

1. $V$ is defined as the non-empty set of a primary context provider $P_i$ and its corresponding complementary context providers $\mathbb{C}_i^P$:

$$V := \{P_i\} \cup \mathbb{C}_i^P$$

2. $R$ is the defined as the set of relations $r$ that exist between the context providers contained in $V$ where $r$ is defined as the tuple

$$r \in \{P_i\} \times \mathbb{C}_i^P \cup \mathbb{C}_i^P \times \mathbb{C}_i^P$$

3. $V \cap R = \varnothing$

4. $\alpha: R \to V$ and $\omega: R \to V$ are projections where $\alpha(r)$ is the starting or direct *predecessor* context provider and $\omega(r)$ the ending or direct *successor* context provider of the relation $r$. $\alpha(r)$ and $\omega(r)$ are adjacent to each other.

5. $\forall r \in R: \alpha(r) \neq \omega(r)$

With this definition, orchestration trees can be represented in an adjacency tableau as depicted in Figure 5 that represents the relations $r_i$ together with preceding and adjacent context providers $P_i \in \mathbb{P}$ and $C_j \in \mathbb{C}$.

The set $V_i^O \in O_i^P$ contains the primary context provider $P_i \in \mathbb{P}$ as well as all compatible complementary context providers orchestrated in the orchestration tree $O_i^P$. Therefore, let $\mathbb{C}_i^P$ be the set of complementary context provider $C_j \in \mathbb{C}$, which are orchestrated in an orchestration tree $O_i^P$ where $\mathbb{C}_i^P \in \mathbb{C}$. Hence, $V_i^O \in O_i^P$ can be defined as the union of the primary context provider $P_i \in \mathbb{P}$ and the set of compatible context providers $C_j \in \mathbb{C}_i^P$:

$$V_i^O := \{P_i\} \cup \mathbb{C}_i^P \text{ where } \mathbb{C}_i^P \subseteq \mathbb{C} \qquad (1.2)$$

As a consequence, the set of all complementary context providers $\mathbb{C}$ can also be defined as the union of the partial sets of complementary context providers $\mathbb{C}_i^P$ orchestrated in an orchestration tree $O_i^P$ and the set of providers $C_k$ that are also deployed on a system but not part of any orchestration tree $O_i^P$:

$$\mathbb{C} := \bigcup_{i=1}^{|\mathbb{P}|} \mathbb{C}_i^P \cup C_k \text{ with } \bigcap_{i=1}^{|\mathbb{P}|} \mathbb{C}_i^P = \varnothing \qquad (1.3)$$

Since an orchestration tree $O_i^P$ contains only distinct elements, where a context provider $P_i$ or $C_j$ can only be part of one orchestration tree, we can state that all sets $V_i^O$ and $V_j^O$ are pairwise disjunct:

$$\forall V_i^O \in O_i^P, V_j^O \in O_j^P, i \neq j : V_i^O \cap V_j^O = \varnothing$$
$$(1.4)$$

Further, let $\mathbb{O}$ denote the union of all orchestration trees $O^P_i$ with $P_i \in \mathbb{P}$ and $|\mathbb{O}| = |\mathbb{P}|$, and $V^O_i$ the set of all context providers orchestrated within an orchestration tree $O^P_i$. Therefore, $V^O_i$ can be defined as:

$$\mathbb{O} := \bigcup_{i=1}^{n} O^P_i \text{ and } \bigcap_{i=1}^{n} V^O_i = \varnothing \text{ where } n = |\mathbb{P}|$$
(1.5)

For building an orchestration tree, the orchestration framework analyzes the data description of each context provider and builds an adjacency matrix. The adjacency matrix represents a graph $G = (V, E)$ with $V = \{v_1, \ldots, v_n\}, n = |\mathbb{A}|$ where vertices $v_i$ represent context providers orchestrated in an orchestration tree. Then, $E \leftrightarrow R$ can be represented in a $n \times n$-matrix, where

$$a_{ij} = \begin{cases} 1 \rightarrow \{v_i, v_j\} \in R \\ 0 \rightarrow \text{ otherwise} \end{cases}$$
(1.6)

The adjacency matrix servers as a basis for building the orchestration trees $O^P_i$ of primary context providers $P_i \in \mathbb{P}$ and associated complementary context providers $C_j \in \mathbb{C}^P_i$. It contains a separate row and column for every element $P_i \in \mathbb{P}$ and $C_j \in \mathbb{C}$. Each element in the matrix corresponds to a distinct tuple $(P_i, C_j)$ or $(C_j, C_k)$ from the cartesian products $\mathbb{P} \times \mathbb{C}$ and $\mathbb{C} \times \mathbb{C}$. In case there is a relation between $P_i$ and $C_j$ or $C_j$ and $C_k$, the matrix coefficient is set to 1 at the corresponding position. All other coefficients remain at 0.

Figure 6 depicts an example of an adjacency matrix containing three primary context providers $P_{1,\ldots,3}$ and five complementary context provider $C_{1,\ldots,5}$. Each row represents a particular context provider where the matrix coefficients in each column indicate a relationship to the context provider given in each column in case its value is set to 1. Likewise, 0 indicates no relationship r between the context provider of the current row and the context provider of the current column. Primary context providers cannot be connected among each other, i.e., primary context providers always have complementary context providers as successors. Complementary context providers in contrast can be connected among each other in a non-cyclic way.

As an example, let there be three primary context providers $P_{1,\ldots,3}$ and five complementary context providers $C_{1,\ldots,5}$ deployed within the framework. By analyzing the data descriptions exposed by each context provider, the adjacency matrix depicted in Figure 6 can be created based on the compatibility indicators computed by the orchestration framework. The first three rows represent the adjacent complementary context providers $C_j \in \mathbb{C}$ of $P_1, P_2,$ and $P_3$ that take their context models as input. For instance, $P_1$ has the two complementary context providers $C_1$ and $C_2$ as direct successors that are able to refine and augment the context model delivered by $P_1$.

The values in row 4 show the direct successors of the first complementary context provider $C_1$. Since the matrix coefficients in the entire row are set to 0, there exist no relationships to other complementary context providers. $C_2$ in line 5 for instance has two adjacent context providers $C_3$ and $C_4$, which both do not have any further successors that take their context models as input for their acquisition tasks. By analyzing the given adjacency matrix, the orchestration trees $O^P_i$ depicted in Figure 7 can be derived for each primary context provider $P_{1,\ldots,3}$. Those orchestration trees represent atomic units of context acquisition workflow schemes.

## Context Model Aggregation

The following paragraph provides an overview of the context acquisition workflows from a data-centric view and illustrates the process of building a global context model, the so-called *context configuration* from the context providers' context models.

Let $M^P_i$ be the context model created by a primary context provider $P_i$, and $M^C_j$ the context model created by a complementary context provider $C_j$. Since there is a 1:1 relation between a context provider and the context model it creates, we can define a projection $A_k \mapsto M^A_k$ with $A_k \in \mathbb{A}$ between any context provider and its context model. A context model $M$ is represented as an RDF graph consisting of a finite set of resources represented by their URIs and denoted as $\mathbb{U}$, a finite set of blank nodes $\mathbb{B}$, and a finite set of literals $\mathbb{L}$, where no distinction is made between plain and typed literals (cf. Klyne & Carroll, 2004). Those elements are the building blocks of an RDF statement which is represented as a triple T = ($\Sigma$, $\Pi$, O) consisting of a *subject* $\Sigma$ which is either a URI reference or a blank node, a *predicate* $\Pi$ which is a URI reference, and an *object* O that can either be a URI reference, a blank node, or a literal. A context model $M$ itself can be defined as an element from the power set $\mathbb{M}$ that is defined as the cartesian product of the union of Y and B, Y, and the union of the sets Y, B, and $\Lambda$:

$$M \in \mathbb{M} \quad \text{where} \quad \mathbb{M} := \mathcal{P}\left(\mathbb{U} \cup \mathbb{B} \times \mathbb{U} \times \mathbb{U} \cup \mathbb{L} \cup \mathbb{B}\right) \tag{1.7}$$

An orchestration tree $O^P_i$ for a primary context provider $P_i$ is considered as an atomic unit, which—when all acquisition tasks of the contained context providers have finished—provides an aggregated context model $M^O_i$ that is composed of the context models of contained context providers. Since $O^P_i$ denotes the orchestration tree of a primary context provider $P_i \in \mathbb{P}$, let $M^O_i$ denote the compounded context model acquired by the context providers orchestrated within the orchestration tree $O^P_i$. As there is a 1:1 relationship between an orchestration tree $O^P_i \in \mathbb{O}$ and the compounded context model $M^O_i$ acquired by its

containing context providers, we can define the projection $O^P_i \mapsto M^O_i$. Further, let $\mathbb{M}^C_i$ be the set of context models acquired by all $C_j \in \mathbb{C}^P_i$. Hence, the compound context model $M^O_i$ of orchestration tree $O^P_i$ is defined as

$$M^O_i := \left\{M^P_i\right\} \cup \bigcup_{j=1}^{|\mathbb{C}^P_i|} M^C_j \in \mathbb{M}^C_i \tag{1.8}$$

When the context models $M^A_k \in \mathbb{M}$ of all context providers being part of all orchestration trees $O^P_i \in \mathbb{O}$ were acquired successfully, the context configuration $CC$ can be created. Therefore, let $CC$ be the union of all compounded context models $M^O_i$ of the orchestration trees $O^P_i$ where $i = |\mathbb{O}|$. Then, we can define $CC$ as the result of a function $f(x)$ that processes the compounded context models of all orchestration trees $O^P_i \in \mathbb{O}$ and infers additional information by applying reasoning techniques.

$$CC := f\left(M^O_i \mid i = 1 \ldots |\mathbb{O}|\right) \tag{1.9}$$

The orchestration tree $O^P_i$ of a primary context provider $P_i \in \mathbb{P}$ delivers an aggregated context model $M^O_i$ that is compounded of the context models of the constituting context providers $M^P_i$ and $\mathbb{M}^C_i$. The context configuration is then built by the context dispatcher by integrating, aggregating, and consolidating those compounded context models.

## Processing Workflow

For processing contextual information represented as RDF graphs, the authors apply concepts from graph theory and distributed transaction management to maintain context accuracy, consistency, and completeness while guaranteeing computational efficiency in terms of required resources and processor load. An orchestration tree $O^P_i$ is executed within a *context acquisition workflow* denoted as $WF^P_i$ that serves as a workflow scheme for the execution of the contained context providers and monitors acquisition progress. At any point in time, multiple context acquisition workflows can be active while there is a 1:1:1-relationship between a primary context provider $P_i$, the corresponding orchestration tree $O^P_i$, and its workflow scheme $WF^P_i$. Every workflow scheme is considered an atomic unit where the containing context providers acquire their context models independently from context providers contained in other workflow schemes. These concepts resemble the idea of the atomicity and isolation properties defined in the *ACID paradigm* for database transactions (cf. Gray & Reuter, 1992). In that sense, the context models acquired within the course of their corresponding acquisition workflows are considered as a single transaction where only the corresponding compound model $M^O_i \subseteq CC$ is updated as a whole in the context configuration $CC$.

Information about adjacent providers $C_j \in \mathbb{C}^P_i$ is requested from the orchestration framework that returns a context provider's direct descendants. Those complementary context providers are executed in self-contained, dedicated, and independent control threads called *context acquisition control threads* denoted as $PCT(C_j)$ that control and coordinate acquisition tasks and set the corresponding entries in the model manager when the control thread finishes, i.e., when a context update has been acquired. Context acquisition control threads are instantiated on demand in separate threads whenever the context of a preceding context provider has changed. The acquisition processes of the involved context providers are completely decoupled and independently executed in form of a *single atomic transaction* (Gray & Reuter, 1992).

Algorithm 1 describes the relevant steps in the acquisition workflow carried out by the context dispatcher for processing and aggregating context descriptions to build a global context model. The context dispatcher regularly checks whether there are new context descriptions available. A context description $D^A_k$ for a context provider $A_k$ can be defined as the quadruple $D = (M, \rho, \sigma, \tau)$ where $M$ represents the context model $M^A_k$ of the corresponding context provider $A_k$, $\rho$ and $\sigma$ represent status information and a status code, and $\tau$ a time stamp when the context model was created. Therefore, a context description $D^A_k$ can be considered a projection of a context model $M^A_k$ for a given context provider $A_k$ at a given time $\tau : M^A_k \mapsto D^A_k(\tau)$.

In case a new context description $D^A_k$ is available, the context dispatcher retrieves the new context description and stores it in the model manager. For each context description $D^A_k$, the corresponding context provider $A_k$ is requested from the registry. The context dispatcher then checks the context description $D^A_k$ and calculates a unique value that reflects certain characteristics and the current status (see Table 2) of the corresponding context provider $A_k$. Therefore, the authors have defined some basic conditions derived from the requirements of the proposed mobile context framework (see Overview). Each condition is assigned a unique value of the form $2^n$ where n refers to a certain condition. This allows to represent unique states by simply summarizing the distinct property values that match for a given provider. For instance, if a context provider is primary typed and has adjacent providers but no acquisition workflow has been initiated yet, its status value is 34 ($0 * 2^0 + 1 * 2^1 + 0 * 2^2 + 0 * 2^3 + 0 * 2^4 + 1 * 2^5 + 0 * 2^6$). If a condition evaluates to true, the corresponding value is added to the provider's status value that determines how a context provider as well as its context model and adjacent providers are to be processed. In accordance to this value, the corresponding processing steps (see Table 3) will be executed; if the context description $D^A_k$ was sent by a pri-

*Algorithm 1. Algorithm for processing context descriptions in pseudocode*

**Input:** ContextDescription $D_i$
**Output:** -

**while** *ContextDispatcher is running* **do**
  **if** *all acquisition workflows have finished* **then**
    collect and aggregate context models $M_j^A$ where $1 \leq j \leq |\mathbb{A}|$ ;
    create ContextConfiguration $CC$ ;
    notify ReplicationManager ;
    reset ModelManager ;
  **end**
  **if** *ContextDescriptionQueue contains new ContextDescriptions* **then**
    **forall the** *ContextDescriptions $D_i \in$ ContextDescriptionQueue* **do**
      obtain context description $D_i$ from ContextDescriptionQueue ;
      retrieve corresponding context provider $A_j \leftrightarrow D_i$ ;
      calculate status value for $D_i$ ;
      **if** $A_j \in \mathbb{P}$ *and* $M_j^A \notin$ *ModelManager* **then**
        store $M_j^A$ in ModelManger ;
        **if** *hasSuccessor($A_j$)* **then**
          $WF_j^A \leftarrow$ new ContextAcquisitionWorkflowThread ;
          **forall the** *Successors $C_k \in \mathbb{C}_j^A$* **do**
            $PCT(C_k) \leftarrow$ new ContextAcquisitionControlThread ;
            $PCT(C_k)$.start() ;
          **end**
        **end**
      **end**
      **else if** $A_j \in \mathbb{C}$ *and* $A_j \in WF_x^P$ **then**
        store $M_j^A$ in ModelManger ;
        **if** *hasSuccessor($A_j$)* **then**
          **forall the** *Successors $C_k \in \mathbb{C}_j^A$* **do**
            $PCT(C_k) \leftarrow$ new ContextAcquisitionControlThread ;
            $PCT(C_k)$.start() ;
          **end**
        **end**
      **end**
      **else if** $A_j \in \mathbb{P}$ *and* $M_j^A \in$ *ModelManager and exists($WF_x^P, x \neq j$)* **then**
        replace $M_j^A$ in ModelManager ;
        **if** *hasSuccessor($A_j$)* **then**
          $WF_j^A \leftarrow$ new ContextAcquisitionWorkflowThread ;
          **forall the** *Successors $C_k \in \mathbb{C}_j^A$* **do**
            $PCT(C_k) \leftarrow$ new ContextAcquisitionControlThread ;
            $PCT(C_k)$.start() ;
          **end**
        **end**
      **end**
    **end**
  **end**
**end**

*Table 2. Conditions for determining a context provider's state*

| Condition | Value | Description |
|-----------|-------|-------------|
| A | $2^0$ | Checks whether a context model has been sent by the context provider and this is already contained int hem odel manager. |
| B | $2^1$ | Indicates whether the context provider has adjacent complementary context providers that are orchestrated within a context acquisition workflow. |
| C | $2^2$ | Indicates whether an acquisition workflow thread that coordinates the execution of the context providers orchestrated within that workflow has already been initiated for the given primary context provider. |
| D | $2^3$ | Indicates whether the given complementary context provider is part of an already instantiated acquisition workflow. |
| E | $2^4$ | Indicates whether the corresponding entry for a given context provider in the model manager is locked, meaning that no onctext model propagation is possible until the corresponding workflow thread has finished. |
| F | $2^5$ | Checks whether the given context provider is an active, that is, primary context provider. |
| G | $2^6$ | Checks whether the corresponding context acquisition workflow or control thread has been finished depending on the context provider's type. |

mary context provider $P_i$ whose context model $MP_i$ has not been stored in the model manager yet, it will be extracted from the context description DP i and stored in the model manager. If $P_i$ has complementary context providers $C_j \in \mathbb{C}_i^P, 1 \leq i \leq \left|\mathbb{C}_i^P\right|$ associated to it in an orchestration tree $O_i^P$, a new instance of a context acquisition workflow $WF_i^P$ is created. The direct descendants of $P_i$ (cf. the adjacency matrix depicted in Figure 6) are requested from the orchestration framework and the context acquisition control threads $PCT(C_j)$ which executes and controls the acquisition tasks will be created and initiated.

If $A_i$ is a primary context provider with $A_i \rightarrow P_i$ that has already committed its context model $M_i^A \rightarrow M_i^P$ but has no direct descendants, that is, no relations to complementary context providers and there are other context acquisition workflows $WF_k^P$ where $i \neq k$ running in parallel, its already stored context model $M_i^P$ can be replaced by the new one in the model manager without violating consistency requirements. The same applies to primary context providers $P_k$ whose context acquisition workflows $WF_k^P$ are finished that is, all

associated complementary context providers $C_j \in \mathbb{C}_k^P, 1 \leq j \leq \left|\mathbb{C}_k^P\right|$ have delivered their context models, but other context acquisition workflows $WF_l^P$ with $l \neq k$ are still running. In this case, a new context acquisition workflow $WF_k^P$ for the primary context provider $P_k$ will be initiated and the previous context models $M_j^C \in \mathbb{M}_k^P$ with $1 \leq j \leq \left|\mathbb{C}_k^P\right|$ can be replaced by new (updated) models.

In case a context acquisition workflow $WF_i^A$ has finished, it notifies the workflow manager, which then checks whether there exist further context acquisition workflows $WF_j^A$ where $j \neq k$. If all acquisition workflows have finished, the workflow manager notifies the context dispatcher that a new context configuration $CC$ can be built. The context dispatcher then starts the merging and consolidation process by collecting all updated and unaltered context models $M_k^A$. Updated context models are retrieved from the model manager whereas unaltered context models are retrieved from the context description queue. Those models will then be aggregated (merged) where consolidation and reasoning rules

*Table 3. Excerpt of the calculation and processing matrix for context providers including descriptions and rules*

| Condition | A | B | C | D | E | F | G | | |
|---|---|---|---|---|---|---|---|---|---|
| Value | $2^0$ | $2^1$ | $2^2$ | $2^3$ | $2^4$ | $2^5$ | $2^6$ | Sum | Description & Rule |
| #1 | 0 | 1 | 0 | 0 | 0 | 1 | 0 | 34 | This is a primary context provider with adjacent context providers but no context acquisition workflow has been created yet; **Rule**: *create a new context acquisition workflow for the given context provider* |
| #2 | 0 | 1 | 0 | 1 | 0 | 0 | 0 | 10 | This complementary context provider is part of a context acquisition workflow and has adjacent complementary context providers attached to it that further augment its context model. **Rule:***retrieve complementary context providers and initiate acquisition tasks* |
| #3 | 0 | 1 | 0 | 0 | 0 | 0 | 0 | 2 | This is a complementary context provider that has further complementary context providers attached to it as successors that take its context model as input data for their acquisition tasks. **Rule:***retrieve complementary context providers and initiate acquisition tasks* |
| #4 | 0 | 0 | 0 | 1 | 0 | 0 | 0 | 8 | This is a complementary context provider that is part of a context acquisition workflow. It has no adjacent providers attached to it and its context model isn't stored in the model manager yet. **Rule:***store model in model manager* |
| #5 | 1 | 1 | 1 | 0 | 1 | 1 | 0 | 55 | This is the status of a primary context provider with adjacent complementary context providers that has already delivered a context model stored in the model manager. The corresponding context acquisition workflow is still in progress since not every complementary context provider contained in the corresponding orchestration tree has finished its acquisition task yet. **Rule:***refuse the context provider's updated context model* |
| #6 | 1 | 1 | 0 | 0 | 1 | 1 | 1 | 115 | A primary context provider whose context acquisition workflow has already finished sends an updated context model. Since there are other acquisition workflows currently running, its updated context model can be accepted and a new context acquisition workflow can be initiated. **Rule:***reset corresponding entries in model manager, store context model and initiate new context acquisition workflow* |
| #7 | 1 | 0 | 0 | 0 | 0 | 1 | 1 | 97 | A primary context provider without adjacent complementary context providers as sent an updated context model. **Rule:***update corresponding entries and store context model in model manager* |

are applied by a lightweight rule reasoner. After the context configuration is built, it will be forwarded to the replication manager to initiate the replication tasks of the deployed data providers.

# IMPLEMENTATION

## The Google Android Platform

Android is an open software stack for mobile devices consisting of a Linux-based operating system kernel, a middleware, key applications, and a set of API libraries for accessing native system functions and natively deployed sensors (*Android*

*Developer's Guide*, 2010). It was released in 2007 under the auspices of the Open Handset Alliance[7], a coalition of 79 technology and mobile companies and has gained noticeable attentions since it represents a new generation of mobile application development platform and software development kit (Rogers et al., 2009). One of the key architectural features of the Android platform is the open communication infrastructure where application can reuse functionality and exchange data in a controlled and flexible way. Android includes a highly specialized virtual machine, called Dalvik VM that was designed for low-powered handheld devices as those devices *"lag behind their desktop counterparts in memory and speed by eight to ten years"* (Hashimi, Komatineni, & MacLean, 2010). In contrast to conventional Java virtual machines which use a stack-based architecture for data storage, the Dalvik VM is built upon a register-based architecture and transforms generated Java classes into a performance and memory optimized Dalvik specific file format.

Android includes a lightweight and powerful relational SQLite database that offers dedicated libraries for its utilization within application and services to store data persistently. Such data can be shared across applications in a controlled manner using Android's inter-process communication model. In contrast to the hard-wired application models of desktop operating systems, Android offers an intend-based application model that allows an application to specify a certain kind of functionality it requires for data processing where the operating system chooses the application that best matches. Access to core-system libraries and functions is offered via native APIs that can be used by both native and non-native applications. Android employs an equal, non-prioritized execution policy for native and non-native applications that are executed in the same runtime, and offers a complete multithreading environment where applications can place extensive computational tasks in separate threads (Meier, 2010).

This environment brings the following options for deploying a framework that provides the desired functionality necessary for an efficient acquisition, management, storage, and dissemination of context information:

- A service can constantly scan the environment for exploitable context sensors and ubiquitous devices and integrate them dynamically in a context framework.
- A context framework can be realized as a background service that is executed transparently and autonomously from the user while not affecting any of their tasks nor requiring explicit user attention.
- The concept of content providers can be deployed for the controlled and fine-grained provision of replicated data as well as for the provision and utilization of context data.
- Broadcast receivers can notify the user about relevant events in a non-disruptive manner while providing specialized configuration views in case explicit user inputs are necessary, e.g., for integrating a recently discovered ubiquitous sensor.
- The open and uniform architecture of Android allows for a broad utilization of locally deployed sensors where native APIs offer uniform and controlled access to peripheral hardware.

## Context and Data Providers

All the basic data of context and data providers such as name, identifier, data description, and status is held in the AbstractBasicProvider class and specified by the IBasicProvider interface (see Figure 8). The AbstractContextProvider class implements methods regarding the communication with the framework as well as for the creation and management of the context providers' models. Primary context providers must be inherited from ActiveContextProvider, whereas

Figure 8. Structured class diagram for context and data providers

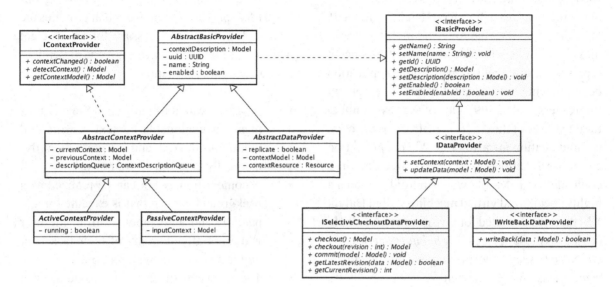

complementary context providers are deduced from the abstract class PassiveContextProvider. Both specify the abstract methods detectContextImpl(), which contains the concrete acquisition logic, and buildRdfModelImpl(), which holds the code for building the RDF context models. In contrast to primary context providers, which acquire context-relevant data proactively, passive context providers initiate their context acquisition tasks whenever they are notified by the context dispatcher about the availability of a required context input model.

The generic functionality of data providers is implemented in the abstract class AbstractDataProvider, which contains member variables for storing the context configuration and specifies the two abstract method declarations updateContextImpl() and updateDataImpl() that must be implemented by individual data providers. The authors decided to decouple the replication process so that a data provider can perform additional processing before it initiates its replication workflow. Two specialized data provider types, ISelectiveCheckoutDataProvider and IWriteBackDataProvider, define behavior for more complex data replication tasks, e.g., for replicating only subsets

of data sets (Schandl, 2010) or bidirectional synchronization of replicated data.

So far, a number of concrete context providers were implemented:

- GPSContextProvider is an active (primary) context provider that utilizes the device's internal GPS module for retrieving GPS coordinates. It can further be configured to use a WiFi-based location ascertainment method. The GPS context provider returns its data as latitude and longitude coordinates using the W3C WGS84 vocabulary[8] (see Figure 10).

- GeonamesContextProvider is a passive (complementary) context provider that is able to interpret any geographical coordinates expressed using the W3C WGS84 vocabulary and requests the *GeoNames. org* web service in order to resolve the GPS coordinates to a concrete geographical location identified by a GeoNames URI. Therefore it is an ideal complement to the GPSContextProvider, whose output is enriched with logical location concepts in addition to physical GPS coordinates;

*Figure 10. Context model as returned by the GPSContextProvider (Turtle notation)*

```
1   @prefix mobisem: <http://www.mobisem.org/2009/01/context#> .
2   @prefix geo:     <http://www.w3.org/2003/01/geo/wgs84_pos#> .
3
4   [] a mobisem:Context ;
5      mobisem:currentLocation [ geo:lat 48.175443 ; geo:long 16.375493 . ] .
```

however, it can in general be used to process any GPS coordinate information regardless of its source.

- GoogleCalendarContextProvider is an active (primary) context provider that retrieves appointments from the user's calendar within a configurable time span (e.g., 72 hours) starting from the current time. For all events, a set of metadata are extracted and added to the context model, including the event's names and start/end times, location descriptions associated to the events, and information about each event's attendees. It uses terms from the popular Dublin Core vocabulary[9] as well as the NEPOMUK calendar ontology[10] for modeling extracted metadata.

- LocalAddressbookContextProvider is an active (primary) context provider that exposes the contacts found in the user's local address book to make them available for relevant data providers. Properties like name, email addresses, phone numbers, postal addresses, and instant messaging contact IDs are extracted and inserted into the context model, where they are expressed using the popular FOAF vocabulary[11].

Typically, context providers use any kind of input data and convert it to an RDF model that represents this data in machine-processable form. Ideally, context providers use standardized, publicly available, and community-accepted vocabularies for this purpose in order to enable interoperability between components from different

manufacturers. As an example, Figure 9 shows how the GPSContextProvider builds a simple RDF model representing containing GPS coordinates obtained from the device's location service. In this example, the context provider registers itself as a location listener in the Android location subsystem; in consequence, the onLocationChanged() method is called every time the device's location changes. In this case the context provider buffers the most recent Location object. In turn, the detectContextImpl() method is regularly invoked by the ActiveContextProvider class. It creates a resource of type mobisem:Context and uses the property mobisem:currentLocation to associate the context with another resource, representing the actual location. This second resource is described with two properties for the latitude and longitude values, as well as one property for the type of the used location provider, if applicable.

In our motivating example (cf. Section Motivating Example) a component that provides location information can be used to update information while John is on travel. It can be used for a variety of purposes, including the provisioning of general information about his location. It could be combined with information about John's personal interests in order to suggest points to visit in his spare time, or it could be combined with his calendar information in order to suggest public transport routes to the address where his next meeting takes place.

Data providers are built upon the abstract classes and interfaces of context provides AbstractBasicProvider and IBasicProvider; they actually expose very generic and common interfaces for

*Figure 9. Code snippet of GPSContextProvider, converting location data into an RDF-based context model*

```
1  // handle location updates
2  // (called by Android location manager)
3  public void onLocationChanged(Location location) {
4      this.location = location;
5  }
6
7  @Override
8  // detect context model
9  // (called by ActiveContextProvider)
10 protected void detectContextImpl() {
11     // read data from the location object (location object is provided by )
12     double lat = this.location.getLatitude();
13     double lon = this.location.getLongitude();
14     String gpsProviderType = this.location.getProvider();
15
16
17     // clear context model
18     this.getCurrentContextModel().removeAll();
19
20     // create new context resources
21     Resource context = getCurrentContextModel().createResource();
22     context.addProperty(RDF.type, Vocabulary.Mobisem.CONTEXT);
23     Resource location = getCurrentContextModel().createResource();
24     context.addProperty(Vocabulary.Mobisem.CURRENT_LOCATION, location);
25
26     // attach coordinates to location resource
27     location.addProperty(Vocabulary.Geo.LAT, lat);
28     location.addProperty(Vocabulary.Geo.LONG, lon);
29     if (gpsProviderType != null)
30         location.addProperty(Vocabulary.Mobisem.LOCATION_PROVIDER_TYPE, gpsProviderType);
31 }
```

managing their basic data as depicted in Figure 8. Data providers are responsible for handling all data replication tasks. They retrieve the context configuration from the replication manager which signals them to initiate their replication tasks based on their analysis of the relevant entities contained in the context configuration. Each data provider wraps a specific data source and replicates data to the local SQLite database deployed on the Android platform. The framework does not expose any restrictions on the data sources to be wrapped by data providers; they can wrap a remote file system object, a remote database, web services and web applications respectively, or generate data themselves. For instance, a data provider may act upon changes of the current location and retrieve

information about nearby points of interest. Each data provider is assigned a named graph under which it stores its data in the triple store. The only requirement exposed is that replicated data must be available as RDF. The following concrete data providers were implemented so far:

- DBpediaLocationDataProvider is able to process geo coordinates as well as Geonames ids of location resources (which are taken from the global context model). For all resources of type geonames:Feature, the data provider retrieves relevant location information from DBpedia.org by querying the DBpedia SPARQL endpoint for relevant resources, based on their label. For all returned resources, descriptive triples are stored on the mobile device.

- MobiSemDataProvider is the default data provider for replicating data from a (versioned) MobiSem repository (Schandl, 2010) to the mobile device. Through this client-/server-based infrastructure, any data source that can be expressed as RDF triples (e.g., file systems, relational data bases, etc.) can be exposed as versioned RDF graph repository, and can be replicated to a mobile device. For this purpose, a REST-style protocol has been developed that is used to specify additional metadata (like context information or the current graph version) in the form of additional HTTP headers. In addition to serving and versioning RDF graphs, the MobiSem repository is able to perform ranking and selection of RDF triples based on a context model that may be sent from the client alongside a checkout request. In this case, only parts of RDF graphs are replicated to the mobile device, saving transmission and storage capacity (Schandl, 2010). The SelectiveHomebaseDataProvider, in addition to providing support for such partial replicas, enables the client to write changes to replicated data back to the remote repository, which then takes care of merging and conflict detection (Schandl, 2010).
- SindiceTopResultsDataProvider extracts possible search terms from the global context model by analyzing all literals found therein. It tokenizes the literals and constructs a list of most frequent keywords, which are in turn sent to the Sindice.com Semantic Web indexing service. From there it retrieves the top results and converts them into an RDF model, which is then stored on the mobile device. Optionally, the data provider is able to retrieve more information about the top results by de-referencing the resource URIs, according to

Linked Data principles. Any data retrieved via this option is also stored on the mobile device's data store.

To illustrate the basic functionality of a data provider, the authors describe the DBpediaLocationDataProvider in more detail (Figure 11). It consists of two main methods: the first one, updateContextImpl(), is called by the data provider's implementation base class (AbstractDataProvider) whenever it receives an updated context model from the context dispatcher. In this method, the data provider analyzes the context model and iterates over all resources of type geonames:Feature; i.e., all logical locations. For all these resources it extracts the labels and stores them in a list of resource labels. In the second method, updateDataImpl() (which is again called by the implementation base class whenever the data provider is requested to update data) a SPARQL DESCRIBE query is constructed that incorporates all resource labels (in this example the authors assume the user is interested only in information in English). Additionally it restricts the query to resources of type dbpedia-owl:Place. This query is sent to the DBpedia query endpoint, and the results are read into the targetModel variable, which is subsequently processed by the context dispatcher.

Since DBpedia provides a large amount of information about locations (including names, descriptions, geographic and statistical data, as well as links to related persons, buildings, and events) such a data provider can be of great use in the case of our motivating example. John will be enabled to browse relevant information about the places that he will visit during his next trip without the need to actually search for it. By matching location information with John's interests (which could be derived from his Web browsing history, or tags he has used to annotate his photos) the system could recommend points of interests to him; by combining location data with publicly available personal profiles (e.g., from FOAF data) John could be notified of people that

*Figure 11. Code snippet of DBpediaLocationProvider, querying DBpedia for data about location resources*

```
1   // analyze the current context model (stored in this.contextModel)
2   @Override
3   protected void updateContextImpl() {
4       this.currentResourceLabels = new ArrayList<String>();
5
6       // iterate over all geonames features in the context model
7       StmtIterator si1 = this.contextModel.listStatements(null, RDF.type, GEONAMES_Feature);
8       while(si1.hasNext()) {
9           Resource featureResource = si1.nextStatement().getSubject();
10
11          // iterate over all statements of these features
12          StmtIterator si2 = this.contextModel.listStatements(featureResource, null, (RDFNode) null);
13          while(si2.hasNext()) {
14              // check if a label property is attached
15              Statement s = si2.nextStatement();
16              if( s.getObject().isLiteral() && ! this.currentResourceLabels.contains(s.getString())) {
17                  this.currentResourceLabels.add(s.getString());
18              }
19          }
20      }
21  }
22
23  // update data from the remove data source
24  @Override
25  protected void updateDataImpl(Model targetModel) {
26      // construct DESCRIBE query for all location resources
27      StringBuffer queryBuffer = new StringBuffer();
28      queryBuffer.append("DESCRIBE ?concept WHERE { \n");
29      for(String featureLabel: this.currentResourceLabels) {
30          queryBuffer.append("{" +
31              "?concept rdfs:label \"" + featureLabel + "\"@en . " +
32              "?concept rdf:type dbpedia-owl:Place . " +
33          "} UNION \n");
34      }
35      queryBuffer.append("{} }");
36
37      // send query to DBpedia
38      HttpClient client = new DefaultHttpClient();
39      String url = this.serviceUri + "?query=" + URLEncoder.encode(queryBuffer.toString());
40      HttpGet method = new HttpGet(url);
41      method.addHeader("Accept", "text/plain");    // accept only N-TRIPLES
42      try {
43          // execute request
44          HttpResponse response = client.execute(method);
45
46          // read model into targetModel (which is then further processed by the context framework)
47          targetModel.read(response.getEntity().getContent(), "N-TRIPLE");
48      }
49      catch (Exception e) {
50          // error handling
51      }
52  }
```

*Figure 12. Example of storing an RDF graph using the RDFContentProvider's insert()method*

```
1  ContentValues values = new ContentValues();
2  values.put(RDFContentProvider.ContentValues.NAME, GRAPH_NAME);
3  values.put(RDFContentProvider.ContentValues.MODEL, serializeModelAsByteArray(model));
4  URI uri = this.mResolver.insert(RDFContentProvider.CONTENT_URI_GRAPH, values);
```

live in cities John is going to visit, and based on their interests it could suggest meetings with potential customers or cooperation partners.

## Data Access and Provision

Replicated data are provided to other applications through an RDF content provider, which is automatically instantiated and started by the Android when the provider is first requested. Content providers are treated as *singleton classes* but can communicate with multiple ContentResolver-objects across system processes and applications. This is the reason why the operations a content provider offers must be implemented in a thread-safe manner to avoid side effects and data inconsistencies.

Content providers usually expose their data using a table-based data model, where columns represent specific types of data and rows a single record. Each data set a content provider exposes is identified via a unique and public-accessible URI called the *content URI* that follows a well-defined scheme. A content URI always starts with the content:// scheme to signal the system that the data is exposed by a content provider followed by an authority part that specifies the content provider, e.g., org.mobisem.rdfprovider, a path segment that can be used to address specific types of data in case a content provider exposes multiple data sets, e.g., /binary for a binary serialization, and an optional ID segment that allows for addressing specific records, e.g., the URI or name of a data replica.

Figure 12 provides a example of storing an RDF graph to the database using the RDFContentProvider's insert() method, which supports three different types of insertions: (1) inserting a complete graph, (2) adding a partial graph to an existing one, and (3) adding single triples. The URI determines which insert operation is to be performed; for instance, by specifying the URI content://org.mobisem.rdfprovider/graphs, a complete RDF graph is to be stored in the database. If only a partial graph is to be added to an existing graph, the content://org.mobisem.rdfprovider/graph URI can be used instead. In case a graph already exists under the given URI, the existing graph will be overwritten. The authors assume that the new graph is an updated replica that results from a context update. All the data to be stored in the database must be wrapped in a µJena model and serialized as a byte array.

Line 2 stores the graph's URI (indicated by the GRAPH_NAME variable) in the ContentValues using the name key (indicated by the RDFContentProvider.ContentValues.NAME constant). In the next line, the RDF graph is serialized as a byte array before it is added to the ContentValues. Finally, the insert() method of the RDFContentProvider is requested using the URI for adding a complete graph together with the graph data stored in the ContentValues object. In case storage was successful, the graph's URI is returned.

The query() method allows to query for data replicas stored in the local database. Depending on the content URI, the RDFContentProvider either returns a collection of all data replicas' names together with the amount of triples included in each graph, an N-Triple-based representation of the triples contained in a specific replica, or a serialized µJena model of a replica that can be parsed for further processing. Figure 13 depicts a code fragment for retrieving a byte array serial-

*Figure 13. Example of transforming a data replica to a workable in-memory RDF model*

```
1  Cursor c = this.mResolver.query(
2     RDFContentProvider.CONTENT_URI_BINARY, null, REPLICA_UUID, null, null);
3  assertTrue(c.moveToFirst());
4  Model m = transformToRdfModel(c.getBlob(0));
```

ized data replica that can directly transformed to an in-memory RDF model.

The data replica retrieved from the triple store is stored in a cursor-object that allows for iterating over the data contained in the result set. Therefore, the cursor is moved to the first entry (line 3) and the byte array-based serialization of the requested data replica is passed to a transformation method that parses the byte stream and creates an in-memory model of the data replica (line 4).

In general, the RDFContentProvider also allows for adding and removing triples (update)[12] as well as deleting all replicas stored in the database, a specific replica, or all the triples of a specific replica's resource.

## FUTURE WORK

Although this framework demonstrates that semantic technologies can make substantial contributions in realizing a mobile context-aware infrastructure assisting users in satisfying their mobile information needs, there are still some open issues that need to be addressed in future research. For instance, the integration of dynamically discovered context sources is a challenge most context-management frameworks face, especially in mobile and ubiquitous environments. The authors therefore suggest to direct future research towards finding and integrating such sources and develop common interfaces and protocols for this purpose. For this, Semantic Web technologies are a candidate because they have been proven to be an appropriate infrastructure for the explicit representation and exchange of data semantics. Those semantics can be used not only for describing contextual information, but also for sensor interfaces to facilitate discovery and integration in running systems, as well as the exchange of contextual information. The authors therefore plan to investigate additional methods for dynamic context source discovery and integration as well as heuristics for transforming sensorial data into qualitative context descriptions.

Further, our framework could be extended to include feedback loops that would allow for adjusting context acquisition and aggregation tasks according to data provisioning needs. Further, more sophisticated reasoning capabilities are missing, which the authors plan to address in future work. A context framework as such should be aware of the operating system conditions to adapt itself to changing conditions. Finally, support for efficient and complex inferencing techniques is still an open issue on mobile systems and should be addressed in future research.

## CONCLUSION

In this chapter, the authors have introduced the concepts of Semantic Web-enhanced context-aware computing for mobile systems and presented the MobiSem context framework that provides an infrastructure for intelligently assisting mobile users by selectively replicating RDF data from remote data sources to a mobile device according to their current and future information needs and the different contexts they are operating in. Since research in context-aware computing and mobile systems shows that context and context awareness should be central parts of future mobile information systems, the authors drafted an application scenario that outlines how Semantic Web-enhanced context-aware computing can support mobile users in fulfilling their information needs that also serves as a motivating example.

This chapter provides an overview on how context is used in information system and discussed problems and limitations of current context-aware computing approaches in information systems. Areas were presented where the introduction of concepts and technologies from the Semantic Web can make substantial contributions in representing, processing, and managing contextual information. For this purpose, concepts from graph theory, distributed transaction management, and the Semantic Web were adopted in order to demonstrate that a synthesis of those fields can serve as an architectural infrastructure for the efficient acquisition, management, and processing of contextual information directly on a mobile device. In consequence, dependencies to external systems can be reduced, and security and privacy issues are reflected since private data do not need to be transferred outside the mobile system.

The proposed architecture allows for a controlled acquisition, aggregation, and consolidation of contextual information taking into account mobile operating system peculiarities. At the same time it guarantees consistency, accurateness, and completeness among contextual data and acquisition workflows. It employs a loose coupling between context acquisition and data provisioning components, which is gained by applying semantic technologies (data models, vocabularies, inference) to interpret and process context information. Practical insights into implementation details of context and data providers were discussed, and concrete examples were shown on how context-relevant data can be acquired from locally deployed sensors, represented using semantic vocabularies, and utilized by external applications.

## REFERENCES

Abadi, D. J., Marcus, A., Madden, S. R., & Hollenbach, K. (2009). Sw-store: A vertically partitioned DBMS for Semantic Web data management. *The VLDB Journal, 18*(2), 385–406. doi:10.1007/s00778-008-0125-y

Abowd, G. D., Dey, A. K., Brown, P. J., Davies, N., Smith, M., & Steggles, P. (1999). Towards a better understanding of context and context-awareness. In *Huc '99: Proceedings of the 1st International Symposium on Handheld and Ubiquitous Computing* (pp. 304-307). London, UK: Springer-Verlag.

Anagnostopoulos, C. B., Ntarladimas, Y., & Hadjiefthymiades, S. (2007). Situational computing: An innovative architecture with imprecise reasoning. *Journal of Systems and Software, 80*(12), 1993–2014. doi:10.1016/j.jss.2007.03.003

Anagnostopoulos, C. B., Tsounis, A., & Hadjiefthymiades, S. (2007). Context awareness in mobile computing environments. *Wireless Personal Communications: An International Journal, 42*(3), 445–464. doi:10.1007/s11277-006-9187-6

*Android developer's guide.* (2010). Retrieved December 13, 2010, from http://developer.android.com/guide/index.html

Auer, S., Bizer, C., Kobilarov, G., Lehmann, J., & Zachary, I. (2007). DBpedia: A nucleus for a Web of open data. In *Proceedings of the 6th International Semantic Web Conference ISWC* (pp. 11-15).

Baader, F., Calvanese, D., McGuinness, D. L., Nardi, D., & Patel-Schneider, P. F. (Eds.). (2003). *The description logic handbook: Theory, implementation, and applications*. Cambridge University Press.

Baldauf, M., Dustdar, S., & Rosenberg, F. (2007). A survey on context-aware systems. *International Journal of Ad Hoc and Ubiquitous Computing, 2*(4), 263–277. doi:10.1504/IJAHUC.2007.014070

Becker, C., & Bizer, C. (2008). *DBpedia mobile: A location-enabled linked data browser*. In Workshop on linked data on the Web (LDOW 2008).

Becker, C., & Bizer, C. (2009a). Exploring the geospatial semantic web with dbpedia mobile. *Journal of Web Semantics, 7*(4), 278–286. doi:10.1016/j.websem.2009.09.004

Becker, C., & Bizer, C. (2009b). *Marbles*. Retrieved from http://marbles.sourceforge.net/

Becket, D. (2004). *RDF/XML syntax specication* (W3C recommendation 10 February 2004). Retrieved November 19, 2010, from http://www.w3.org/TR/2004/REC-rdf-syntax-grammar-20040210

Beckett, D., & Berners-Lee, T. (2008). *Turtle - Terse RDF triple language* (W3C Team Submission 14 January 2008). Retrieved November 25, 2010, from http://www.w3.org/TeamSubmission/turtle/

Bergman, M. K. (2005). *Untapped assets: The $3 trillion value of U.S. enterprise documents*. BrightPlanet Corporation White Paper, July 2005, 42 pp.

Berners-Lee, T., Fielding, R., & Masinter, L. (2005). *Uniform resource identier (URI): Generic syntax (RFC 3986)*. Retrieved 11th December, 2010, from http://www.faqs.org/rfcs/rfc3986.html

Berners-Lee, T., Hendler, J., & Lassila, O. (2001). The Semantic Web. *Scientific American, 284*(5), 34–43. doi:10.1038/scientificamerican0501-34

Biegel, G., & Cahill, V. (2004). A framework for developing mobile, context-aware applications. In *Proceedings of the Second IEEE International Conference on Pervasive Computing and Communications (PerCom'04)* (pp. 361-365). Washington, DC, USA.

Bizer, C. (2009). The emerging Web of linked data. *IEEE Intelligent Systems, 24*, 87–92. doi:10.1109/MIS.2009.102

Bizer, C., Heath, T., & Berners-Lee, T. (2009). Linked data - The story so far. *International Journal on Semantic Web and Information Systems, 5*(3), 1–22. doi:10.4018/jswis.2009081901

Boardman, R., & Sasse, M. A. (2004). Stuff goes into the computer and doesn't come out: A cross-tool study of personal information management. In *Proceedings of Sigchi Conference on Human Factors in Computing Systems* (pp. 583-590). New York, NY: ACM.

Boehm, S., Koolwaaij, J., & Luther, M. (2008). Share whatever you like. *Electronic Communication of the EASST (ECEASST)*, 11.

Boehm, S., Koolwaaij, J., Luther, M., Souville, B., Wagner, M., & Wibbels, M. (2008). Introducing IYOUIT. In *The Semantic Web* (pp. 804–817). ISWC.

Bolchini, C., Curino, C. A., Quintarelli, E., Schreiber, F. A., & Tanca, L. (2007). A data-oriented survey of context models. *SIGMOD Record, 36*(4), 19–26. doi:10.1145/1361348.1361353

Boley, H., Hallmark, G., Kifer, M., Paschke, A., Polleres, A., & Reynolds, D. (2010). *RIF core dialect* (W3C recommendation 22 June 2010). Retrieved from http:// www.w3.org/TR/rif-core/

Bradley, N. A., & Dunlop, M. D. (2005). Toward a multidisciplinary model of context to support contextaware computing. *Human-Computer Interaction, 20*(4), 403–446. doi:10.1207/s15327051hci2004_2

Brickley, D., & Guha, R. (2004). *RDF vocabulary description language 1.0: RDF schema* (W3C Recommendation February 10, 2004).

Brickley, D., & Miller, L. (2007). *FOAF vocabulary specication 0.91*.

Broekstra, J., Kampman, A., & van Harmelen, F. (2002). Sesame: A generic architecture for storing and querying RDF and RDF schema. In I. Horrocks & J. Hendler (Eds.), *Proceedings of the First International Semantic Web Conference* (pp. 54-68). Springer Verlag.

Cai, G., & Xue, Y. (2006). Activity-oriented context-aware adaptation assisting mobile geospatial activities.In *IUI '06: Proceedings of the 11th International Conference on Intelligent User Interfaces* (pp. 354-356). New York, NY: ACM.

Carroll, J. J., Bizer, C., Hayes, P., & Stickler, P. (2005). Named graphs. *Journal of Web Semantics, 3*(4), 247–267. doi:10.1016/j.websem.2005.09.001

Carton, A., Clarke, S., Senart, A., & Cahill, V. (2007). Aspect-oriented model-driven development for mobile context-aware computing. In *SEP-CASE '07: Proceedings of the 1st International Workshop on Software Engineering for Pervasive Computing Applications, Systems, and Environments*. Washington, DC: IEEE Computer Society.

Chen, H. (2004). *An intelligent broker architecture for pervasive context-aware systems.* Unpublished doctoral dissertation, University of Maryland, Baltimore County.

Chen, H., & Joshi, A. (2004). An ontology for context-aware pervasive computing environments. *Special Issue on Ontologies for Distributed Systems. The Knowledge Engineering Review, 18*(3), 197–207. doi:10.1017/S0269888904000025

Cheng, G., & Qu, Y. (2009). Searching linked objects with falcons: Approach, implementation and evaluation. *International Journal on Semantic Web and Information Systems, 5*(3), 49–70. doi:10.4018/jswis.2009081903

Coutaz, J., Crowley, J. L., Dobson, S., & Garlan, D. (2005). Context is key. *Communications of the ACM- Special Issue. The Disappearing Computer, 48*(3), 49–53.

Dey, A. K. (2000). *Providing architectural support for building context-aware applications.* Unpublished doctoral dissertation, Georgia Institute of Technology.

Dey, A. K. (2001). Understanding and using context. *Personal and Ubiquitous Computing, 5*(1), 4–7. doi:10.1007/s007790170019

Dey, A. K., Abowd, G. D., Pinkerton, M., & Wood, A. (1997). Cyberdesk: A framework for providing selfintegrating ubiquitous software services. In *ACM Symposium on User Interface Software and Technology* (pp. 75-76).

Dey, A. K., Abowd, G. D., & Salber, D. (2001). A conceptual framework and a toolkit for supporting the rapid prototyping of context-aware applications. *Human-Computer Interaction, 16*(2), 97–166. doi:10.1207/S15327051HCI16234_02

Diaper, D. (1990). *Task analysis for human-computer interaction.* Upper Saddle River, NJ: Prentice Hall PTR.

Dourish, P. (2004). What we talk about when we talk about context. *Personal and Ubiquitous Computing, 8*(1), 19–30. doi:10.1007/s00779-003-0253-8

Egenhofer, M. J. (2002). Toward the semantic geospatial Web. In Voisard, A., & Chen, S.-C. (Eds.), *ACM-GIS* (pp. 1–4). ACM.

Erling, O., & Mikhailov, I. (2007). RDF support in the Virtuoso DBMS. In S. Auer, C. Bizer, C. Müller, & A. V. Zhdanova (Eds.), *1st Conference on Social Semantic Web,* (vol. 113, pp. 59-68). Leipzig, Germany.

Euzenat, J. (2005). *Alignment infrastructure for ontology mediation and other applications (Vol. 168*, pp. 81–95). Mediate.

Euzenat, J., Pierson, J., & Ramparany, F. (2008). Dynamic context management for pervasive applications. *The Knowledge Engineering Review, 23*(1), 21–49. doi:10.1017/S0269888907001269

Fielding, R., Gettys, J., Mogul, J., Frystyk, H., Masinter, L., Leach, P., et al. (1999). *Hypertext transfer protocol - HTTP/1.1 (RFC 2616).*

Forman, G. H., & Zahorjan, J. (1994). The challenges of mobile computing. *Computer, 27*(4), 38–47. doi:10.1109/2.274999

Fournier, D., Mokhtar, S. B., Georgantas, N., & Issarny, V. (2006). Towards ad hoc contextual services for pervasive computing. In *MW4SOC '06: Proceedings of 1st Workshop on Middleware for Service Oriented Computing (MW4SOC 2006)* (pp. 36-41). New York, NY: ACM.

Franz, T., Ansgar, S., & Staab, S. (2009). Are semantic desktops better? Summative evaluation comparing a semantic against a conventional desktop. *In Proceedings of Fifth International Conference on Knowledge Capture* (pp. 1-8). New York, NY: ACM

Gellersen, H. W., Schmidt, A., & Beigl, M. (2002). Multi-sensor context-awareness in mobile devices and smart artifacts. *Mobile Networks and Applications, 7*(5). doi:10.1023/A:1016587515822

Gray, J., & Reuter, A. (1992). *Transaction processing: Concepts and techniques* (1st ed.). San Francisco, CA: Morgan Kaufmann Publishers Inc.

Grosz, B. J., & Kraus, S. (1996). Collaborative plans for complex group action. *Artificial Intelligence, 86*(2), 269–357. doi:10.1016/0004-3702(95)00103-4

Harris, S., & Gibbins, N. (2003). 3store: Efficient bulk RDF storage. In *Practical and Scalable Semantic Systems, Proceedings of the First International Workshop on Practical and Scalable Semantic Systems.*

Hashimi, S. Y., Komatineni, S., & MacLean, D. (2010). *Pro Android 2.* Apress. doi:10.1007/978-1-4302-2660-4

Heath, T., Dzbor, M., & Motta, E. (2005). *Supporting user tasks and context: Challenges for Semantic Systems* (pp. 433-442). New York, NY: ACM.

Heath, T., & Motta, E. (2008). Revyu: Linking reviews and ratings into the Web of data. *Journal of Web Semantics: Science. Services and Agents on the World Wide Web, 6,* 266–273. doi:10.1016/j.websem.2008.09.003

Heath, T., Motta, E., & Dzbor, M. (2005). Context as a foundation for a semantic desktop. In S. Decker, J. Park, D. Quan, & L. Sauermann (Eds.), *Proceedings of the Semantic Desktop Workshop at the ISWC,* Galway, Ireland, (vol. 175).

Henricksen, K., Indulska, J., McFadden, T., & Balasubramaniam, S. (2005). *Middleware for distributed context-aware systems.* In OTM Conference.

Henricksen, K., Indulska, J., & Rakotonirainy, A. (2002). Modeling context information in pervasive computing systems. In *Pervasive '02: Proceedings of First International Conference on Pervasive Computing* (pp. 167-180). London, UK: Springer-Verlag.

Hertel, A., Broekstra, J., & Stuckenschmidt, H. (2008). *RDF storage and retrieval systems.* Online. Retrieved from http://ki.informatik.uni-mannheim.de/fileadmin/publication/Hertel-08RDFStorage.pdf

Hofer, T., Schwinger, W., Pichler, M., Leonhartsberger, G., Altmann, J., & Retschitzegger, W. (2003). Context-awareness on mobile devices - The Hydrogen approach. In *HICSS '03: Proceedings of 36th Annual Hawaii International Conference on System Sciences (HICSS'03).* Washington, DC, USA: IEEE Computer Society.

Hu, D. H., Dong, F., & Wang, C.-L. (2009). A semantic context management framework on mobile device. In *ICESS 2009: Proceedings of the 2009 International Conference on Embedded Software and Systems* (pp. 331-338). Washington, DC: IEEE Computer Society.

Huebscher, C., & McCann, A. (2005). An adaptive middleware framework for context-aware applications. *Personal and Ubiquitous Computing, 10*(1), 12–20. doi:10.1007/s00779-005-0035-6

Indulska, J., & Sutton, P. (2003). Location management in pervasive systems. In *ACSW Frontiers '03: Proceedings of Australasian Information Security Workshop Conference on ACSW Frontiers 2003* (pp. 143-151). Darlinghurst, Australia: Australian Computer Society, Inc.

Kaenampornpan, M., & Ay, B. B. (2004). *An intergrated context model: Bringing activity to context*. In Workshop on Advanced Context Modelling, Reasoning and Management - UBICOMP.

Kamvar, M., & Baluja, S. (2006). A large scale study of wireless search behavior: Google mobile search. In *CHI '06: Proceedings of Sigchi Conference on Human Factors in Computing Systems* (pp. 701-709).

Kearney, A. T. (2001). *Network publishing: Creating value through digital content*. ATKearney White Paper, April 2001, 32 pp.

Kim, S. W., Park, S. H., Lee, J., Jin, Y. K., Park, H.-M., & Chung, A. (2004). Sensible appliances: Applying context-awareness to appliance design. *Personal and Ubiquitous Computing, 8*(3-4), 184–191. doi:10.1007/s00779-004-0276-9

Klyne, G., & Carroll, J. J. (2004). *Resource description framework (RDF): Concepts and abstract syntax* (W3C recommendation 10 February 2004).

Korpipää, P., Malm, E., Salminen, I., Rantakokko, T., Kyllönen, V., & Känsälä, I. (2005). Context management for end user development of context-aware applications. In *MDM '05: 6th International Conference on Mobile Data Management* (pp. 304-308). ACM.

Korpipää, P., Mantyjarvi, J., Kela, J., Keranen, H., & Malm, E. (2003). Managing context information in mobile devices. *Pervasive Computing, 2*(3), 42–51. doi:10.1109/MPRV.2003.1228526

Krogstie, J., Lyytinen, K., Opdahl, A., Pernici, B., Siau, K., & Smolander, K. (2004). Research areas and challenges for mobile Information Systems. *International Journal of Mobile Communications, 2*(3), 220–234.

Kuutti, K. (1995). *Activity theory as a potential framework for human-computer interaction research*. Cambridge, MA: Massachusetts Institute of Technology.

Lehmann, J., Bizer, C., Kobilarov, G., Auer, S., Becker, C., & Cyganiak, R. (2009). DBpedia - A crystallization point for the Web of data. *Journal of Web Semantics, 7*(3), 154–165. doi:10.1016/j.websem.2009.07.002

Lewis, R. (2007). *Dereferencing HTTP URIs*. Retrieved January 9, 2010, from http://www.w3.org/2001/tag/doc/httpRange-14/2007-05-31/HttpRange-14

Luther, M., Fukazawa, Y., Wagner, M., & Kurakake, S. (2008). Situational reasoning for task-oriented mobile service recommendation. *The Knowledge Engineering Review, 23*(1), 7–19. doi:10.1017/S0269888907001300

Luther, M., Mrohs, B., Wagner, S., Steglich, M., & Kellerer, W. (2005). Situational reasoning - A practical OWL use case. In *Proceedings of 7th International Symposium on Autonomous Decentralized Systems (ISADS2005)* (pp. 96-103). Chengdu, China.

McGuinness, D. L., & van Harmelen, F. (Eds.). *OWL Web ontology language overview* (W3C recommendation). World Wide Web Consortium. Retrieved January 10, 2010, from http://www.w3.org/TR/2004/REC-owl-features-20040210/

Meier, R. (2010). *Professional Android 2 application development*. Wiley Publishing.

Mihalic, K., & Tscheligi, M. (2007). 'Divert: Mother-in-law': Representing and evaluating social context on mobile devices. In Mobilehci '07: *9th International Conference on Human Computer Interaction with Mobile Devices & Services* (pp. 257-264). ACM.

Miller, E., Shen, D., Liu, J., & Nicholas, C. (2000). Performance and scalability of a large-scale n-gram based information retrieval system. *Journal of Digital Information*, 1.

Motik, B., Patel-Schneider, P. F., & Parsia, B. (2009, October). *OWL 2 Web ontology language structural specication and functional-style syntax* (W3C recommendation 27 October 2009). Retrieved from http://www.w3.org/TR/owl2-syntax/

Nardi, B. A. (Ed.). (1995). *Context and consciousness: Activity theory and human-computer interaction*. Cambridge, MA: Massachusetts Institute of Technology.

New York, NY: ACM.

Oren, E., Delbru, R., Catasta, M., Cyganiak, R., Stenzhorn, H., & Tummarello, G. (2008). Sindice.com: A document-oriented lookup index for Open Linked data. *International Journal of Metadata, Semantics and Ontologies, 3*(1).

Pawar, P., van Halteren, A. T., & Sheikh, K. (2007, March). Enabling context-aware computing for the nomadic mobile user: A service oriented and quality driven approach. In *IEEE Wireless Communications and Networking Conference WCNC 2007* (pp. 2531-2536). IEEE Communication Society.

Pietriga, E., Bizer, C., Karger, D., & Lee, R. (2006). Fresnel: A browser-independent presentation vocabulary for RDF. In *Proceedings of the 5th International Semantic Web Conference ISWC 2006* (Vol. 4273, pp. 158-171). Springer-Verlag.

Preuveneers, D., den Bergh, J. V., Wagelaar, D., Georges, A., Rigole, P., Clerckx, T., et al. (2004). Towards an extensible context ontology for ambient intelligence. In P. Markopoulos, B. Eggen, E. Aarts, & J. L. Crowley (Eds.), *Second European Symposium on Ambient Intelligence* (vol. 3295, pp. 148-159). Eindhoven, The Netherlands: Springer.

Prud'hommeaux, E., & Seaborne, A. (2008*). SPARQL query language for RDF* (W3C recommendation January 15, 2008). Retrieved from http://www.w3.org/TR/rdf-sparql-query/

Ramparany, F., Euzenat, J., Broens, T. H. F., Bottaro, A., & Poortinga, R. (2006, April). *Context management and semantic modelling for ambient intelligence* (Technical Report No. TR-CTIT-06-52). Enschede.

Raptis, D., Tselios, N., & Avouris, N. (2005). *Context-based design of mobile applications for museums: A survey of existing practices*. In MobileHCI '05: 7th International Conference on Human Computing Interaction with Mobile Devices & Services. ACM.

Redish, J., & Wixon, D. (2003). In Jacko, J. A., & Sears, A. (Eds.), *The human-computer interaction handbook* (pp. 922–940). Hillsdale, NJ: L. Erlbaum Associates Inc.

Rogers, R., Lombardo, J., Mednieks, Z., & Meike, B. (2009). *Android application development: Programming with the Google SDK*. Beijing, China: O'Reilly.

Ryan, N. S., Pascoe, J., & Morse, D. R. (1998). Enhanced reality fieldwork: The context-aware archaeological assistant. In V. Ganey, M. van Leusen, & S. Exxon (Eds.), *Computer applications in archaeology 1997*. Oxford, UK: Tempus Reparatum. Retrieved from http://www.cs.kent. ac.uk/pubs/ 1998/616

Sauermann, L., Bernardi, A., & Dengel, A. (2005). Overview and outlook on the semantic desktop. In S. Decker, J. Park, D. Quan, & L. Sauermann (Eds.), *Proceedings of the 1ˢᵗ Workshop on the Semantic Desktop at the ISWC 2005 Conference* (vol. 175, pp. 1-18). CEUR-WS.

Sauermann, L., & Cyganiak, R. (2008). *Cool URIs for the Semantic Web*. W3C Interest Group Note. Retrieved from http://www.w3.org/TR/cooluris/

Schandl, B. (2010). Replication and versioning of partial RDF graphs. In *Proceedings of 7ᵗʰ European Semantic Web Conference (ESWC 2010)*.

Schilit, B., & Theimer, M. (1994, Sep/Oct). Disseminating active map information to mobile hosts. *Network, 8*(5), 22–32.

Schmidt, A., Beigl, M., & Gellersen, H.-W. (1998). There is more to context than location. *Computers & Graphics, 23*, 893–901. doi:10.1016/S0097-8493(99)00120-X

Schraefel, M. C., Smith, D. A., Owens, A., Russell, A., Harris, C., & Wilson, M. (2005). The evolving mspace platform: Leveraging the Semantic Web on the trail of the memex. In *Hypertext 2005: Proceedings of Sixteenth ACM Conference on Hypertext and Hypermedia* (pp. 174-183). New York, NY: ACM.

Seaborne, A. (2004). *RDQL - A query language for RDF*. W3C member submission. Retrieved from http://www.w3.org/Submission/2004/SUBMRDQL-20040109/

Sohn, T., Li, K. A., Griswold, W. G., & Hollan, J. D. (2008). A diary study of mobile information needs. In *CHI '08: Proceedings of the Twenty-sixth Annual Sigchi Conference on Human Factors in Computing*

Springer, T., Wustmann, P., Braun, I., Dargie, W., & Berger, M. (2008). A comprehensive approach for situation-awareness based on sensing and reasoning about context. In *UIC '08: Proceedings of 5ᵗʰ International Conference on Ubiquitous Intelligence and Computing* (pp. 143-157). Berlin, Germany: Springer-Verlag.

Strang, T., & Popien, C. L. (2004, September). A context modeling survey. In *Ubicomp 1ˢᵗ International Workshop on Advanced Context modelling, Reasoning and Management* (pp. 31-41). Nottingham.

Teo, H.-S. (2008). An activity-driven model for context-awareness in mobile computing. In *MobileHCI '08: 10th International Conference on Human Computer Interaction with Mobile Devices & Services* (pp. 545-546). New York, NY: ACM.

Volz, J., Bizer, C., Gaedke, M., & Kobilarov, G. (2009). Discovering and maintaining links on the Web of data. In *The Semantic Web - ISWC 2009: 8ᵗʰ International Semantic Web Conference* (pp. 650-656). Chantilly, VA, USA.

Web research. *Proceedings of the ESWC2005: Workshop on End-User Aspects of the Semantic Web (UserSWeb)*.

Weithöner, T., Liebig, T., Luther, M., Böhm, S., Henke, F., & Noppens, O. (2007). Real-world reasoning with OWL. In ESWC2007: *Proceedings of 4ᵗʰ European Conference on the Semantic Web* (pp. 296-310). Berlin, Germany: Springer-Verlag.

Widjaja, I., & Balbo, S. (2005). Spheres of role in Context-awareness. In *OZCHI '05: Proceedings of 17th Australia Conference on Computer-Human Interaction* (pp. 1-4). Narrabundah, Australia: Computer-Human Interaction Special Interest Group (CHISIG) of Australia.

Wilson, M., Russell, A., Smith, D. A., Owens, A., & Schraefel, M. C. (2005). mSpace mobile: A mobile application for the Semantic Web. *End User Semantic Web Workshop, ISWC2005*, (p. 11).

Wojciechowski, M., & Xiong, J. (2006). Towards an open context infrastructure. *Proceedings of Workshop on Context Awareness for Proactive Systems (CAPS'06)* (pp.125-136).

Xia, F., Yakovlev, A. V., Clark, I. G., & Shang, D. (2002). Data communication in systems with heterogeneous timing. *IEEE Micro, 22*, 58–69. doi:10.1109/MM.2002.1134344

Zander, S., & Schandl, B. (2010). A framework for context-driven RDF data replication on mobile devices. In *Proceedings of the 6th International Conference on Semantic Systems (I-Semantics)*. Graz, Austria.

Zander, S., & Schandl, B. (2011). Context-driven RDF data replication on mobile devices. *Semantic Web Journal- Interoperability, Usability, Applicability, 1*(1).

## ADDITIONAL READING

Abowd, G. D., Mynatt, E. D., & Rodden, T. (2002). The Human Experience. *IEEE Pervasive Computing / IEEE Computer Society [and] IEEE Communications Society, 1*, 48–57. doi:10.1109/MPRV.2002.993144

Allemang, D., & Hendler, J. A. (2008). *Semantic Web for the Working Ontologist: Modeling in RDF, RDFS and OWL*. Morgan Kaufmann.

Antoniou, G., & van Harmelen, F. (2008). *A Semantic Web Primer*. Cambridge, MA: MIT Press.

Beckett, D. (2004). *RDF/XML Syntax Specification (W3C Recommendation February 10, 2004)*, http://www.w3.org/TR/rdf-syntax-grammar/.

Berners-Lee, T. (1999). *Weaving the Web: The Past, Present and Future of the World Wide Web by its Inventor*. London: Texere.

Berners-Lee, T., Hendler, J., & Lassila, O. (2001). The Semantic Web. *Scientific American, 284*(5), 34–43. doi:10.1038/scientificamerican0501-34

Bizer, C., Cyganiak, R., & Heath, T. (2007). *How to publish Linked Data on the Web*. http://www4.wiwiss.fu-berlin.de/bizer/pub/LinkedDataTutorial/.

Bizer, C., Heath, T., & Berners-Lee, T. (2009). Linked Data - The Story So Far. *International Journal on Semantic Web and Information Systems, 5*(3), 1–22. doi:10.4018/jswis.2009081901

Bu, Y., Tao, X., & Chen, S. (2006). Managing quality of context in pervasive computing. In QSIC 2006: *Proceedings of the Sixth International Conference on Quality Software* (pp. 193-200).

Church, K., Smyth, B., Cotter, P., & Bradley, K. (2007). Mobile information access: A study of emerging search behavior on the mobile internet. *ACM Transactions on the Web, 1*(1), 4. doi:10.1145/1232722.1232726

Dudkowski, D., Weinschrott, H., & Marron, P. J. (2008). Design and implementation of a reference model for context management in mobile ad-hoc networks. *In AINAW '08: Proceedings of the 22nd International Conference on Advanced Information Networking and Applications - Workshops*, (pp. 832-837). Washington, DC, USA, IEEE Computer Society.

Erickson, T. (2002). Some problems with the notion of context-aware computing. *Communications of the ACM - Ontology: Different ways of representing the same concept, 45*(2), 102-104.

Fielding, R., et al. (1999). *Hypertext Transfer Protocol -- HTTP/1.1*. RFC 2616, http://www.w3.org/Pro-tocols/rfc2616/rfc2616.html.

Gehlen, G., Aijaz, F., Sajjad, M., & Walke, B. (2007). A mobile context dissemination middleware. In *Proceedings of the International Conference on Information Technology, ITNG '07* (pp. 155-160). Washington, DC, USA. IEEE Computer Society.

Greenberg, S. (2001). Context as a dynamic construct. *Journal on Human-Computer Interaction*, *16*(2), 257–268. doi:10.1207/S15327051HCI16234_09

Heath, T. (2008). How Will We Interact with the Web of Data? *IEEE Internet Computing*, *12*(5), 88–91. doi:10.1109/MIC.2008.101

Kjaer, K. E. (2007). A survey of context-aware middleware. In SE'07: *Proceedings of the 25th Conference on IASTED International Multi-Conference* (pp. 148-155). Anaheim, CA, USA. ACTA Press.

Lei, H., Sow, D. M., Davis, J. S., Banavar, G., & Ebling, M. R. (2002). The design and applications of a context service. *SIGMOBILE Mobile Computing and Communications Review*, *6*(4), 45–55. doi:10.1145/643550.643554

McGuinness, D. L. (2002). Ontologies Come of Age. In Fensel, D., Hendler, J., Lieberman, H., & Wahlster, W. (Eds.), *Spinning the Semantic Web: Bringing the World Wide Web to Its Full Potential*. MIT Press.

Preuveneers, D., & Berbers, Y. (2008). Encoding Semantic Awareness in Resource-Constrained Devices. *IEEE Intelligent Systems*, *23*(2), 26–33. doi:10.1109/MIS.2008.25

Schmidt, H., Flerlage, F., & Hauck, F. J. (2009). A generic context service for Ubiquitous environments. In *Proceedings of the 2009 IEEE International Conference on Pervasive Computing and Communications* (pp. 1-6). Washington, DC, USA. IEEE Computer Society.

Schmohl, R., & Baumgarten, U. (2008). A generalized context-aware architecture in heterogeneous mobile computing environments. In ICWMC '08: *Proceedings of the Fourth International Conference on Wireless and Mobile Communications* (pp. 118-124). Washington, DC, USA. IEEE Computer Society.

Shadbolt, N., Berners-Lee, T., & Hall, W. (2006). The Semantic Web Revisited. *IEEE Intelligent Systems*, *21*(3), 96–101. doi:10.1109/MIS.2006.62

Trajcevski, G., & Scheuermann, P. (2009). Managing context evolution in pervasive environments. In *Proceedings of the 2nd International Conference on Pervasive Technologies Related to Assistive Environments, PETRA '09* (pp.19:1-19:2). New York, NY, USA. ACM.

## KEY TERMS AND DEFINITIONS

**Context:** Context can be defined as any information that characterizes the situation of an entity and gives meaning to something. An entity can be a person, place, or object being considered relevant with respect to the interaction between a user and an application, including the user and applications themselves.

**Context Awareness:** Context awareness describes a system's capability to conceive aspects of the physical and virtual environment it is operating in and tailor its behavior and responses according to the computational analysis of such aspects.

**Contextual Information:** Contextual information are constituting parts of the surrounding context that may be used for describing relevant aspects of the situation a user or device is currently operating in. Contextual Information is acquired through a variety of different technologies and protocols and can be considered as computational abstractions over real-world aspects that have a specific realationship to the user's current situation and tasks at hand.

**Linked Data:** Linked Data denotes the practice of publishing data on the Web according to simple core principles. Its basic idea is to denote resources exclusively using HTTP URIs, and to allow data consumers to directly de-reference these URIs to retrieve structured information about a resource including links to other relevant resources.

**Ontology:** An Ontology in information science is a formal and explicit specification of a shared conceptualization to model the entities of a specific domain as well as the relationships among them using shared vocabularies and logic-based relations. Ontologies are used to reason about the entities of a domain – the universe of discourse.

**Resource Description Framework (RDF):** RDF represents a family of W3C standards for the formal description of meta information about resources identified through unique URIs on the Web. RDF is a core technology of the Semantic Web and consists of a simple triple-based structure that forms an RDF graph.

**Semantic Web:** The Semantic Web refers to the idea of expressing rich, machine-processable knowledge using the Web infrastructure where such knowledge is published in a structured format using vocabularies that follow a well-defined semantics. It is envisioned to serve as an underlying information-provisioning infrastructure for information-centric applications, whereas the information processing is performed semi-automatically by computer programs.

## ENDNOTES

[1] http://esw.w3.org/SweoIG/TaskForces/CommunityProjects/LinkingOpenData

[2] http://www.w3.org/TR/owl-time

[3] http://www.w3.org/2003/01/geo

[4] http://www.w3.org/Mobile/CCPP

[5] http://www.foaf-project.org

[6] mSpace: http://research.mspace.fm/mspace

[7] Open Handset Alliance: http://www.open-handsetalliance.com/

[8] W3C WGS84 vocabulary: http://www.w3.org/2003/01/geo/

[9] Dublin Core vocabulary: http://purl.org/dc/terms/

[10] NEPOMUK ontology: http://www.semanticdesktop.org/ontologies/2007/04/02/ncal/

[11] FOAF vocabulary: http://xmlns.com/foaf/0.1/

[12] Updates to RDF graphs usually consist of two sets: one set that contains the triples to be added to an RDF graph and another set that contains the triples to be removed from a graph.

# Section 2
# Applications and Cases

# Chapter 4
# Applications of Graph Theory Algorithms in Mobile Ad hoc Networks

**Natarajan Meghanathan**
*Jackson State University, USA*

## ABSTRACT

*Various simulators (e.g., ns-2 and GloMoSim) are available to implement and study the behavior of the routing protocols for Mobile Ad hoc Networks (MANETs). However, students and investigators who are new to this area often get perplexed in the complexity of these simulators and lose the focus in designing and analyzing the characteristics of the network and the protocol. Most of the time is spent in learning the existing code modules of the simulator and the logical flow between the different code modules. The purpose of this chapter is to illustrate the applications of Graph Theory algorithms to study, analyze, and simulate the behavior of routing protocols for MANETs. Specifically, the chapter focuses on the applications of Graph Theory algorithms to determine paths, trees, and connected dominating sets for simulating and analyzing respectively unicast (single-path and multi-path), multicast, and broadcast communication in mobile ad hoc networks (MANETs). The chapter discusses the (i) Dijkstra's shortest path algorithm and its modifications for finding stable paths and bottleneck paths; (ii) Prim's minimum spanning tree algorithm and its modification for finding all pairs smallest and largest bottleneck paths; (iii) Minimum Steiner tree algorithm to connect a source node to all the receivers of a multicast group; (iv) A node-degree based algorithm to construct an approximate minimum Connected Dominating Set (CDS) for sending information from one node to all other nodes in the network; and (v) Algorithms to find a sequence of link-disjoint, node-disjoint, and zone-disjoint multi-path routes in MANETs.*

DOI: 10.4018/978-1-4666-0080-5.ch004

# INTRODUCTION

A Mobile Ad hoc Network (MANET) is a dynamically changing infrastructureless and resource-constrained network of wireless nodes that may move arbitrarily, independent of each other. The transmission range of the wireless nodes is often limited, necessitating multi-hop routing to be a common phenomenon for communication between any two nodes in a MANET. Various routing protocols for unicast, multicast, multi-path and broadcast communication have been proposed for MANETs. The communication structures that are often determined include: a path (for unicast – single-path and multi-path routing), a tree (for multicast routing) and a Connected Dominating Set – CDS (for broadcast routing). Within a particular class, it is almost impossible to find a single routing protocol that yields an optimal communication structure with respect to different route selection metrics and operating conditions.

Various simulators such as ns-2 (Fall & Varadhan, 2001) and GloMoSim (Zeng, Bagrodia, & Gerla, 1998) are available to implement and study the behavior of the routing protocols. But, students and investigators who are new to this area often get perplexed in the complexity of these simulators and lose the focus in designing and analyzing the characteristics of the network and the protocol. Most of the time would be spent in learning the existing code modules of the simulator and the logical flow between the different code modules. The purpose of this chapter would be to illustrate the applications of Graph Theory algorithms to study, analyze and simulate the behavior of routing protocols for MANETs. Applications of Graph Theory algorithms for unicast (single-path and multi-path), multicast and broadcast communication in MANETs will be discussed.

An ad hoc network is often approximated as a unit disk graph (Kuhn, Moscibroda, & Wattenhofer, 2004). In this graph, the vertices represent the wireless nodes and an edge exists between two vertices $u$ and $v$ if the normalized Euclidean distance (i.e., the physical Euclidean distance divided by the transmission range) between $u$ and $v$ is at most 1. Two nodes can communicate only if each node lies within (or on the edge of) the unit disk of the other node. The unit disk graph model neatly captures the behavior of many practical ad hoc networks and would be used in the rest of this chapter for discussing the algorithms to simulate the MANET routing protocols.

Most of the contemporary routing protocols proposed in the MANET literature adopt a Least Overhead Routing Approach (LORA) according to which a communication structure (route, tree or CDS) discovered through a global flooding procedure would be used as long as the communication structure exist, irrespective of the structure becoming sub-optimal since the time of its discovery in the MANET. This chapter will also adopt a similar strategy and focus only on discovering a communication structure on a particular network graph taken as a snapshot during the functioning of the MANET. Such a graph snapshot would be hereafter referred to as a 'Static Graph' and a sequence of such static graphs over the duration of the MANET simulation session would be called a 'Mobile Graph'. A communication structure determined on a particular static graph would be then validated for its existence in the subsequent static graphs and once the structure breaks, the appropriate graph algorithm can be invoked on the static graph corresponding to that particular time instant and the above procedure would be continued for the rest of the static graphs in the mobile graph. The big-O notation is used to express the theoretical worst-case run-time complexity of the algorithms discussed in this paper. Given a problem size $x$, where $x$ is usually the number of items, one can say $f(x) = O(g(x))$, when there exists positive constants $c$ and $k$ such that $0 \leq f(x) \leq cg(x)$, for all $x \geq k$ (Cormen et al. 2001).

The following are the learning objectives of this chapter with regards to the applications of Graph Theory algorithms in Mobile Ad hoc Networks:

1. Learn the application of the classical minimum-weight Dijkstra shortest path algorithm and its extensions to find stable paths, smallest and largest bottleneck paths.
2. Learn the application of the classical minimum spanning tree Prim's algorithm and its modification to find all pairs smallest and all pairs largest bottleneck paths.
3. Learn the application of the classical Steiner tree algorithm for multicast communication.
4. Learn the application of a minimum connected dominating set for broadcast communication.
5. Learn the algorithms to determine a set of link-disjoint, node-disjoint and zone-disjoint multi-path routes and the relationships between these multi-paths.

The rest of the chapter is organized as follows: The next section reviews related works on Unicast, Multicast, Broadcast and Multi-path communication in MANETs. In the subsequent sections, the chapter discusses Graph Theory algorithms for Unicast communication, the Tree-based algorithms for Multicast communication, a maximum density-based CDS algorithm for broadcast communication and multi-path algorithms for determining link-disjoint, node-disjoint and zone-disjoint routes in MANETs. After concluding the chapter, future research directions in this area are discussed. Throughout the chapter, the terms 'route' and 'path', 'link' and 'edge', 'message' and 'packet' are used interchangeably. They mean the same.

## BACKGROUND

### Unicast Communication in MANETs

There are two broad classifications of unicast routing protocols: minimum-weight based routing and stability-based routing. Routing protocols under the minimum-weight category have been primarily designed to optimize the hop count of source-destination (*s-d*) routes. Some of the well-known minimum-hop based routing protocols include the Dynamic Source Routing (DSR) protocol (Johnson, Maltz, & Broch, 2001) and the Ad hoc On-demand Distance Vector (AODV) routing protocol (Perkins & Royer, 1999). The stability-based routing protocols aim to minimize the number of route failures and in turn reduce the number of flooding-based route discoveries. Some of the well-known stability-based routing protocols include the Flow-Oriented Routing Protocol (FORP; Su, Lee, & Gerla, 2001) and the Node Velocity-based Stable Path (NVSP) routing protocol (Meghanathan, 2009). In previous work, Meghanathan (2008; 2010) has observed that there exists a stability-hop count tradeoff and it is not possible to simultaneously optimize both the hop count as well as the number of route discoveries.

The DSR protocol is a source routing protocol that requires the entire route information to be included in the header of every data packet. However, because of this feature, intermediate nodes do not need to store up-to-date routing information in their routing tables. Route discovery is by means of the broadcast query-reply cycle. The *Route Request* (RREQ) packet reaching a node contains the list of intermediate nodes through which it has propagated from the source node. After receiving the first RREQ packet, the destination node waits for a short time period for any more RREQ packets, then chooses a path with the minimum hop count and sends a *Route Reply* packet (RREP) along the selected path. Later, if any new RREQ is received through a path with hop count less than that of the selected path, another RREP would be sent on the latest minimum hop path discovered.

The AODV protocol, like DSR, is also a shortest path based routing protocol. However, it is table-driven. Upon receiving an unseen RREQ packet (with the highest sequence number seen so far), an intermediate node records the upstream node (sender) of the RREQ packet in its routing table entry for the source-destination route. The intermediate node then forwards the RREQ packet by incrementing the hop count of the path from the source node. The destination node receives RREQ packets on several routes and selects that RREQ packet that traversed on the minimum-hop path to the destination node. The RREP packet is then sent on the reverse of this minimum-hop path towards the source node. The destination node includes the upstream node from which the RREQ was received as the downstream node on the path from the destination node to the source node. An intermediate node upon receiving the RREP packet will check whether it has been listed as the downstream node ID. In that case, the intermediate node processes the RREP packet and completes its routing table by including the sender of the RREP packet as the next hop node on the path from the source node towards the destination node. The intermediate node then replaces its own ID in the RREP downstream node entry with the ID of the upstream node that it has in its routing table for the path from the source node to the destination node.

The FORP protocol has been observed to discover the sequence of most stable routes among the contemporary stable path routing protocols (Meghanathan, 2008). FORP utilizes the mobility and location information of the nodes to approximately predict the expiration time (LET) of a wireless link. The minimum of LET values of all wireless links on a path is termed as the Route Expiration Time (RET). The route with the maximum RET value is selected as the desired route. Each node is assumed to be able to predict the LET values of the links with its neighboring nodes based on the information regarding the current position of the nodes, velocity, the direction of movement, and transmission range. FORP assumes the availability of location-update mechanisms like Global Positioning System (GPS; Hofmann-Wellenhof, Lichtenegger, & Collins, 2004) to identify the location of the nodes and also requires each node to periodically broadcast its location and mobility information to its neighbors through beacons.

The NVSP protocol is the only beaconless routing protocol that can discover long-living stable routes without significant increase in the hop count per path. FORP discovers routes that have a significantly larger hop count than the minimum value. NVSP only requires each intermediate node to include its velocity in the RREQ packets propagated via flooding from the source node to the destination node. With flooding, each intermediate node forwards the RREQ packet exactly once, the first time the node sees the packet as part of a particular route discovery session. The destination node receives the RREQ packets through several paths and determines the bottleneck velocity of each of those paths. The bottleneck velocity of a path is the maximum among the velocities of the intermediate nodes on the path. The destination node chooses the path with the minimum bottleneck velocity and sends a RREP packet along that path. In case of a tie, the destination node chooses the path with the lowest hop count and if the tie could not be still broken, the destination node chooses an arbitrary path among the contending paths.

## Multicast Communication in MANETs

Multicast communication refers to sending messages from one source node to a set of receiver nodes in a network. The receiver nodes form the multicast group and one typically finds a tree that connects the source node to the multicast group members such that there is exactly one path from the source node to each receiver node. The tree could be constructed based on either one of the following two objectives: (i) Shortest path tree – the tree would have the minimum hop count paths from the source node to each receiver node and (ii) Steiner tree – the tree would have the minimum number of links spanning the source node and the multicast group members. Both these trees cannot be simultaneously built and there would always be a tradeoff between the above two objectives (Meghanathan, 2010). The Multicast extension of the Ad hoc On-demand Distance Vector (MAODV) protocol and the Bandwidth Efficient Multicast Routing Protocol (BEMRP) are respectively examples of the minimum hop and minimum link based multicast protocols.

MAODV (Royer & Perkins, 1999) is the multicast extension of the AODV unicast routing protocol. Here, a receiver node joins the multicast tree through a member node that lies on the minimum-hop path to the source node. A potential receiver node wishing to join the multicast group broadcasts a RREQ message. If a node receives the RREQ message and is not part of the multicast tree, the node broadcasts the message in its neighborhood and also establishes the reverse path by storing the state information consisting of the group address, requesting node id and the sender node id in a temporary cache. If a node receiving the RREQ message is a member of the multicast tree and has not seen the RREQ message earlier, the node waits to receive several RREQ messages and sends back a RREP message on the shortest path to the receiver node. The member node also informs in the RREP message, the number of hops from itself to the source node. The potential receiver node receives several RREP messages and selects the member node which lies on the shortest path to the source node. The receiver node sends a *Multicast Activation* (MACT) message to the selected member node along the chosen route. The route from the source node to the receiver node is set up when the member node and all the intermediate nodes in the chosen path update their multicast table with state information from the temporary cache.

According to BEMRP (Ozaki, Kim, & Suda, 2001), a newly joining node to the multicast group opts for the nearest forwarding node in the existing tree, rather than choosing a minimum-hop count path from the source node of the multicast group. As a result, the number of links in the multicast tree is reduced leading to savings in the network bandwidth. Multicast tree construction is receiver-initiated. When a node wishes to join the multicast group as a receiver node, it initiates the flooding of *Join control* packets targeted towards the nodes that are currently members of the multicast tree. On receiving the first *Join control* packet, the member node waits for a certain time before sending a *Reply* packet. The member node sends a *Reply* packet on the path, traversed by the *Join* control packet, with the minimum number of intermediate forwarding nodes. The newly joining receiver node collects the *Reply* packets from different member nodes and would send a *Reserve* packet on the path that has the minimum number of forwarding nodes from the member node to itself.

## Broadcast Communication in MANETs

Broadcast communication refers to sending a message from one node to all the other nodes in the network. Since MANET topology is not fully connected as nodes operate with a limited transmission range, multi-hop communication is a common phenomenon in routing. As a result, a message has to be broadcast by more than one node (in its neighborhood) so that the message can reach all the nodes in the network. An extreme case of broadcasting is called flooding wherein each node broadcasts the message among its neighbors, exactly once, when the message is seen for the first time. This ensures that the message is received by all the nodes in the network. However, flooding would cause unnecessary retransmissions, exhausting the network bandwidth and the energy reserves at the nodes.

Connected Dominating Sets (CDS) are considered to be very efficient for broadcasting a message from one node to all the nodes in the network. A CDS is a sub graph of a given undirected connected graph such that all nodes in the graph are included in the CDS or directly attached to a node (i.e., covered by a node) in the CDS. A

Minimum Connected Dominating Set (MCDS) is the smallest CDS (in terms of the number of nodes in the CDS) for the entire graph. For a virtual backbone-based route discovery, the smaller the size of the CDS, the smaller is the number of unnecessary retransmissions. If the RREQ packets of a broadcast route discovery process get forwarded only by the nodes in the MCDS, one will have the minimum number of retransmissions. Unfortunately, the problem of determining the MCDS in an undirected graph, like that of the unit disk graph considered for modeling MANETs, is NP-complete. Alzoubi and his colleagues (2002) and Butenko and his colleagues (2003a ; 2003b) have proposed efficient algorithms to approximate the MCDS for wireless ad hoc networks. A common thread among these algorithms is to give preference to nodes with high neighborhood density (i.e., a larger number of uncovered neighbors) for inclusion in the MCDS.

## Multi-Path Communication in MANETs

MANET routing protocols incur high route discovery latency and also incur frequent route discoveries in the presence of a dynamically changing topology. Recent research has started to focus on multi-path routing protocols for fault tolerance and load balancing. Multi-path on-demand routing protocols tend to compute multiple paths, at both the traffic sources as well as at intermediary nodes, in a single route discovery attempt. This reduces both the route discovery latency and the control overhead as a route discovery is needed only when all the discovered paths fail. Spreading the traffic along several routes could alleviate congestion and bottlenecks. Multi-path routing also provides a higher aggregate bandwidth and effective load balancing as the data forwarding load can be distributed over all the paths.

Multi-paths can be of three types: link-disjoint, node-disjoint and zone-disjoint. For a given source node *s* and destination node *d*, the set of

link-disjoint *s-d* routes comprises of paths that have no link present in more than one constituent *s-d* path. Similarly, the set of node-disjoint *s-d* routes comprises of paths that have no node (other than the source node and destination node) present in more than one constituent *s-d* path. A set of zone-disjoint *s-d* routes comprises of paths such that an intermediate node in one path is not a neighbor node of an intermediate node in another path. Multi-path on-demand routing protocols tend to compute multiple paths between a source-destination (*s-d*) pair, in a single route discovery attempt. A new network-wide route discovery operation is initiated only when all the *s-d* paths fail. The Split Multi-path Routing (SMR) protocol (Lee & Gerla, 2001), the AODV-Multi-path (AODVM) protocol (Ye, Krishnamurthy, & Tripathi, 2003) and the Zone-Disjoint multi-path extension to the DSR (ZD-DSR) protocol (Javan & Dehghan, 2007) are respectively well-known examples for link-disjoint, node-disjoint and zone-disjoint multi-path routing protocols.

In SMR, the intermediate nodes forward RREQs that are received along a different link and with a hop count not larger than the first received RREQ. The destination node selects the route on which it received the first RREQ packet (which will be a shortest delay path), and then waits to receive more RREQs. The destination node then selects the path which is maximally disjoint from the shortest delay path. If more than one maximally disjoint path exists, the tie is broken by choosing the path with the shortest hop count.

In AODVM, an intermediate node does not discard duplicate RREQ packets and records them in a RREQ table. The destination node responds with an RREP for each RREQ packet received. An intermediate node, on receiving the RREP, checks its RREQ table and forwards the packet to the neighbor that lies on the shortest path to the source node. The neighbor entry is then removed from the RREQ table. Also, whenever a node hears a neighbor node forwarding the RREP packet, the node removes the entry for the neighbor node in its RREQ table.

The Zone-Disjoint Multi-Path extension of the Dynamic Source Routing (ZD-MPDSR) protocol proposed for an omni-directional system works as follows: Whenever a source node has no route to send data to a destination node, the source node initiates broadcast of the RREQ messages. The number of active neighbors for a node indicates the number of neighbor nodes that have received and forwarded the RREQ message during a route discovery process. The RREQ message has an *ActiveNeighborCount* field and it is updated by each intermediate node before broadcasting the message in the neighborhood. When an intermediate node receives the RREQ message, it broadcasts a 1-hop RREQ-query message in its neighborhood to determine the number of neighbors who have also seen the RREQ message. The number of RREQ-query-replies received from the nodes in the neighborhood is the value of the *ActiveNeighborCount* field updated by a node in the RREQ message. The destination node receives several RREQ messages and selects the node-disjoint paths with lower *ActiveNeighborCount* values and sends the RREP messages to the source node along these paths. Even though the selection of the zone-disjoint paths with lower number of active neighbors will lead to reduction in the end-to-end delay per data packet, the route acquisition phase will incur a significantly longer delay as RREQ-query messages are broadcast at every hop (in addition to the regular RREQ message) and the intermediate nodes have to wait to receive the RREQ-query and reply messages from their neighbors. This will significantly increase the control overhead in the network.

## GRAPH THEORY ALGORITHMS FOR MANET UNICAST COMMUNICATION

In a graph theoretic context, this chapter illustrates that the minimum-weight (minimum-hop) based

routing protocols could be simulated by running the shortest-path Dijkstra algorithm (Cormen et al. 2001) on a mobile graph (i.e. a sequence of static graphs). Similarly, the chapter illustrates that the NVSP and FORP protocols could be simulated by respectively solving the smallest bottleneck and the largest bottleneck path problems – each of which could be implemented as a slight variation of the shortest path Dijkstra algorithm. In addition, the chapter also illustrates that the Prim's minimum spanning tree algorithm and its modification to compute the maximum spanning tree can be respectively used to determine the 'All Pairs Smallest Bottleneck Paths' and 'All Pairs Largest Bottleneck Paths' in a weighted network graph.

## Shortest Path Problem

Given a weighted graph $G = (V, E)$, where $V$ is the set of vertices and $E$ is the set of weighted edges, the shortest path problem is to determine a minimum-weight path between any two nodes (identified as source node $s$ and destination node $d$) in the graph. Execution of the Dijkstra algorithm (pseudo code in Figure 1) on a weighted graph starting at the source node $s$ results in a shortest path tree rooted at $s$. In other words, the Dijkstra algorithm will actually return the minimum-weight paths from the source vertex $s$ to every other vertex in the weighted graph. If all the edge weights are 1, then the minimum-weight paths are nothing but minimum-hop paths.

Dijkstra algorithm proceeds in iterations. To begin with, the weights of the minimum-weight paths from the source vertex to every other vertex is assumed to be $+\infty$ (an estimate value, indicating that the paths are actually not known) and from the source vertex to itself is assumed to be 0. During each iteration, the algorithm determines the shortest path from the source vertex $s$ to a particular vertex $u$, which would be the vertex with the minimum weight among the vertices that have been not yet optimized (i.e. for which the shortest path has not been yet determined). The

*Figure 1. Pseudo code for Dijkstra's shortest path algorithm*

**Begin** Algorithm *Dijkstra-Shortest-Path* (*G*, *s*)
1   **For** each vertex *v* ∈ *V*
2       *weight* [*v*] ← ∞ // an estimate of the minimum-weight path from *s* to *v*
3   **End For**
4   *weight* [*s*] ← 0
5   *S* ← Φ // set of nodes for which the minimum-weight path from *s* is known
6   *Q* ← *V* // set of nodes for which only an estimate of the minimum-weight path from *s* is known
7   **While** *Q* ≠ Φ
8       *u* ⌐ EXTRACT-MIN (*Q*)
9       *S* ← *S* ∪ {*u*}
10      **For** each vertex *v* such that (*u*, *v*) ∈ *E*
11          **If** *weight* [*v*] > *weight* [*u*] + *weight* (*u*, *v*) **then**
12              *weight* [*v*] ← *weight* [*u*] + *weight* (*u*, *v*)
13              Predecessor (*v*) ← *u*
14          **End If**
15      **End For**
16  **End While**
17  **End** *Dijkstra-Shortest-Path*

*Figure 2. Example to illustrate the working of the Dijsktra's shortest path algorithm*

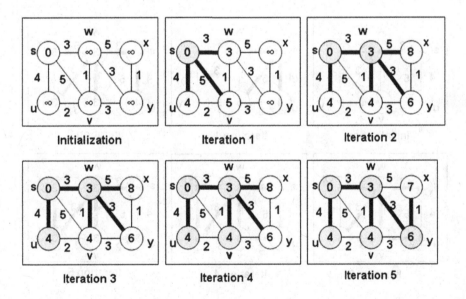

algorithm then explores the neighbors of *u* and determines whether any of the neighbor vertices, say *v*, can be reached from *s* through *u* on a path with weight less than the estimated weight of the current path known from *s* to *v*. If one could find such a neighbor *v*, then vertex *u* is set to be the predecessor of vertex *v* on the shortest path from *s* to *v*. This step is called the relaxation step and is repeated over all iterations. The darkened edges shown in the working example of Figure 2 are the edges that are part of the shortest-path tree rooted at the source vertex *s*. The run-time complexity of the Dijkstra's shortest path algorithm is O(|*V*|²).

*Figure 3. Pseudo code for the modified Dijkstra's algorithm for the smallest bottleneck path problem*

**Begin** Algorithm *Modified-Dijkstra-Smallest-Bottleneck-Path (G, s)*
1   **For** each vertex $v \in V$
2       *weight* [$v$] $\leftarrow +\infty$ // an estimate of the smallest bottleneck weight path from $s$ to $v$
3   **End For**
4   *weight* [$s$] $\leftarrow -\infty$
5   $S \leftarrow \Phi$ // set of nodes for which the smallest bottleneck weight path from $s$ is known
6   $Q \leftarrow V$ // set of nodes for which an estimate of the smallest bottleneck weight path from $s$ is known
7   **While** $Q \neq \Phi$
8       $u \neg$ EXTRACT-MIN ($Q$)
9       $S \leftarrow S \cup \{u\}$
10      **For** each vertex $v$ such that $(u, v) \in E$
11          **If** *weight* [$v$] > Max (*weight* [$u$], *weight* ($u, v$)) **then**
12              *weight* [$v$] $\leftarrow$ Max (*weight* [$u$], *weight* ($u, v$))
13              Predecessor ($v$) $\leftarrow u$
14          **End If**
15      **End For**
16  **End While**
17 **End** *Modified-Dijkstra-Smallest-Bottleneck-Path*

*Figure 4. Example for the smallest bottleneck path problem*

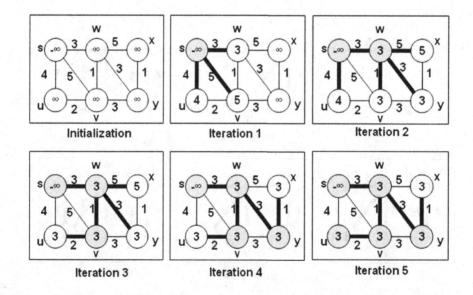

| Initialization | Iteration 1 | Iteration 2 |
|---|---|---|

| Iteration 3 | Iteration 4 | Iteration 5 |
|---|---|---|

## Smallest Bottleneck Path Problem

In the context of the smallest bottleneck path problem, the bottleneck weight of a path $p$ is defined to be the maximum of the weights of the constituent edges, $e \in p$. Given the set of all loop-free paths $P$ between a source node $s$ and destination node $d$, the smallest bottleneck path is the path with the smallest bottleneck weight. One can express this mathematically as
$$\underset{p \in P}{Min}\left[\underset{\forall e \in p}{Max}\left(weight(e)\right)\right].$$

The NVSP protocol can be implemented in a graph theoretic context through a modified ver-

*Figure 5. Pseudo code for the Prim's algorithm to determine a minimum spanning tree*

**Begin** Algorithm *Prim (G, s)*; *s* – is any arbitrarily chosen starting vertex
1    **For** each vertex $v \in V$
2        *weight* [*v*] ← ∞
3    **End For**
4    *weight* [*s*] ← 0
5    $S \leftarrow \Phi$ // set of nodes whose bottleneck weights will not change further
6    $Q \leftarrow V$ // set of nodes whose bottleneck weights are only estimates and the final weight could change
7    **While** $Q \neq \Phi$
8      .   *u* ¬ EXTRACT-MIN (*Q*)
9        $S \leftarrow S \cup \{u\}$
10       **For** each vertex *v* such that $(u, v) \in E$
11           **If** *weight* [*v*] > *weight* (*u, v*) **then**
12              *weight* [*v*] ← *weight* (*u, v*)
13              Predecessor (*v*) ← *u*
14           **End If**
15       **End For**
16   **End While**
17 **End** *Prim*

sion of the Dijkstra's algorithm (pseudo code in Figure 3) that solves the smallest bottleneck path problem. Accordingly, the weight of a link from node *u* to node *v* is the velocity of the downstream node *v*. To start with, the weight of the smallest bottleneck path from the source vertex *s* to every other vertex is estimated to be +∞; whereas the weight of the smallest bottleneck path from the source vertex *s* to itself is set to –∞. At the beginning of an iteration, the vertex (say *u*) with the smallest bottleneck weight among the vertices that have been not yet optimized is now considered to be optimized. As part of the relaxation step, the idea is to check whether the current weight of the smallest bottleneck path to any non-optimized neighbor *v*, i.e. *weight*[*v*], is greater than the maximum of the weight of the recently optimized largest bottleneck path from *s* to *u*, i.e. *weight*[*u*] and the weight of the edge (*u, v*). If this relaxation condition evaluates to true, then the bottleneck weight of the path from *s* to *v* is correspondingly updated (i.e., *weight*[*v*] = Max (*weight*[*u*], *weight*(*u, v*)) and the predecessor of *v* is set to be *u* for the path from *s* to *v*. This step is repeated over all iterations. A working example is

presented in Figure 4. The run-time complexity of the modified Dijkstra algorithm for the smallest bottleneck path problem is the same as that of the original Dijkstra algorithm.

## All Pairs Smallest Bottleneck Paths Algorithm

This section shows that the smallest bottleneck path between any two vertices *u* and $v \in V$ in an undirected weighted graph $G = (V, E)$ is the path between *u* and *v* in the minimum spanning tree of *G*. The Prim's algorithm (Cormen et al., 2001) is a well-known algorithm to determine the minimum spanning tree of weighted graphs and its pseudo code is illustrated in Figure 5. The Prim's algorithm is very similar to the Dijkstra algorithm – the major difference is in the relaxation step.

The Prim's algorithm work as follows: The starting vertex is any arbitrarily chosen vertex (say *s*) in the given undirected weighted graph *G*. To begin with, the weights of the smallest bottleneck paths from the starting vertex to every other vertex is assumed to be +∞ (an estimate value, indicating that the paths are actually not

*Figure 6. Example for the minimum spanning tree – All pairs smallest bottleneck paths*

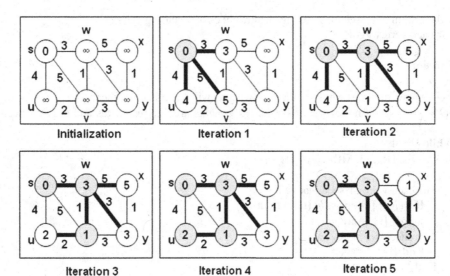

known) and the path from the starting vertex to itself is assumed to be 0. During every iteration, the idea is to determine the smallest bottleneck path from the starting vertex $s$ to a particular vertex $u$, which would be the vertex with the minimum weight among the vertices that have been not yet optimized (i.e. for which the smallest bottleneck path has not been yet determined). The neighbors of $u$ are then explored and checked if one can reach any of the neighbor vertex, say $v$, from $s$ through $u$ on a path with weight less than the estimated bottleneck weight of the current path known from $s$ to $v$. If one could find such a neighbor $v$ as part of the relaxation step, the new estimated bottleneck weight of vertex $v$ is set to be the weight of the edge $(u, v)$ and vertex $u$ is set to be the predecessor of $v$ on the smallest bottleneck path from $s$ to $v$. The darkened edges shown in the working example of Figure 6 are the edges that are part of the smallest bottleneck path tree rooted at the starting vertex $s$. The path between any two vertices in this smallest bottleneck path tree is the smallest bottleneck path between the two vertices in the original graph. The run-time complexity of the Prim's minimum spanning tree algorithm is $O(|V|^2)$.

Note that in both Figures 4 and 6, the algorithm starts with the same initial graph. Since, the relaxation step of the modified Dijkstra algorithm and the Prim's algorithm are different, the sequence of vertices that are optimized in each algorithm is different from one another. However, the final tree rooted at the starting vertex $s$ is the same in both the figures. This example vindicates the argument that the minimum spanning tree contains the smallest bottleneck paths between any two vertices in the original graph. This argument is now formally proven (refer Figure 7 for an illustration of the example) below through the method of Proof by Contradiction.

Let there be a pair of vertices $u \in V_1$ and $v \in V_2$ in $G = (V, E)$ such that the edge $(u, v) \in E$, $V_1 \cup V_2 = V$ and $V_1 \cap V_2 = \Phi$. Assume the edge $(u, v)$ belongs to the minimum spanning tree $T$ of $G$; but the edge is not part of the smallest bottleneck path from $u$ to $v$. Let there exist an alternate path from $u$ to $v$ that is the smallest bottleneck path. Since $u$ and $v$ are in two disjoint vertex partitions, there should be at least one edge (call it $e$) in the path from $u$ to $v$ with endpoint vertices in each partition. But, by definition of a minimum spanning tree, the weight($u$, $v$) ≤ weight(edge $e$);

*Figure 7. Proof by contradiction: Minimum spanning tree contains all pairs smallest bottleneck paths*

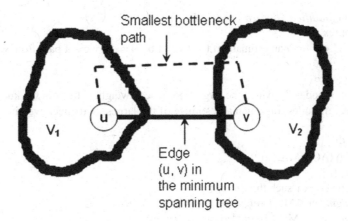

otherwise, a cheaper tree could be obtained by replacing $(u, v)$ with the edge $e$ and $T$ containing $(u, v)$ would not be a minimum spanning tree. Hence, edge $e$ could be replaced by edge $(u, v)$ without increasing the weight of any smallest bottleneck path. Likewise, one can prove that every edge in $T$ would be part of a smallest bottleneck path. Since $T$ is a minimum spanning tree, all its edges constitute the all pairs smallest bottleneck paths for the entire graph.

## Largest Bottleneck Path Problem

In the context of the largest bottleneck path problem, the bottleneck weight of a path $p$ is defined to be the minimum of the weights of the constituent edges, $e \in p$. Given the set of all loop-free paths $P$ between a source node $s$ and destination node $d$, the largest bottleneck path is the path with the largest bottleneck weight. One can express this mathematically as $\underset{p \in P}{Max}\left[\underset{\forall e \in p}{Min}\left(weight(e)\right)\right]$.

The FORP protocol can be simulated using a modified version of the Dijsktra's algorithm (pseudo code in Figure 8) that solves the Largest Bottleneck Path problem on a static graph. The edge weights correspond to the predicted LET values for the corresponding links. To start with, the weight of the largest bottleneck path from the source vertex $s$ to every other vertex is estimated

to be $-\infty$; whereas the weight of the largest bottleneck path from the source vertex $s$ to itself is set to $+\infty$. At the beginning of an iteration, the vertex (say $u$) with the largest bottleneck weight among the vertices that have been not yet optimized is now considered to be optimized (i.e., the largest bottleneck path from the source vertex $s$ to the vertex $u$ is considered to have been determined by now). As part of the relaxation step, the algorithm checks whether the current weight of the largest bottleneck path to any non-optimized neighbor $v$, i.e. $weight[v]$, is lower than the minimum of the weight of the recently optimized largest bottleneck path from $s$ to $u$, i.e. $weight[u]$ and the weight of the edge $(u, v)$. If this relaxation condition evaluates to true, then the bottleneck weight of the path from $s$ to $v$ is correspondingly updated (i.e., $weight[v] = Min(weight[u], weight(u, v))$ and vertex $u$ is set to be the predecessor of vertex $v$ on the path from $s$ to $v$. This step is repeated over all iterations. A working example is presented in Figure 9. The run-time complexity of the modified Dijkstra algorithm for the largest bottleneck path problem is the same as that of the original algorithm for the shortest path problem.

*Figure 8. Pseudo code for the modified Dijkstra's algorithm for the largest bottleneck path problem*

**Begin** Algorithm *Modified-Dijkstra-Largest-Bottleneck-Path (G, s)*
1   **For** each vertex $v \in V$
2       *weight* $[v] \leftarrow -\infty$ // an estimate of the largest bottleneck weight path from $s$ to $v$
3   **End For**
4   *weight* $[s] \leftarrow +\infty$
5   $S \leftarrow \Phi$ // set of nodes for which the largest bottleneck weight path from $s$ is known
6   $Q \leftarrow V$ // set of nodes for which an estimate of the largest bottleneck weight path from $s$ is known
7   **While** $Q \neq \Phi$
8       $u \neg$ EXTRACT-MAX $(Q)$
9       $S \leftarrow S \cup \{u\}$
10      **For** each vertex $v$ such that $(u, v) \in E$
11          **If** *weight* $[v] <$ Min (*weight* $[u]$, *weight* $(u, v)$) then
12              *weight* $[v] \leftarrow$ Min (*weight* $[u]$, *weight* $(u, v)$)
13              Predecessor $(v) \leftarrow u$
14          **End If**
15      **End For**
16  **End While**
17 **End** *Modified-Dijkstra-Largest-Bottleneck-Path*

*Figure 9. Example for the largest bottleneck path problem*

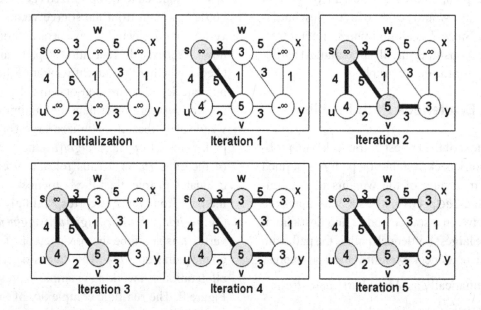

| Initialization | Iteration 1 | Iteration 2 |
| Iteration 3 | Iteration 4 | Iteration 5 |

## All Pairs Largest Bottleneck Paths Algorithm

This section shows that the largest bottleneck path between any two vertices $u$ and $v \in V$ in an undirected weighted graph $G = (V, E)$ is the path

between $u$ and $v$ in the maximum spanning tree of $G$. The maximum spanning tree of a graph can be determined using a slightly modified version of the Prim's algorithm – the modification is in the initialization step and the relaxation condition. In the original Prim's algorithm, the initial weight of

*Figure 10. Pseudo code for the modified Prim's algorithm to determine a maximum spanning tree*

**Begin** Algorithm *Modified-Prim* $(G, s)$; $s$ – is any arbitrarily chosen starting vertex
1    **For** each vertex $v \in V$
2        $weight\ [v] \leftarrow -\infty$
3    **End For**
4    $weight\ [s] \leftarrow 0$
5    $S \leftarrow \Phi$ // set of nodes whose bottleneck weights will not change further
6    $Q \leftarrow V$ // set of nodes whose bottleneck weights are only estimates and the final weight could change
7    **While** $Q \neq \Phi$
8        $u \neg$ EXTRACT-MAX $(Q)$
9        $S \leftarrow S \cup \{u\}$
10       **For** each vertex $v$ such that $(u, v) \in E$
11           **If** $weight\ [v] < weight\ (u, v)$ then
12               $weight\ [v] \leftarrow weight\ (u, v)$
13               Predecessor $(v) \leftarrow u$
14           **End If**
15       **End For**
16   **End While**
17 **End** *Modified-Prim*

all the vertices other than the starting vertex is set to $+\infty$; whereas in the modified Prim's algorithm for the all pairs largest bottleneck path problem, the initial weight of all the vertices other than the starting vertex is set to $-\infty$ (an initial estimate for the largest bottleneck paths, which are actually not known to start with). The weight of the starting vertex, $s$, in both algorithms is 0. The pseudo code of the modified Prim's algorithm is given in Figure 10.

During every iteration, the algorithm determines the largest bottleneck path from the starting vertex $s$ to a particular vertex $u$, which would be the vertex with the maximum weight among the vertices that have been not yet optimized (i.e. for which the largest bottleneck path has not been yet determined). The algorithm explores the neighbors of $u$ and determines whether one can reach any of the neighbor vertices, say $v$, from $s$ through $u$ on a path with weight greater than the estimated bottleneck weight of the current path known from $s$ to $v$. If one could find such a neighbor $v$ as part of the relaxation step, the new estimated bottleneck weight of vertex $v$ is set to the weight of the edge $(u, v)$ and vertex $u$ is also set to be the predecessor

of $v$ on the largest bottleneck path from $s$ to $v$. The darkened edges shown in the working example of Figure 11 are the edges that are part of the largest bottleneck path tree rooted at the starting vertex $s$. The path between any two vertices in this largest bottleneck path tree is the largest bottleneck path between the two vertices in the original graph. The run-time complexity of the modified Prim's algorithm to determine the maximum spanning tree is $O(|V|^2)$, the same as that of the original Prim's algorithm for minimum spanning trees. The correctness of the modified Prim's algorithm for the all pairs largest bottleneck path problem can be proved using the same logic used to prove the correctness of the Prim's algorithm for the all pairs smallest bottleneck path problem.

## GRAPH THEORY ALGORITHMS FOR MANET MULTICAST COMMUNICATION

The original Dijkstra shortest path algorithm described previously can be used to determine

*Figure 11. Example for the maximum spanning tree – All pairs largest bottleneck paths*

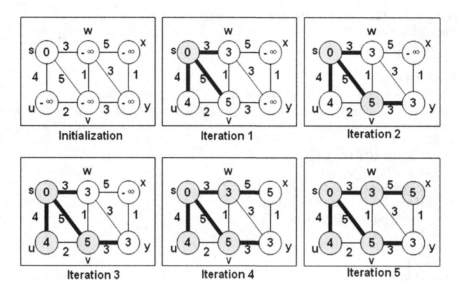

shortest path trees with minimum hop count per source-receiver path. The problem of determining a multicast tree with the minimum number of links is a NP-complete problem for which there is no polynomial-time algorithm yielding an optimistic solution. Hence, algorithms have been proposed in the literature to approximate such multicast trees. The Steiner tree algorithm is the best-known algorithm in the literature to approximate multicast trees with the minimum number of links connecting a source node to the receiver nodes of the multicast group.

Given a static graph, $G = (V, E)$, where $V$ is the set of vertices, $E$ is the set of edges and a subset of vertices (called the multicast group or Steiner points) $MG \subseteq V$, the multicast Steiner tree is the tree with the least number of edges required to connect all the vertices in $MG$. In this chapter, a well-known $O(|V||MG|^2)$ algorithm of Kou and his colleagues (1981) is used, where $|V|$ is the number of nodes in the network graph and $|MG|$ is the size of the multicast group comprising of the source nodes and the receiver nodes, to approximate the minimum edge Steiner tree in graphs representing snapshots of the network topology. In unit disk graphs such as the static graphs used

in this research, Step 5 of the algorithm (pseudo code in Figure 12) is not needed and the minimal spanning tree $T_{MG}$ obtained at the end of Step 4 could be considered as the minimum edge Steiner tree. One could use the Prim's algorithm to find the minimum spanning trees.

The working of the Kou et al's algorithm is illustrated through an example in Figure 13. The vertices {D, G, E, M, N, P} form the multicast group in the vertex set {A, B ... P}. As observed in the example, the sub graph $G_{MG}$ obtained in Step 3 is nothing but the minimal spanning tree $T_{MG}$, which is the output of Step 4. In general, for unit disk graphs, like the static graphs that are used in this chapter, the outputs of both Steps 3 and 4 are the same and it is enough to stop at Step 3 and output the MG-Steiner-tree.

# GRAPH THEORY ALGORITHMS FOR MANET BROADCAST COMMUNICATION

This section describes a Maximum-Density based CDS (MaxD-CDS) graph theoretic algorithm to approximate a Minimum Connected Dominating

*Figure 12. Kou et al's algorithm to find an approximate minimum edge Steiner tree*

**Input:**   A Static Graph $G = (V, E)$
Multicast Group $MG \subseteq V$
**Output:** A *MG-Steiner-tree* for the set $MG \subseteq V$
**Begin** Kou et al Algorithm $(G, MG)$
**Step 1:** Construct a complete undirected weighted graph $G_C = (MG, E_C)$ from $G$ and $MG$ where $\forall (v_i, v_j) \in E_C$, $v_i$ and $v_j$ are in $MG$, and the weight of edge $(v_i, v_j)$ is the length of the shortest path from $v_i$ to $v_j$ in $G$.
    **Step 2:** Find the minimum weight spanning tree $T_C$ in $G_C$ (If more than one minimal spanning tree exists, pick an arbitrary one).
    **Step 3:** Construct the sub graph $G_{MG}$ of $G$, by replacing each edge in $T_C$ with the corresponding shortest path from $G$ (If there is more than one shortest path between two given vertices, pick an arbitrary one).
    **Step 4:** Find the minimal spanning tree $T_{MG}$ in $G_{MG}$ (If more than one minimal spanning tree exists, pick an arbitrary one). Note that each edge in $G_{MG}$ has weight 1.
    **return** $T_{MG}$ as the *MG-Steiner-tree*
**End** *Kou et al Algorithm*

Set (MCDS) in a static graph, taken as a snapshot of a MANET topology.

## Data Structures Used by the MaxD-CDS Algorithm and Breadth First Search

The MaxD-CDS algorithm uses the following principal data structures:

1.  *CDS-Node-List* – includes all nodes that are members of the CDS
2.  *Covered-Nodes-List* – includes all nodes that are in the CDS-Node-List and all nodes that are adjacent to at least one member of the CDS-Node-List.

The CDS formation algorithm first makes sure that the underlying network graph is connected by running the Breadth First Search (BFS) algorithm (Cormen et al., 2001); because, if the underlying network graph is not connected, one would not be able to find a CDS that will cover all the nodes in the network. BFS is run starting with an arbitrarily chosen node in the network graph. If one is able to visit all the vertices in the graph, then the corresponding network is said to be connected.

If the graph is not connected, the procedure is to simply continue with the static graph (snapshot of the network topology) collected at the next time instant and start with the BFS test. The pseudo code for BFS is shown in Figure 14. The run-time complexity of BFS is $O(|V|+|E|)$.

## Maximum Density-Based Algorithm to Approximate a MCDS

The idea of the Maximum Density (MaxD)-based CDS formation algorithm is to select nodes with larger number of uncovered neighbors for inclusion in the CDS. The algorithm forms and outputs a CDS based on a given input graph representing a snapshot of the MANET at a particular time instant. Specifically, the algorithm outputs a list (*CDS-Node-List*) of all nodes that are part of the CDS formed based on the given MANET. The first node to be included in the *CDS-Node-List* is the node with the maximum number of uncovered neighbors (any ties are broken arbitrarily). A CDS member is considered to be "covered", so a CDS member is additionally added to the *Covered-Nodes-List* when it is included in the *CDS-Node-List*. All nodes that are adjacent to a CDS member are also said to be covered, so the

*Figure 13. Construction of a minimum Steiner tree using Kou et al.'s algorithm*

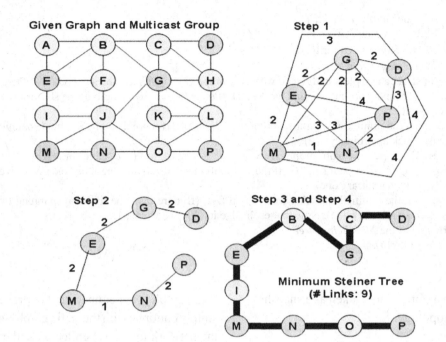

*Figure 14. Pseudo code for the BFS algorithm to determine network connectivity*

**Input:** Graph $G = (V, E)$
**Auxiliary Variables/Initialization:** *Nodes-Explored* ← Φ, *FIFO-Queue* ← Φ (First-In First-Out Queue)
**Begin** Algorithm *BFS* $(G, s)$
  *root-node* ← randomly chosen vertex in *V*
  *Nodes-Explored* ← *Nodes-Explored* ∪ {*root-node*}
  *FIFO-Queue* ← *FIFO-Queue* ∪ {*root-node*}
  **while** ( |*FIFO-Queue*| > 0 ) **do**
    *front-node u* ← Dequeue(*FIFO-Queue*) // extract the first node
    **for** (every edge $(u, v)$ ) **do** // i.e. every neighbor *v* of node *u*
      **if** ($v \notin$ *Nodes-Explored*) **then**
        *Nodes-Explored* ← *Nodes-Explored* ∪ {*v*}
        *FIFO-Queue* ← *FIFO-Queue* ∪ {*v*}
        *Parent* (*v*) ← *u*
      **end if**
    **end for**
  **end while**
  **if** ( | *Nodes-Explored* | = | *V* | ) **then return** Connected Graph - true
  **else return** Connected Graph - false
  **end if**
**End** Algorithm *BFS*

*Figure 15. Pseudo code for the algorithm to construct the maximum density (MaxD)-based CDS*

**Input:** Graph $G = (V, E)$, where $V$ is the vertex set and $E$ is the edge set.
  Source vertex, $s$ – vertex with the largest number of uncovered neighbors in $V$.
**Auxiliary Variables and Functions:** *CDS-Node-List, Covered-Nodes-List, Neighbors(v),* $\forall v \in V$.
**Output:** *CDS-Node-List*
**Initialization:** *Covered-Nodes-List* $\leftarrow$ $\{s\}$, *CDS-Node-List* $\leftarrow$ $\Phi$
**Begin** Construction of *MaxD-CDS*
    **while** ( $|Covered\text{-}Nodes\text{-}List| < |V|$ ) **do**
            Select a vertex $r \in Covered\text{-}Nodes\text{-}List$ and $r \notin CDS\text{-}Node\text{-}List$ such that $r$ has the largest
        number of uncovered neighbors that are not in *Covered-Nodes-List*
                *CDS-Node-List* $\leftarrow$ *CDS-Node-List* $\cup$ $\{r\}$
            **for** all $u \in Neighbors(r)$ and $u \notin Covered\text{-}Nodes\text{-}List$
        *Covered-Nodes-List* $\leftarrow$ *Covered-Nodes-List* $\cup$ $\{u\}$
            **end for**
        **end while**
    **return** *CDS-Node-List*
**End** Construction of *MaxD-CDS*

uncovered neighbors of a CDS member are also added to the *Covered-Nodes-List* as the member is added to the *CDS-Node-List*. To determine the next node to be added to the *CDS-Node-List*, one must select the node with the largest density amongst the nodes that meet the criteria for inclusion into the CDS.

The criteria for CDS membership selection are the following: the node should not already be a part of the CDS (*CDS-Node-List*), the node must be in the *Covered-Nodes-List*, and the node must have at least one uncovered neighbor (at least one neighbor that is not in the *Covered-Nodes-List*). Amongst the nodes that meet these criteria for CDS membership inclusion, the node with the largest density (i.e., the largest number of uncovered neighbors) is selected to be the next member of the CDS. Ties are broken arbitrarily. This process is repeated until all nodes in the network are included in the *Covered-Nodes-List*. Once all nodes in the network are considered to be "covered", the CDS is formed and the algorithm returns a list of nodes in the resulting MaxD-CDS (nodes in the *CDS-Node-List*). The run-time complexity of the MaxD-CDS algorithm is $O(|V|^2+|E|)$ since there would be at most $|V|$ iterations – it would

take $O(|V|)$ time to determine the node with the largest number of uncovered neighbors in each iteration and across all these iterations, a total of $|E|$ edges would be visited. The pseudo code for the MaxD-CDS algorithm is given in Figure 15 and a working example of the algorithm is illustrated in Figure 16. The legend for Figure 16 is shown in Figure 17.

## GRAPH THEORY ALGORITHMS FOR MANET MULTI-PATH COMMUNICATION

Let $P_L$, $P_N$ and $P_Z$ be the set of link-disjoint, node-disjoint and zone-disjoint $s$-$d$ routes respectively. The Dijkstra $O(|V|^2)$ algorithm is used to determine the minimum hop $s$-$d$ path in a graph of $n$ nodes. It is assumed that the $s$-$d$ routes in a multi-path set are used in the increasing order of the hop count. In other words, the $s$-$d$ route with the least hop count is used as long as it exists, then the $s$-$d$ route with the next highest hop count is used as long as it exists and so on. Thus, the determined multi-path set of $s$-$d$ routes is persisted with as long as at least one path in the set exists.

*Figure 16. Example to illustrate the construction of a Maximum Density (MaxD)-based CDS*

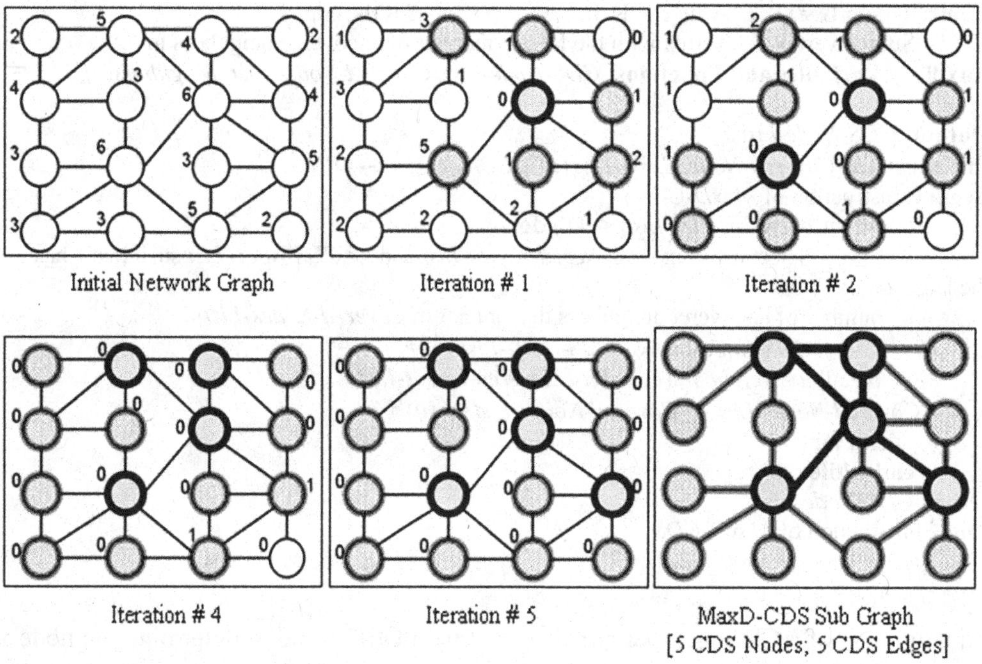

Initial Network Graph          Iteration # 1          Iteration # 2

Iteration # 4          Iteration # 5          MaxD-CDS Sub Graph
[5 CDS Nodes; 5 CDS Edges]

*Figure 17. Legend for Figure 16*

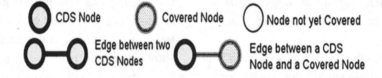

CDS Node    Covered Node    Node not yet Covered

Edge between two CDS Nodes    Edge between a CDS Node and a Covered Node

## Algorithm to Determine Link-Disjoint Paths

To determine the set of link-disjoint paths, $P_L$, (refer Figure 18), all the links that were part of minimum hop $s$-$d$ path $p$ from the graph $G$ are removed to obtain a modified graph $G^L$ ($V, E^L$). A new minimum hop $s$-$d$ path is then determined in the modified graph $G'$, added to the set $P_L$ and the links that were part of this path are removed to get a new updated $G^L$ ($V, E^L$). This procedure is repeated until there exist no more $s$-$d$ paths in the network. The set $P_L$ is now said to have the link-disjoint $s$-$d$ paths in the original network graph $G$ at the given time instant. Figure 19 illustrates a

working-example of the algorithm to find the set of link-disjoint paths on a static graph.

## Algorithm to Determine Node-Disjoint Paths

To determine the set of node-disjoint paths, $P_N$, (refer Figure 20), all the intermediate nodes (nodes other than the source vertex $s$ and destination vertex $d$) that were part of the minimum hop $s$-$d$ path $p$ in the original graph $G$ are removed to obtain the modified graph, $G^N$ ($V^N, E^N$). The minimum hop $s$-$d$ path in the modified graph $G^N$ ($V^N, E^N$) is determined and added to the set $P_N$; the intermediate nodes part of this $s$-$d$ path are removed

*Figure 18. Algorithm to determine the set of link-disjoint s-d paths in a network graph*

**Input:** Graph $G$ $(V, E)$, source vertex $s$ and destination vertex $d$
**Output:** Set of link-disjoint paths $P_L$
**Auxiliary Variables:** Graph $G^L$ $(V, E^L)$
**Initialization:** $G^L$ $(V, E^L) \leftarrow G$ $(V, E)$, $P_L \leftarrow \Phi$
**Begin** *Algorithm Link-Disjoint-Paths*
1      **While** ( $\exists$ at least one *s-d* path in $G^L$)
2         $p \leftarrow$ Minimum hop *s-d* path in $G^L$.
3         $P_L \leftarrow P_L \cup \{p\}$
4         $\forall$    $G^L$ $(V, E^L) \leftarrow G^L$ $(V, E^L$ -$\{e\})$
      *edge,* $e \in p$
5      **end While**
6      **return** $P_L$
**End** *Algorithm Link-Disjoint-Paths*

to get a new updated $G^N$ $(V^N, E^N)$. This procedure is repeated until there exist no more *s-d* paths in the network. The set $P_N$ is now said to contain the node-disjoint *s-d* paths in the original network graph $G$. Figure 21 illustrates a working example of the algorithm to find the set of node-disjoint paths on a static graph.

## Algorithm to Determine Zone-Disjoint Paths

To determine the set of zone-disjoint paths, $P_Z$, (refer Figure 22), all the intermediate nodes (nodes other than the source vertex *s* and destination vertex *d*) that were part of the minimum hop *s-d* path *p* and also all their neighbor nodes from the original graph $G$ are removed to obtain the modified graph $G^Z$ $(V^Z, E^Z)$. The minimum hop *s-d* path in the modified graph $G^Z$ is then determined and added to the set $P_Z$; the intermediate nodes part

*Figure 19. Example to illustrate the working of the algorithm to Find Link-Disjoint Paths*

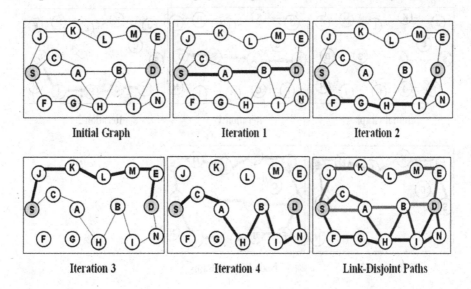

Initial Graph        Iteration 1        Iteration 2

Iteration 3        Iteration 4        Link-Disjoint Paths

*Figure 20. Algorithm to determine the set of node-disjoint s-d paths in a network graph*

**Input:** Graph $G$ $(V, E)$, source vertex $s$ and destination vertex $d$
**Output:** Set of node-disjoint paths $P_N$
**Auxiliary Variables:** Graph $G^N$ $(V^N, E^N)$
**Initialization:** $G^N$ $(V^N, E^N) \leftarrow G$ $(V, E)$, $P_N \leftarrow \Phi$
**Begin** *Algorithm Node-Disjoint-Paths*
1   **While** ( $\exists$ at least one *s-d* path in $G^N$)
        $p \leftarrow$ Minimum hop *s-d* path in $G^N$.
2       $P_N \leftarrow P_N \cup \{p\}$
3       $\forall_{\substack{vertex, v \in p \\ v \neq s, d \\ edge, e \in Adj-list(v)}}$ $G^N$ $(V^N, E^N) \leftarrow G^N$ $(V^N - \{v\}, E^N - \{e\})$

4   **end While**
5   **return** $P_N$
**End** *Algorithm Node-Disjoint-Paths*

of this *s-d* path and all their neighbor nodes are removed to obtain a new updated graph $G^Z$ $(V^Z, E^Z)$. This procedure is repeated until there exist no more *s-d* paths in the network. The set $P_Z$ is now said to contain the set of zone-disjoint *s-d* paths in the original network graph $G$. Note that when a node *v* is removed from a network graph, all the links associated with the node (i.e., links belonging to the adjacency list *Adj-list(v)*) are removed; whereas, when a link is removed from

a graph, no change occurs in the vertex set of the graph. Figure 23 illustrates a working example of the algorithm to find the set of zone-disjoint paths on a static graph.

## FUTURE RESEARCH DIRECTIONS

Graph theory algorithms form the backbone for research on communication protocols for wire-

*Figure 21. Example to illustrate the working of the algorithm to find node-disjoint paths*

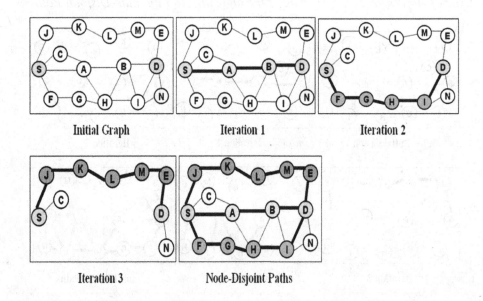

Initial Graph        Iteration 1        Iteration 2

Iteration 3        Node-Disjoint Paths

*Figure 22. Algorithm to determine the set of zone-disjoint s-d paths in a network graph*

**Input:** Graph $G(V, E)$, Source vertex $s$ and Destination vertex $d$
**Output:** Set of Zone-Disjoint Paths $P_Z$
**Auxiliary Variables:** Graph $G^Z(V^Z, E^Z)$
**Initialization:** $G^Z(V^Z, E^Z) \leftarrow G(V, E)$, $P_Z \leftarrow \Phi$
**Begin** *Algorithm Zone-Disjoint-Paths*
1   **While** ($\exists$ at least one $s$-$d$ path in $G^Z$)
2       $p \leftarrow$ Minimum hop $s$-$d$ path in $G^Z$
3       $P_Z \leftarrow P_Z \cup \{p\}$
4       $\displaystyle\mathop{\forall}_{\substack{vertex, u \in p, u \neq s, d \\ edge, e \in Adj-list(u)}}$   $G^Z(V^Z, E^Z) \leftarrow G^Z(V^Z - \{u\}, E^Z - \{e\})$
5       $\displaystyle\mathop{\forall}_{\substack{vertex, u \in p, u \neq s, d \\ v \in Neighbor(u), v \neq s, d \\ edge, e' \in Adj-list(v)}}$   $G^Z(V^Z, E^Z) \leftarrow G^Z(V^Z - \{v\}, E^Z - \{e'\})$
6   **end While**
7   **return** $P_Z$
**End** *Algorithm Zone-Disjoint-Paths*

less ad hoc networks and sensor networks. This chapter lays the foundation for use of several simplistic graph theoretic algorithms (taught at the undergraduate and graduate level) to simulate the behavior of the complex MANET routing protocols. The next step of research in this direction would involve implementing these graph theoretic algorithms in a centralized environment using offline traces of the mobility profiles of the nodes (under a particular mobility model) to generate the mobile graph (i.e., sequence of static graphs representing snapshots of the network topology at different time instants) and compare the performance metrics obtained for the communication structures with that of those obtained for the actual routing protocols when simulated in a discrete-event simulator such as ns-2, Glo-MoSim etc. Some of the performance metrics that could be directly compared are the hop count per source-destination path (for unicasting), hop count per source-receiver path (for multicasting), number of links per multicast tree, lifetime per path, lifetime per multicast tree, time between two consecutive route discoveries for link-disjoint,

node-disjoint and zone-disjoint routes, number of nodes per CDS, hop count of a source-destination path per CDS etc. It is conjectured that the results obtained for the above performance metrics from the centralized graph theory implementations will serve as the optimal benchmarks to which the results obtained from the actual routing protocols in a discrete-event simulator environment would be actually bounded under. This is because the centralized implementations would assume an ideal Medium-Access Control (MAC) layer that would not offer any interference to constrain the communication.

If the simulations could be conducted in more than one discrete-event simulator, then the results for the performance metrics obtained from the different simulators could be compared to the optimal benchmarks obtained with the theoretical algorithms presented in this chapter and could be helpful in identifying the simulator that gives performance closest to the optimum for a particular communication problem (unicast, multicast, broadcast, multi-path) under specific operating conditions. The proposed approach of using graph

*Figure 23. Example to illustrate the working of the algorithm to find zone-disjoint paths*

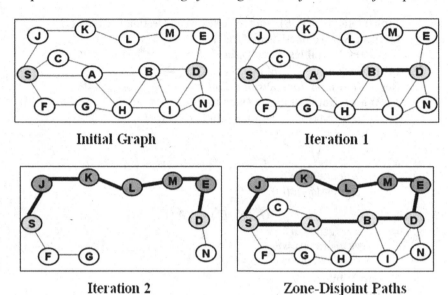

| Initial Graph | Iteration 1 |
|---|---|
| Iteration 2 | Zone-Disjoint Paths |

theory algorithms to study the MANET routing protocols could also be extended to wireless sensor networks, wherein one can use the tree and CDS construction algorithms to study the data gathering protocols.

## CONCLUSION

The high-level contribution of this chapter is the idea of using traditional graph theory algorithms, which have been taught in academic institutions at undergraduate and graduate level, to simulate and study the behavior of the complex routing protocols for unicast, multicast, broadcast and multipath communication in MANETs. The section on Background work provides an exhaustive set of information on the routing protocols that have been proposed for the above different communication problems. In the subsequent sections, one or more graph theoretic algorithms are described for studying each of these communication problems. The Dijkstra algorithm is chosen as the core algorithm for shortest path routing and is also meticulously modified and/or adopted to (i) find a solution for

the largest bottleneck path and smallest bottleneck path problems, which could be used to determine a sequence of stable routes as well as to (ii) find a set of link-disjoint, node-disjoint or zone-disjoint routes for multi-path communication. The use of Prim's algorithm for minimum spanning tree is illustrated to determine the 'All Pairs Smallest Bottleneck Paths' and it is also shown how the modified version of the Prim's algorithm to determine maximum spanning tree can be used to solve the 'All Pairs Largest Bottleneck Paths' problem. A proof is also illustrated to show that the path between any two nodes in the minimum spanning tree is the smallest bottleneck path between those two nodes in the original graph. The chapter also discusses a maximum density-based algorithm to approximate minimum connected dominating sets (MCDS) for MANETs such that the MCDS can be used as a backbone towards broadcast communication for route discoveries and other global communication needs. In addition, the chapter discusses the Steiner tree problem and the Kou et al.'s algorithm to approximate a multicast tree that has the minimum number of links connecting the source nodes to the members of a multicast group. Each of the graph theoretic algorithms discussed in this chapter presents the optimal solutions or the best approximations for the appropriate problems they have been suggested for and all of them could be implemented in the most efficient manner using the pseudo code presented and executed in a polynomial run-time.

As a concluding remark, the proposed idea in this chapter could go a long way in avoiding the significant loss of time faced by student researchers to understand and modify the simulation code for even conducting simple experimental studies. This idea could be adopted to facilitate undergraduate student research in the area of MANETs without requiring the students to directly work on the complex discrete-event simulators. An example for the successful implementation of this idea is the Research Experiences for Undergraduates (REU) site in the areas of Wireless Ad hoc Networks and Sensor Networks, hosted by the Department of Computer Science at Jackson State University, Jackson, MS, USA. The REU site is currently being funded by the U.S. National Science Foundation (NSF) and is accessible through http://www.jsums.edu/cms/reu.

# REFERENCES

Alzoubi, K. M., Wan, P-J., & Frieder, O. (2002). Distributed heuristics for connected dominating set in wireless ad hoc networks. *IEEE / KICS Journal on Communication Networks, 4*(1), 22-29.

Butenko, S., Cheng, X., Du, D.-Z., & Pardalos, P. M. (2003a). On the construction of virtual backbone for ad hoc wireless networks. In Butenko, S., Murphey, R., & Pardalos, P. M. (Eds.), *Cooperative control: Models, applications and algorithms* (pp. 43–54). Berlin, Germany: Springer Publishers.

Butenko, S., Cheng, X., Oliviera, C., & Pardalos, P. M. (2003b). A new heuristic for the minimum connected dominating set problem on ad hoc wireless networks. In Butenko, S., Murphey, R., & Pardalos, P. M. (Eds.), *Recent developments in cooperative control and optimization* (pp. 61–73). Berlin, Germany: Springer Publishers. doi:10.1007/978-1-4613-0219-3_4

Cormen, T. H., Leiserson, C. E., Rivest, R. L., & Stein, C. (2001). *Introduction to algorithms.* Cambridge, MA: MIT Press.

Fall, K., & Varadhan, K. (2001). *NS notes and documentation.* Unpublished Manual – The VINT Project at LBL, Xerox PARC, UCB, and USC/ISI.

Hofmann-Wellenhof, B., Lichtenegger, H., & Collins, J. (2004). *Global positioning system: Theory and practice.* Berlin, Germany: Springer Publishers.

Javan, N. T., & Dehghan, M. (2007). Reducing end-to-end delay in multi-path routing algorithms for mobile ad hoc networks. In Zhang, H., Olariu, S., Cao, J., & Johnson, D. B. (Eds.), *Proceedings of the International Conference on Mobile Ad hoc and Sensor Networks, Lecture Notes in Computer Science: Vol. 4864* (pp. 703-712). Berlin, Germany: Springer Publishers.

Johnson, D. B., Maltz, D. A., & Broch, J. (2001). *DSR: The dynamic source routing protocol for multi-hop wireless ad hoc networks. Ad Hoc Networking* (pp. 139–172). Boston, MA: Addison Wesley.

Kou, L., Markowsky, G., & Berman, L. (1981). A fast algorithm for Steiner trees. *Acta Informatica, 15*(3), 141–145. doi:10.1007/BF00288961

Kuhn, F., Moscibroda, T., & Wattenhofer, R. (2004). Unit disk graph approximation. In Basagni, S., & Phillips, C. (Eds.), *Proceedings of the Joint Workshop on Foundations of Mobile Computing* (pp. 3201-3205). Philadelphia, PA: ACM.

Lee, S., & Gerla, M. (2001). Split multi-path routing with maximally disjoint paths in ad hoc networks. In *Proceedings of the IEEE International Conference on Communications, Vol. 10* (pp. 3201-3205). Helsinki, Finland: IEEE.

Meghanathan, N. (2008). Exploring the stability-energy consumption-delay-network lifetime tradeoff of mobile ad hoc network routing protocols. *Journal of Networks, 3*(2), 17–28. doi:10.4304/jnw.3.2.17-28

Meghanathan, N. (2009). A beaconless node velocity-based stable path routing protocol for mobile ad hoc networks. In *Proceedings of the Sarnoff Symposium Conference*. Princeton, NJ: IEEE.

Meghanathan, N. (2010). Benchmarks and tradeoffs for minimum hop, minimum edge and maximum lifetime per multicast tree in mobile ad hoc networks. *International Journal of Advancements in Technology, 1*(2), 234–251.

Ozaki, T., Kim, J.-B., & Suda, T. (2001). Bandwidth-efficient multicast routing for multi-hop, ad hoc wireless networks. In *Proceedings of the International Conference on Computers and Communication Networks,* vol. 2 (pp. 1182-1192). Anchorage, AK, USA: IEEE.

Perkins, C. E., & Royer, E. (1999). Ad hoc on-demand distance vector routing. In *Proceedings of the 2nd Annual International Workshop on Mobile Computing Systems and Applications* (pp. 90-100). New Orleans, LA: IEEE.

Royer, E., & Perkins, C. E. (1999). Multicast operation of the ad-hoc on-demand distance vector routing protocol. In *Proceedings of the 5th Annual Conference on Mobile Computing and Networking* (pp. 207-218). Seattle, WA: ACM.

Su, W., Lee, S.-J., & Gerla, M. (2001). Mobility prediction and routing in ad hoc wireless networks. *International Journal of Network Management, 11*(1), 3–30. doi:10.1002/nem.386

Ye, Z., Krishnamurthy, S. V., & Tripathi, S. K. (2003). A framework for reliable routing in mobile ad hoc networks. In *Proceedings of the International Conference on Computer Communications, Vol. 1* (pp. 270-280). San Francisco, CA: IEEE.

Zeng, X., Bagrodia, R., & Gerla, M. (1998). GloMoSim: A library for parallel simulation of large-scale wireless networks. In *Proceedings of the 12th Workshop on Parallel and Distributed Simulations* (pp. 154). Banff, Canada: ACM.

## ADDITIONAL READING

Agarwal, S., Ahuja, A., Singh, J. P., & Shorey, R. (2000). Route Lifetime Assessment Based Routing Protocol for Mobile Ad hoc Networks. In *Proceedings of the International Conference on Communications* (pp. 1697-1701). New Orleans, LA, USA: IEEE.

Bae, S., Lee, S., Su, W., & Gerla, M. (2000). The Design, Implementation and Performance Evaluation of the On-demand Multicast Routing Protocol in Multi-hop Wireless Networks. *IEEE Network, 4*(1), 70–77.

Bettstetter, C., Hartenstein, H., & Perez-Costa, X. (2004). Stochastic Properties of the Random-Way Point Mobility Model. *Wireless Networks, 10*(5), 555–567. doi:10.1023/B:WINE.0000036458.88990.e5

Bianchi, G. (2000). Performance Analysis of the IEEE 802.11 Distributed Coordination Function. *IEEE Journal on Selected Areas in Communications, 18*(3), 535–547. doi:10.1109/49.840210

Bononi, L., & Felice, M. D. (2006). Performance Analysis of Cross-layered Multi-path Routing and MAC Layer Solutions for Multi-hop Ad Hoc Networks. In *Proceedings of the International Workshop on Mobility Management and Wireless Access* (pp. 190-197). Terromolinos, Spain: ACM.

Broch, J., Maltz, D. A., Johnson, D. B., Hu, Y. C., & Jetcheva, J. (1998). A Performance Comparison of Multi-hop Wireless Ad Hoc Network Routing Protocols. In *Proceedings of the 4ᵗʰ International Mobile Computing and Networking Conference* (pp. 85-97). Dallas, TX, USA: ACM.

Chen, Y. P., & Liestman, A. L. (2002). Approximating Minimum Size Weakly-Connected Dominating Sets for Clustering Mobile Ad hoc Networks. In *Proceedings of the International Symposium on Mobile Ad hoc Networking and Computing*. Lausanne, Switzerland: ACM.

Creixell, W., & Sezaki, K. (2007). Routing Protocol for Ad hoc Mobile Networks using Mobility Prediction. *International Journal of Ad Hoc and Ubiquitous Computing, 2*(3), 149–156. doi:10.1504/IJAHUC.2007.012416

Das, S. K., Manoj, B. S., & Murthy, C. S. R. (2002). Weight-based Multicast Routing Protocol for Ad hoc Wireless Networks. In *Proceedings of the Global Telecommunications Conference* (pp. 117-121). Taipei, Taiwan: IEEE.

Fasolo, E., Zanella, A., & Zorzi, M. (2006). An Effective Broadcast Scheme for Alert Message Propagation in Vehicular Ad hoc Networks. In *Proceedings of the International Conference on Communications, Vol. 9* (pp. 3960-3965). Istanbul, Turkey: IEEE.

Feeney, L. M. (2001). An Energy Consumption Model for Performance Analysis of Routing Protocols for Mobile Ad hoc Networks. *Journal of Mobile Networks and Applications, 3*(6), 239–249. doi:10.1023/A:1011474616255

Gil, H.-R., Yoo, J., & Lee, J.-W. (2003). An On-demand Energy-efficient Routing Algorithm for Wireless Ad hoc Networks. *Lecture Notes in Computer Science, 2713*, 302–311. doi:10.1007/3-540-45036-X_31

Gunes, M., Sorges, U., & Bouazizi, I. (2002). ARA: The Ant Colony based Routing Algorithm for MANETs. In *Proceedings of the ICPP Workshop on Ad Hoc Networks* (pp. 79-85). Vancouver, Canada: IEEE.

Keiss, W., Fuessler, H., & Widmer, J. (2004). Hierarchical Location Service for Mobile Ad hoc Networks. *ACM Mobile Computing and Communications Review, 8*(4), 47–58. doi:10.1145/1052871.1052875

Lim, G., Shin, K., Lee, S., Yoon, H., & Ma, J. S. (2002). Link Stability and Route Lifetime in Ad hoc Wireless Networks. In *Proceedings of the International Conference on Parallel Processing Workshops* (pp. 116-123). Vancouver, Canada: IEEE.

Lin, X., & Stojmenovic, I. (2003). Location-based Localized Alternate, Disjoint and Multi-path Routing Algorithms for Wireless Networks. *Journal of Parallel and Distributed Computing, 63*, 22–32. doi:10.1016/S0743-7315(02)00037-0

Meghanathan, N. (2007). Impact of Broadcast Route Discovery Strategies on the Performance of Mobile Ad Hoc Network Routing Protocols. In *Proceedings of the International Conference on High Performance Computing, Networking and Communication Systems* (pp. 144-151). Orlando, FL, USA: Promote Research.

Meghanathan, N. (2007). Path Stability based Ranking of Mobile Ad Hoc Network Routing Protocols. *ISAST Transactions Journal on Communications and Networking, 1*(1), 66–73.

Meghanathan, N. (2009). Survey and Taxonomy of Unicast Routing Protocols for Mobile Ad hoc Networks. *The International Journal on Applications of Graph Theory in Wireless Ad hoc Networks and Sensor Networks, 1*(1), 1-21.

Meghanathan, N. (2009). A Location Prediction Based Reactive Routing Protocol to Minimize the Number of Route Discoveries and Hop Count per Path in Mobile Ad hoc Networks. *The Computer Journal, 52*(4), 461–482. doi:10.1093/comjnl/bxn051

Meghanathan, N. (2010). An Algorithm to Determine Minimum Velocity-based Stable Connected Dominating Sets for Ad hoc Networks. *Communications in Computer and Information Science Series, 94*, 206–217. doi:10.1007/978-3-642-14834-7_20

Meghanathan, N., & Sugumar, M. (2009). A Beaconless Minimum Interference Based Routing Protocol for Mobile Ad hoc Networks. *Communications in Computer and Information Science Series, 40*, 58–69. doi:10.1007/978-3-642-03547-0_7

Mueller, S., Tsang, R. P., & Ghosal, D. (2004). Multi-path Routing in Mobile Ad Hoc Networks: Issues and Challenges. *Lecture Notes in Computer Science, 2965*, 209–234. doi:10.1007/978-3-540-24663-3_10

Ni, S., Tseng, Y., Chen, Y., & Sheu, J. (1999). The Broadcast Storm Problem in a Mobile Ad Hoc Network. In *Proceedings of the International Conference on Mobile Computing and Networking* (151-162). Seattle, WA, USA: ACM.

Sheu, P., Tsai, H., Lee, Y., & Cheng, J. (2009). On Calculating Stable Connected Dominating Sets Based on Link Stability for Mobile Ad Hoc Networks. *Tamkang Journal of Science and Engineering, 12*(4), 417–428.

Son, D., Helmy, A., & Krishnamachari, B. (2004). The Effect of Mobility-Induced Location Errors on Geographic Routing in Mobile Ad hoc Sensor Networks: Analysis and Improvement using Mobility Prediction. *IEEE Transactions on Mobile Computing, 3*(3), 233–245. doi:10.1109/TMC.2004.28

Toh, C. K., Guichal, G., & Bunchua, S. (2000). ABAM: On-demand Associatvity-based Multicast Routing for Ad hoc Mobile Networks. In *Proceedings of the 52nd VTS Fall Vehicular Technology Conference, Vol. 3* (pp. 987-993). Boston, MA, USA: IEEE.

Viswanath, K., Obraczka, K., & Tsudik, G. (2006). Exploring Mesh and Tree-based Multicast Routing Protocols for MANETs. *IEEE Transactions on Mobile Computing, 5*(1), 28–42. doi:10.1109/TMC.2006.11

Wang, H., Ma, K., & Yu, N. (2005). Performance Analysis of Multi-path Routing in Wireless Ad Hoc Networks. In *Proceedings of the International Conference on Wireless Communications, Networking and Mobile Computing, Vol. 2* (pp. 723-726). Maui, HI, USA: IEEE.

Ye, Z., Krishnamurthy, S. V., & Tripathi, S. K. (2003). A Framework for Reliable Routing in Mobile Ad Hoc Networks. In *Proceedings of the International Conference on Computer Communications* (pp. 270-280). San Francisco, CA, USA: IEEE.

## KEY TERMS AND DEFINITIONS

**Algorithm:** A set procedure for solving a given problem in a finite sequence of steps.

**Beacon:** Each node in the network periodically broadcasts a small control message called 'beacon' to all the nodes within its transmission range to announce its presence in the neighborhood.

**Broadcast:** Communication initiated from one source node and targeted to all the other nodes in the network.

**Euclidean Distance:** The Euclidean distance between two points $p\,(p_1, p_2, ..., p_n)$ and $q\,(q_1, q_2, ..., q_n)$ in an n-dimensional plane is the length of the line segment connecting the two points. It is given by: $\sqrt{\sum_{i=1}^{n}(q_i - p_i)^2}$

**Mobile Ad hoc Network:** A dynamically changing infrastructureless and resource-constrained network of wireless nodes moving arbitrarily, independent of each other.

**Multicast:** Communication between a single source node and multiple receiver nodes.

**Multi-Path:** Unicast communication involving more than one path, with the paths often independent of each other (either link-disjoint or node-disjoint or zone-disjoint).

**Routing Protocol:** A protocol used for communication between any two nodes or a set of nodes in a network through one or more hops.

**Simulation:** The process of imitating the working of a network through a software program that implements one or more interaction models between the different entities (nodes, links and etc).

**Unicast:** Communication between a single source node and a single destination node.

**Unit Disk Graph:** A collection of vertices (nodes) and edges (links); the vertices represent the wireless nodes and an edge exists between two vertices $u$ and $v$ if the normalized Euclidean distance (i.e., the physical Euclidean distance divided by the transmission range) between $u$ and $v$ is at most 1.

# Chapter 5
# Experiences from Integrating Collaborative Filtering in a Mobile City Guide

**Wolfgang Woerndl**
*Technische Universitaet Muenchen, Germany*

**Korbinian Moegele**
*Technische Universitaet Muenchen, Germany*

**Vivian Prinz**
*Technische Universitaet Muenchen, Germany*

## ABSTRACT

*This chapter presents an approach to extend a real world mobile tourist guide running on personal digital assistants (PDAs) with collaborative filtering. The system builds a model of item similarities based on explicit and implicit ratings. This model is then utilized to generate recommendations in several ways. The approach integrates the current user location as context. Experiences gained in two field studies are reported. In the first one, 30 participants – real tourists visiting Prague – used the recommender function and were asked to fill out a questionnaire with promising results. In a second field study analyzing usage log files, an improvement of recommendations based on the collaborative filter in comparison to the pure location-based filter used before was discovered. In addition, recommendations based on implicit ratings derived from audio playback duration outperformed the model based on explicit ratings.*

## INTRODUCTION

Mobile devices such as smartphones and Personal Digital Assistants (PDAs) are becoming more and more powerful and are increasingly used for tasks other than just making phone calls. Examples include mobile Web browsing, Gaming, Personal Information Management (PIM) and audio and video playback. Tourism and travel are prime application areas for mobile applications. An increasing number of services is offered to support a traveler not only before and after the travel, but also while sightseeing (Ricci, 2010). The development of technologies such as Global

DOI: 10.4018/978-1-4666-0080-5.ch005

Positioning System (GPS) positioning and ubiquitous availability of wireless communication services has fostered this development.

However, mobile information access still suffers from limited resources regarding input capabilities, displays, network bandwidth and other limitations of small mobile devices. In addition, mobile applications must consider mobile user constraints such as limited attention span while moving, changing locations and contexts, and expectations of quick and easy interactions (Subramanya & Yi, 2007). Furthermore, as the amount of information and online services increases, it becomes more and more difficult for users to find the right information needed to complete a particular task (Ricci, 2010). Therefore, it is desirable to tailor information access to the current user needs (Brusilovsky, Kobsa, & Nejdl, 2007).

Recommender systems are a well-established technique to counter this problem of information overload (Borchers, Herlocker, Konstan, & Riedl, 1998; Kantor, Ricci, Rokach, & Shapira, 2010; Jannach, Zanker, Felfernig, & Friedrich, 2010). These systems suggest items like products, restaurants or other Points-Of-Interests (POIs) based on explicit user ratings or implicitly observed user behavior. In a mobile setting, good and precise recommendations are even more important because of the explained intrinsic obstacles of mobile usage environments. Moreover, the context such as the current user position plays a major role in mobile recommendations.

The considered scenario is a real world mobile city guide running on PDAs with GPS positioning capabilities. The guide is currently available for the Czech city of Prague (http://www.voxcity.de). The mobile application plays audio, video, pictures and (HTML) text of tourist attractions based on the current position. This mobile guide has been extended with a collaborative filtering recommender system based on user ratings. The solution exchanges ratings for POIs among PDAs, computes matrices of item similarity and utilizes them to generate recommendations. The basic idea

for taking the current user position into account is to use a weighted combination of the collaborative filtering score with a location score function.

In this chapter, the design and implementation issues to put this mobile recommender system into practice is outlined. In addition, experiences gained in two user studies are presented. The first aimed at the usability of the city guide and the potential benefits of the recommender system from a user perspective. The second field study was performed to investigate the recommendation quality of the approach in more detail.

The rest of the chapter is organized as follows. The next section describes the existing real world mobile city guide and also specifies requirements. Afterwards, fundamentals of recommender systems including the used item-based collaborative filtering algorithm, and properties of context-aware and mobile recommendation are explained. Related work including relevant surveys and systems is then covered. This is followed by a discussion of the system design for the integration of a collaborative recommender in the city guide. The next sections present the findings from the two user studies. Then, some related application areas for context-aware recommender systems in mobile scenarios are explained. Finally, the chapter concludes with a brief summary and outlook on future work.

## THE MOBILE CITY GUIDE: BACKGROUND AND REQUIREMENTS FOR EXTENSION

### The Existing Mobile City Guide (MCG)

The Mobile City Guide (MCG) was developed by the companies voxcity s.r.o. and jomedia s.r.o. (http://www.voxcity.de). The concept is to rent out mobile devices with GPS positioning capabilities to support tourists (Figure 1). The system runs on Microsoft Windows Mobile devices, version

*Figure 1. Using the mobile city guide in Prague*

5 or later. The guide is currently available for the Czech city of Prague and can be rented at several tourist information centers, some major sights and hotels in Prague. The mobile application plays audio, video, pictures and (HTML) text of tourist attractions based on the current position and also traveling direction.

The central idea behind the MCG is to support the user with information about sights while still allowing an individual tour of the city. The goal was not to provide pre-determined routes but letting the user explore the city on his/her own. The mobile guide provides multimedia content of nearby attractions on demand. The current version of the mobile guide mainly features audio files with background information about the sights. Figure 2 shows screenshots of the mobile guide. Users can see a map that is centered on the current location (blue marker). In addition, the sights of the surrounding area are marked. Users can now select the content of the nearest POI by pressing the "play" button. It is also possible to manually

select a POI by using the button in the lower right corner of the screen and entering a number. When a sight is selected, the system shows an image of a selected POI (Figure 2, right), displays some text if available, and plays the corresponding audio (or video) file.

The existing system utilizes the current location of the user as context for selecting a point-of-interest. The goal of the work was to extend this system with a recommender system to improve the user experience. Thus, not only the nearest POI would be selected but recommended ones in the vicinity. The system should determine whether a POI that is located farer away but suits the user's supposed interests better might be more interesting for his/her in the current situation. Additional requirements are discussed in the next subsection.

*Figure 2. Screenshots of the mobile city guide*

## Requirements for Extending the Mobile City Guide with POI Recommendations

The most important functional requirement for the extended mobile city guide was to enhance the system with a recommendation method for POIs. The authors want the system to capitalize on past experiences other users have made in a similar situation with minimal user interaction. Therefore, a collaborative recommender system seems to be preferable. More properties of different recommender algorithms and their suitability for the mobile scenario with PDAs are discussed in the next chapter.

Non-functional requirement include minimal changes in the current system. The recommender part should be independent from the rest of the MCG, since the current MCG is a commercial application. Any malfunction of the recommender part should better not interfere with the other parts of the MCG. Another requirement was the usability of the mobile guide. The application is used by tourists who do not want to have to spend

any time to learn the handling of the mobile application. Therefore, the user interface should be kept as simple as possible. A tiresome dialogue to acquire user interests and preferences is out of the question. In addition, the recommender algorithm should not delay response time of the guide. It is important to allow fast interactions with the system. Otherwise the users might quickly lose interest in using the MCG.

Another important aspect is that the mobile devices do not feature a permanent network connection in this scenario. This is due to the fact that networks connection is not free and would complicate the whole process from a technical and also business point of view. Therefore, all the information needed for recommendations have to be stored on the device and all computation has to be performed locally. To summarize, the recommender system in the mobile city guide should support the tourists by selecting more interesting POIs for audio playback on the device without utilizing a network connection and without any significant changes or drawbacks in the overall user experience.

## RECOMMENDER SYSTEMS BACKGROUND

### Recommender Systems Fundamentals

The basic idea of recommender systems is to recommend products such as books, restaurants and other items for an active user (Kantor et. al., 2010). To do so, the system computes the chance that a user likes an item. This is based on information about the user (e.g. preferred restaurant types and other preferences), the items (e.g. restaurant type or price level) and possibly other data such as contextual information (e.g. the current location) (Woerndl, Muehe, & Prinz, 2009). As a result, systems generally present the user with a ranked list of recommended items new to the user.

In principle, authors distinguish between individual and collaborative recommender systems (Anand & Mobasher, 2003). Individual recommenders determine fitting items utilizing the profile of the active user only. Thereby, the system matches explicitly entered or implicitly observed user preferences and interests with items' meta data. This type of recommender system is often called content-based recommender systems. A number of ways to implement individual recommenders have been proposed (Burke, 2007). Content-based recommenders treat recommendation as a user-specific classification problem and learn a classifier for the user's likes and dislikes based on product features. Other examples for individual recommender algorithms include demographic recommenders providing recommendations using the demographic profile of the user, and knowledge-based systems. A knowledge-based recommender suggests products based on inferences about a user's needs and preferences. This knowledge can contain explicit knowledge about how certain product features meet user needs (Burke, 2007).

The second main category of recommender systems is based on Collaborative Filtering (CF) (Schafer, Frankowski, Herlocker, & Sen, 2007). CF utilizes the ratings of other users for items. Ratings can be captured on an n-ary scale of 1 to n, a binary scale ("good/bad") or on an unary scale. An example for the latter case is a system that exploits information about users clicking on Web links or purchasing products. Ratings can be obtained explicitly or implicitly. Users can often submit explicit ratings in E-Commerce Web sites via forms. Because of the limitations of the user interface in a mobile environment (see "INTRODUCTION"), implicitly observed user behavior may be preferred on mobile devices. In this case, a rating is derived from user actions such as clicking on Web links or usage analysis of other programs such as media players. The system utilizes playback duration of audio files as a source for implicit ratings in the system. The evaluation compares recommendations based on explicit versus implicit ratings, see below.

Memory-based collaborative filtering algorithms operate on ratings only. The raw data of user ratings of items is stored in a user-item matrix (Figure 3, left). One element (u, i) in this matrix represents the rating R a user u gave to an item i. We can differentiate between two variants of CF, user- and item-based collaborative filtering. The recommendation process of user-based CF basically consists of two steps. First, neighborhood creation: Determine a set of k users that have rated similarly to the active user in the past. Secondly, recommendation of new items for the active user. For the neighborhood creation, the rating vector of the active user is compared to the vector of all other users. To do so, different metrics have been proposed in the literature, for example Euclidean distance, Cosine similarity or Pearson-Spearman correlation (Adomavicius & Tuzhilin, 2005). In the second step, the algorithm selects recommended items, which the active user has not rated yet. Recommended items are items that have been rated positively in the neighborhood of the active user. Different metrics can be applied to implement this step. It is necessary to

*Figure 3. User-item and item-item matrices (Woerndl, Muehe, & Prinz, 2009)*

| | $item_1$ | $item_2$ | $\cdots$ | $item_i$ |
|---|---|---|---|---|
| $user_1$ | R | R | $\cdots$ | R |
| $user_2$ | R | - | $\cdots$ | - |
| $\vdots$ | $\vdots$ | $\vdots$ | $\ddots$ | $\vdots$ |
| $user_{u-1}$ | - | - | $\cdots$ | R |
| $user_u$ | - | R | $\cdots$ | R |

| | $item_1$ | $item_2$ | $\cdots$ | $item_i$ |
|---|---|---|---|---|
| $item_1$ | 1 | | $\cdots$ | |
| $item_2$ | S | 1 | $\cdots$ | |
| $\vdots$ | $\vdots$ | $\vdots$ | $\ddots$ | |
| $item_i$ | S | S | $\cdots$ | 1 |

adjust the rating predictions with regard to users who always rate on the extreme ends of the scale or only choose very good ratings, for example.

Item-based CF does not consider the similarity of users, but of items (Sarwar, Karypis, Konstan, & Riedl, 2001; Deshpande & Karypis, 2004). Thus, the user-item matrix is thus not analyzed line by line, but column by column. One significant difference to user-based CF is the independence from the active user: the item similarities can be pre-computed to build an item-item matrix of pair wise item similarity (Figure 3, right). This item-item matrix is the model of the algorithm. Therefore, this type of recommendation algorithm is also called model-based collaborative filtering. One element (i,j) of the item-item matrix expresses the similarity S between items i and j, determined from the users' ratings. The rating of the active user is then used to recommend items that are similar to those positively rated by the active user in the past. It is important to note that item-based CF has little in common with individual, content-based filtering. This is because the users' ratings are solely used for computing the item similarity. Meta data or other information of items is irrelevant. Item-based CF algorithms have successfully been applied in commercial systems such as Amazon.com (Linden, Smith, & York, 2003).

Item-based CF has an advantage over user-based CF with regard to the complexity of the computation because the item-item matrix can be calculated as an intermediate result, independently from the active user. In addition, at the time of the recommendation, the algorithm also needs the rating vector of the active user only, not all ratings. This is promising for improving the privacy of personal user data. Therefore, item-based CF appears to be well suited for implementation on PDAs (Woerndl, Muehe, & Prinz, 2009).

## PocketLens

The basic motivation behind the development of PocketLens is that characteristics of item-based recommenders are well suited for decentralized adoption (Miller, Konstan, & Riedl, 2004). Though, some problems arise from utilizing the item-item matrix. In particular, it is required to recompute the matrix when a new rating is inserted. The algorithm has to evaluate all ratings associated with the concerning items in this process, because only the computed similarity is stored in the matrix. In addition, the recalculation of the matrix every time a new rating is added is not very efficient and increases hardware demands. It is preferable to make use of the item similarity values calculated so far as intermediate results (Woerndl, Muehe, & Prinz, 2009).

PocketLens modifies the basic item-based CF algorithm to compute the item-item matrix. Thereby, all the concerning elements of the matrix do not have to be completely recalculated. This is achieved by storing the intermediate results with every item similarity pair in the item-item matrix. PocketLens uses cosine similarity and stores the dot products and the lengths of the rating vectors for every item pair. Thus, it is possible to easily update the item-item matrix with new ratings: the new rating just has to be added to the dot products and the vector length. The whole data model does not have to be recomputed. Especially reapplying all rating vectors used so far is not necessary. Hence, the PocketLens algorithm allows to delete the rating vector of other users after integrating their ratings into the model (Woerndl, Muehe, Rothlehner, & Moegele, 2009).

The PocketLens approach is well suited for the requirements of a mobile city guide (see above). First of all, the model-based approach offers good performance when generating recommendations. In addition, the model can be updated offline with new ratings of other users. Since the used mobile devices are not permanently connected to a server, the system could recompute the model when a user returns a device to a renting station. Afterwards, all the information – the item-item matrix and ratings of the active user – is locally available on the mobile device. Aside from PocketLens, Berkovsky, Kuflik and Ricci (2007) propose a distributed recommender system that partitions the user-item matrix based on domain-specific item categories. However, additional information about the application domain of the items is needed.

## Context in Recommender Systems

As already mentioned in the Introduction, context plays a major role in mobile recommender systems (Ricci, 2010). Following the definition of Dey, Abowd and Salber (2001), context "is any information that can be used to characterize the situation of entities (i.e. whether a person, place or subject) that are considered relevant to the interaction between a user and an application, including the user and the application themselves". In the considered scenario, context mainly is the current user location. Yet, other context attributes such as the time of the day are possible.

It is important to distinguish between user profile information (e.g. a rating of an item) and context (Woerndl & Schlichter, 2008). While the

user profile is rather static and somewhat longer lasting, context is highly dynamic and transient. Depending on the application area, user models can also be highly dynamic, but a user profile contains information such as preferences, interests or knowledge (e.g. the user's favourite type of restaurant or a rating), whereas the context model describes the current environment in which the user operates in (e.g. his/her current location). Profile information can be implicitly observed or explicitly provided by the user. In contrast, context information is neither stored permanently nor manually entered by the user. In the considered scenario, the context is the current user position and this location is captured and analyzed when a user wants to select an item.

Adomavicius and Tuzhilin (2010) have recently introduced three different algorithmic paradigms for incorporating contextual information into the recommendation process: contextual pre-filtering, post-filtering, and modeling (Adomavicius & Tuzhilin, 2010). In contextual pre-filtering (Figure 4, left) the contextualization is performed prior to applying the recommender function. Information about the current context is used for selecting or constructing the relevant set of data records (i.e., ratings) for that specific context. This is indicated by the symbol "C" in Figure 4. Then, ratings can be predicted using any traditional 2-dimensional recommender algorithm and thus taking the user dimension "U" into account. Contextual post-filtering initially (Figure 4, middle) ignores context information. The contextual filtering is done on the result set of the recommender system. The context modeling paradigm directly uses context information as part of the recommender algorithm (Adomavicius & Tuzhilin, 2010). To do so, one option is the multidimensional approach by Adomavicius et. al. (2005). They propose a reduction-based algorithm with the goal to reduce the dimensions.

*Figure 4. Paradigms for incorporating context in Recommender Systems, adopted from (Adomavicius & Tuzhilin, 2010)*

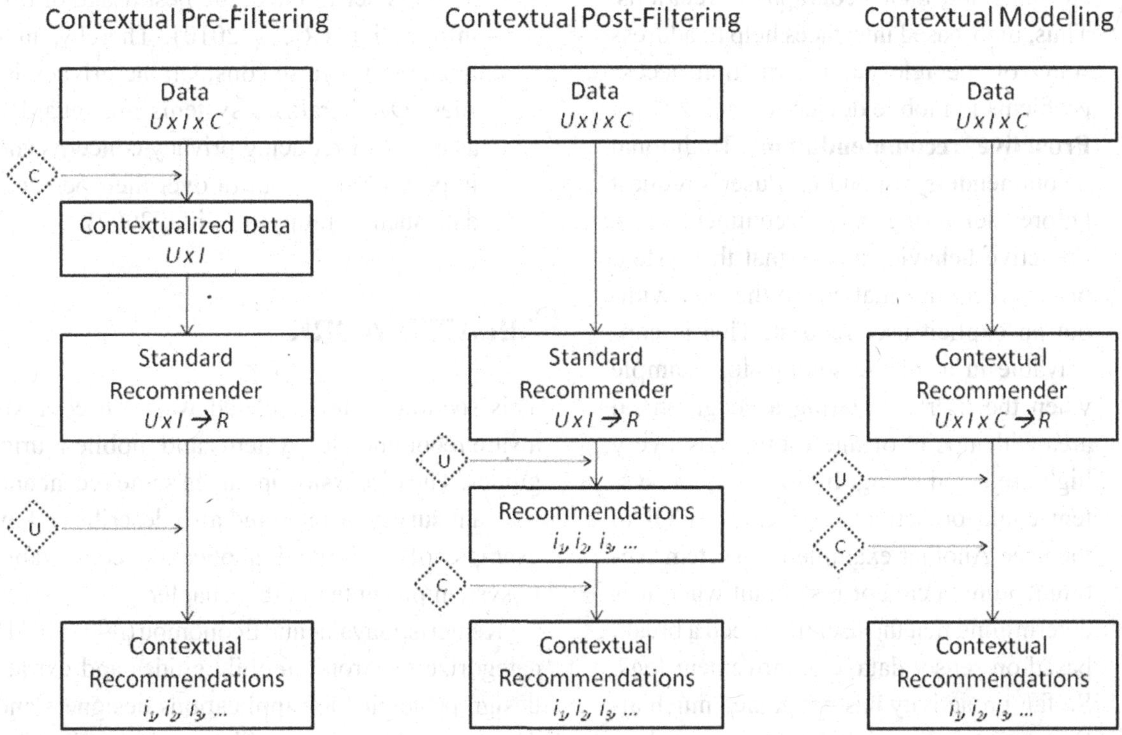

## Other Characteristics of Mobile Recommender Systems

Context-awareness may be the most important property of a mobile recommender system. Moreover, the following characteristics and challenges of mobile recommenders can be noticed (Ricci, 2010):

- **Preciseness of recommendations**: User can not browse through many search results on a mobile device. Therefore, the first single recommended item or the first few items on a list have to be precise and good recommendations. In other words, ranking of items is more important than precision or recall (Herlocker, Konstan, Terveen, & Riedl, et. al., 2004).

- **User interface**: Recommendation sessions on small screen devices can be difficult and frustrating for end-users. Users can actually read and understand information offered by small interfaces, but the size of the display can impact on users' performance (Jones & Marsden, 2005). In addition, mobile devices offer limited input and interaction capabilities. To counter these problems, some techniques that have been exploited in mobile recommender systems, for example starfield displays or map-based interfaces (Ricci, 2010). A starfield display is a mapping of selected attributes of a multidimensional information space onto a two-dimensional representation (Schneiderman, 1994). Several of the mobile guide applications covered in the "Related Work" section apply maps and map-based interfaces as primary access method to visualize the recommended items and their geographical relations. Thus, map-based interfaces help to address some of the relevant information access problems in mobile devices (Ricci, 2010).

- **Proactive recommendation**: Traditional recommenders respond to a user's request before delivering any recommendation. Proactive behavior means that the system pushes recommendations to the user without an explicit user request. This is conceivable in a mobile setting, for example when the user is entering a geographical area with a point-of-interest that has a very high predicted rating. In this case, the system could proactively suggest the POI to the user. Another example is a system recommending a café or restaurant when it is determining that the user may need a break, based on sensor data, e.g. movement logs. So far, proactivity has not gained much attention in recommender system research or has been put into practice. Proactive recommendations may play a larger role in mobile recommender systems in the future.
- **Portable and decentralized user profiles and systems**: Users in practice are using a variety of mobile devices (e.g. phone, laptop and digital camera). Activity performed and services obtained with these devices should be integrated. The user must be recognized and served appropriately whatever device he/she is using (Ricci, 2010). Decentralized recommender systems can be used without a permanent connection to a server and thus improve the portability of the system (Miller, Konstan, & Riedl, 2004).
- **Privacy**: Mobile systems utilize a user's position and other context data, in addition to profile information such as ratings or preferences. An application example is functionality related to support personal

memories. These systems help the user to remember personal facts and tasks, and help his/her to make the best usage of this information (Ricci, 2010). Thereby, mobile systems raise considerable privacy issues. Decentralized systems are regarded as a way of reducing privacy concerns and improving user control over their personal data such as ratings (Kobsa, 2007).

## RELATED WORK

This section reviews related work on context-aware recommender systems and mobile tourist guides. The discussion includes some recent and relevant survey articles and also describes a few examples of mobile guide applications comparable to system presented in this chapter.

Kenteris, Gavalas and Economou (2008; 2011) categorize electronic mobile guides and extract design principles for applications designers and developers. The criteria with regard to application designers are composed of questions regarding the information model (e.g. collaborative filtering or other techniques), input/output modalities and what unique services were designed and implemented. Criteria with regard to technology developers include the type of positioning technologies. The majority of surveyed projects used a centralized approach, i.e. a connection of some sort to feed information to networked mobile devices. This means costly wireless connections for tourists or maybe even more expensive metropolitan installations. In contrast, the approach explained in this chapter stores all the information locally on the PDAs while still being able to utilize ratings of other users modeled in the item-item matrix.

Ricci (2010) is a recent article surveying the field of mobile recommender systems. In the application domain of tourist guides, the article identifies the following three functions: finding relevant attractions, finding relevant services and exploring a city. As an application example for supporting the exploration of a city, (Hinze & Buchanan, 2006) present a mobile infrastructure for cooperating information services. This infrastructure is demonstrated through the example of a Tourist Information Provider (TIP) system. The TIP delivers context-sensitive information from a variety of services to the user. The approach uses an event-based communication layer to support continually changing information.

With regard to the user interface, Raubal and Panov (2009) propose a formal conceptual model for automatic mobile map adaptation composed of three components – a context model, a user model, and a task model. The model aims to reduce the user interaction and thus the cognitive load for the user. It can be applied in different application areas such as pedestrian navigation. This approach does not consider collaborative recommendations though.

Krüger et al. (2007) give an overview on mobile guides focusing on adaptation for user needs. The survey outlines mobile guide applications in the domains of museums, navigation systems and shopping assistants. Most related to the scenario of a mobile city guide are navigation systems where users are supported to explore an unknown area. One example is the "LoL@" (Local Location assistant) system (Anegg, Kunczier, Michlmayr, Pospischil, & Umlauft, 2002; Pospischil, Umlauft, & Michlmayr, 2002), mobile tourist guide for the city of Vienna. The system is similar to the scenario of this chapter: tourists are supported with an overview map and can access information about POI. In addition, LoL@ integrates navigation capabilities and a tour diary. However, the application relies on a network connection and does not incorporate collaborative recommendation.

Baldauf, Dustdar and Rosenberg (2007) survey the field of context-aware systems. They present several design principles with regard to architecture and the context model. In the latter case, it is possible to identify some relevant context modeling approaches based on the data structures used for representing and exchanging contextual information in the respective system (Strang & Linnhoff-Popien, 2004). Examples include simple key-value attributes, object-oriented models or more advanced models based on logic or ontologies. Baldauf, Dustdar and Rosenberg (2007) discuss some existing systems including location-aware tourist guides where information dependent to the current location is displayed.

Rasinger, Fuchs and Höpken, (2007) and Fuchs, Rasinger and Höpken (2007) present an overview of mobile tourist guides based on an empirical approach. Their goal was to find out which services should be developed and identify ideal functionality compositions for promising tourist guides. Other surveys of mobile tourist guides include (Costa, Correia, & Moital, 2008), (Grun, Werthner, Proll, Retchitzegger, & Schwinger, 2008), (Kray & Baus, 2003) and (Stroobants, 2006).

One of the earlier mobile guide systems was "Cyberguide" (Abowd, Atkeson, Hong, Long, & Pinkerton, 1997). The main goal of this project was to support rapid prototyping resulting into many separate systems prototyped for outdoor and indoor use. "TellMaris" (Kray, Laakso, & Elting, 2003), was one of the first mobile systems to use 3D map prototypes in combination with 2D maps for the city of Tonsberg in Norway targeting boating tourists in the Baltic Sea area. The project presented 2D and 3D maps on mobile devices in a way in which to provide easier orientation for tourist. The "CRUMPET" project (Poslad, Laamanen, Malaka, Nick, Buckle, & Zipl, 2001) aimed at providing new information delivery services for a heterogeneous tourist population. The services proposed take advantage of integrating four key emerging technology domains and applying them to the tourism domain: location-aware services, personalized user interaction, seamlessly accessible multi-media mobile communication, and smart component-based middleware. As far as personalization is concerned, their filtering process is based on a user profile describing the interests, abilities and characteristics of a user. The user model acquisition mainly relies on implicit user feedback, taking events when walking around into account. For instance, if a user visits a number of old churches, then he/she is probably interested in churches and perhaps also other historic buildings in the town.

Kenteris, Gavalas and Economou (2010) recently presented the "Mytilene" E-guide. This system incorporates a tool for adopting a "web-to-mobile" model into tourism websites. It allows for the adaptation of personalized tourism web content to be transferred to a mobile application. These applications let users then browse the adapted multimedia content with no requirement for constant network connection. However, the approach does not apply collaborative filtering or other recommendation systems based on users' ratings. The system creates an explicit personal profile used by the recommendation system to suggest content that matches user preferences.

"MobyRek" is another system that is designed to support the user on-the-move (Ricci & Nguyen, 2006). The authors present a computational approach for providing travel product recommendations to on-the-move travelers. The system employs a dialogue approach whereby a set of candidate products are proposed and the user is asked to critique the recommended products. When making a critique in the system, the user is supported by the GUI on the mobile device to express the strength of the preference implied in that critique, i.e., as a "must" condition. In Ricci and Nguyen (2007) it is shown that critique-based recommendation methodology is effective in supporting mobile users in product selection decisions. But the approach relies heavily on user interaction.

Jacobsson, Rost and Holmquist (2006) describe a collaborative filtering approach for mobile "media agents". Thereby, additional information and rules in a profile are attached to items such as music files. The items are then supposed to interact as software agents and show CF-like behavior. "Magitti" is a system to recommend leisure activities on mobile devices (Ducheneaut

et. al., 2009). Their approach is to combine and weigh very different recommenders such as CF, a distance model, content preference and others. This is similar to the approach presented in the next section. However, they do not use item-based collaborative filtering, which is well suited for application on a mobile device without network connection.

## INTEGRATING COLLABORATIVE AND CONTEXT-BASED RECOMMENDATIONS IN THE MOBILE CITY GUIDE

### Building the Item-Based Collaborative Filtering Model

A decentralized, item-based CF based on PocketLens has been implemented and tested for Microsoft Windows Mobile PDAs (Woerndl, Muehe, & Prinz, 2009). The idea is that users rate items on mobile devices. The ratings are then used to calculate the item-item matrix. Since the PDAs are not connected to a network when in use, the ratings of other users are exchanged to update the model when the PDAs are returned to the rental station.

The process of generating context-aware recommendations from the item-item matrix is explained below. An explanation how to acquire the ratings necessary to build the model is first. From the users' perspective, the mobile city guide is unchanged apart from one addition in the user interface. The system includes an option to rate items (Figure 5). This dialogue is shown after the user played an audio file. Note that rating items is optional. Users can just return to the map view of the MCG or replay an audio file without performing a rating. This is the same as in the MCG without the collaborative filtering extension. The dialogue offers two possibilities for performing the rating, a variant with three "smilies" ("good/average/bad"), or selecting 1 to 5 stars with 5

*Figure 5. Rating an item*

stars representing the best value. Users can choose the rating method they like better. Internally, the 3-scale ratings are mapped to the 5-scale variants, and only a number of 1 to 5 is stored as a user's rating for an item.

Since one of the requirements was to make the user interaction as simple as possible (see "Requirements for Extending the Mobile City Guide with POI Recommendations"), the system intended not to rely on these explicit user ratings only. Therefore, the system includes a method to capture implicit user ratings. Analyzing the users' media usage does this. Since the MCG mostly contains audio files about POIs at this time, the system utilizes information about the playback duration. If a user listens to the audio file to the end, the system assumes that the user liked the items and assigns a rating value of 5, the best grade. The other values are determined accordingly; the second best grade of 4 is used when the playback duration is between 60% and 80% of total length. The exact mapping of playback dura-

tion to ratings can be easily configured in the application.

Thus user ratings for items required for the CF model with little user distraction can be acquired. The first approach was to use both explicit and implicit ratings to calculate item similarity and build the model (Figure 6, left). However, it is not possible to evaluate whether the implicit rating are as good as the explicit ones in this case. Therefore, explicit and implicit ratings are separated in two models of item similarity (Figure 6). The disadvantage of this approach is that less data is available for the models. However, the second field study collected enough meaningful data (see below) to compare the two methods with regard to recommendation quality.

## Storage Complexity of the Item-Item Matrix

Item-based collaborative filtering has a disadvantage with regard to storage complexity because the item-item matrix grows with the item set. Storage demand is an issue for mobile applications. Therefore, the storage model of PocketLens has been optimized (Woerndl, Muehe, & Prinz, 2009). The goal is to store the item-item matrix as compactly as possible. For that purpose, the properties of the algorithm were analyzed and many duplicate entries in the item-item matrix were found. That means the computed item similarity value was identical for many item pairs. The elements of the matrix in the PocketLens algorithm require at least 20 bytes of memory in a densely populated matrix. Replacing an element with a pointer (4 bytes) to a duplicate entry reduces the storage requirement for this matrix element by 16 bytes (Woerndl et. al., 2009).

The approach has been evaluated with the "MovieLens" data set (Herlocker et. al., 1999) containing 100000 ratings by 943 users for 1682 movies. Storage requirement was reduced from about 45 megabytes to 32 megabytes. Besides the reduction by using pointers, the algorithm additionally modified the definition of duplicate respective identical matrix elements. PocketLens uses float values for item similarity and vector size as matrix elements. If we assume that two elements are duplicates if they are very similar but not quite identical, we can increase the number of duplicates and thus further improve the storage efficiency. The resulting model quality suffers a little, but findings indicated that cutting the float value after some decimal points does indeed reduce the memory requirements while hurting recommendation quality only marginally (Woerndl, Muehe, & Prinz, 2009).

If the approach compares the field elements of the item-item matrix with an inaccuracy of just 0.05, the data model can be further reduced to 20 megabytes. The search for duplicates is done only when a memory threshold is exceeded. The insertion of a rating vector into the model took about 1.5 seconds on average on the test PDAs. The search for duplicates needs the main share of this time, which can be done in the background. The algorithm needs only a fraction of the 1.5 seconds for actually inserting a rating vector into the item-item matrix. This seems very reasonable for the test scenario (Woerndl, Muehe, & Prinz, 2009).

## Context-Aware Recommendation Methods

The discussion so far has been about how to obtain user ratings on the model device and build a storage-optimized model of item similarity. Now this section explains how to generate recommendations based on the model. The general idea of item-based CF is that items are recommended that are similar to items which the active user rated highly, according to the item similarity stored in the item-item matrix. Note that not only the items receiving the highest ratings are recommended in this approach. If users generally rate two items i and j with low grades, the item similarity between i and j will be high. If an active user then evaluates item i with a good rating, item j is a likely

candidate for recommendation. In this case, the predicted rating of j will be high. Predicted rating of an item is called as the (Collaborative Filtering) score. The system calculates the scores for all relevant items and selects the one with the highest score as a recommendation for the user.

Likewise, the algorithm computes a context score based on the current distance of the user to a POI. Selecting the item with the highest context score was the method used in the MCG without collaborative filtering. By applying item-based CF and separating explicit and implicit ratings, the system can now use three options for recommendation: context- or location-based recommendation, collaborative filtering based on explicit ratings and collaborative filtering based on implicit ratings.

Since it is not useful to recommend POIs at the other end of town, the collaborative filtering recommenders use a cut-off distance to the POI. This means, only items within a configurable threshold of distance – e.g. 800 meters – from the current position are considered as input for the collaborative filtering. In terms of Adomavicius and Tuzhilin (2010), the applied CF methods utilize contextual pre-filtering (see above).

In addition, the approach aims at combining the three methods in a hybrid recommender (Burke, 2007). First, explicit and implicit CF scores are combined by arithmetic mean. It is also possible to use other means or maintain a collaborative recommendation model in addition to the explicit and implicit ones. The score for an item is then calculated by a linear combination of the CF score with a weight of 40% and the context score with a weight of 60%. These weights produced reasonable results in preliminary tests. The score recommender implements contextual modeling as recommendation method (see Figure 4). The algorithm uses information about users (ratings), items (their location) and context (the current user position). This recommendation process to calculate a score for an item is summarized in Figure 7.

The MCG applies all four methods including score recommendation. From a user's perspective,

*Figure 6. Explicit and implicit recommendations models*

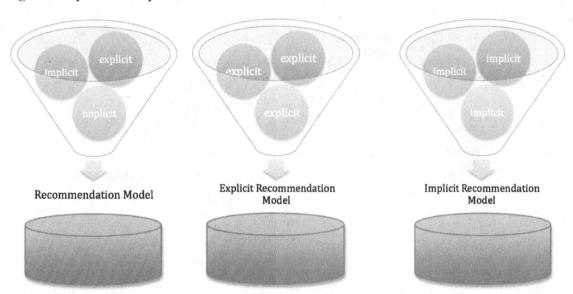

the process of selecting information about a POI is not evident, apart from the fact that not necessarily the nearest POI is selected as before, but the one with the highest score according to the chosen recommendation method. Authors compared the different methods in the second field study (see "Field Study 2: Evaluating Recommendation Quality"). But first, the next section reports on experiences from a user study to investigate the users' acceptance of the recommender system in the MCG in general.

## Technical Details about System Implementation

This chapter provides some details about the implementation of the MCG. As already explained, the approach does not assume a permanent network connection, so all the necessary data and the MCG program itself are stored on the Windows Mobile PDA. The MCG was originally developed using Microsoft Visual Studio 2005 Enterprise Edition and the Windows Mobile 5.0 Developer Resource Kit in C++. The MCG handles maps, POIs with the additional information such as audio files

and text in several languages. The system was extended with the following components respectively classes:

- **Tourist:** Manages the ratings of the active user
- **Recommender:** Implements the explained item-based collaborative filtering algorithm
- **Storage:** Stores the rating vectors of other users which is the groundwork for the model calculation

As far as data storage is concerned, the system uses XML files for the data and an additional Serializer helping class to make the data available for the modules of the system. The basic system loop can be summarized are follows. When the user starts the application, he/she can select a preferred language. The system then retrieves the current position using GPS, displays a maps and sends a request to the Recommender component for a (ranked) list of POIs in the vicinity that are then shown on the map. When a POI is displayed, the Tourist component as an exten-

*Figure 7. Score recommendation*

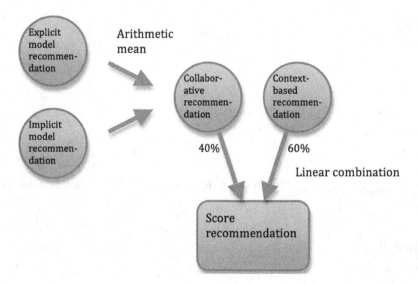

sion to the original MCG application handles the implicit and explicit rating process and also sends a rating to the Recommender module for an update of the model. There is also a Logger component to keep track of all the user actions to be able to later analyze the effectiveness of the system (see "Field Study 2" below). The goal of the design was to encapsulate the recommender functionality from the original system as much as possible. Therefore, only minimal changes in the existing components were necessary. The system design also allows for easy modification of the used recommender algorithms or changes in the user interface.

## FIELD STUDY 1: USABILITY AND USER ASSESSMENT

### Study Setup

The approach was evaluated in a first field study from the users' perspective. Five of the rental devices with the mobile city guide were equipped with the new recommendation function. The users were regular, international tourists in Prague,

Czech Republic. Some tourists were asked to test the new MCG. In return for filling out a questionnaire at the end of the test, the users did not have to pay the standard fee for using the mobile city guide.

### Participants

30 tourists participated in the study. 8 of them were German; the other 22 were from a variety of countries. Figure 8 shows the age distribution. 75% of the test users' age was between 21 and 50. Although the average age of users was probably lower than the average age of all Prague visitors, not only young people used the mobile city guide.

The participants were also asked for the reasons for visiting Prague; participants had the options of culture, relaxing, interest in the country and its people and other (Figure 9). Most of the users selected "culture" (35%).

### Questionnaire Results and Discussion

The next questions were concerned with the overall usability of the MCG. 76% of the participants had

*Figure 8. Age distribution*

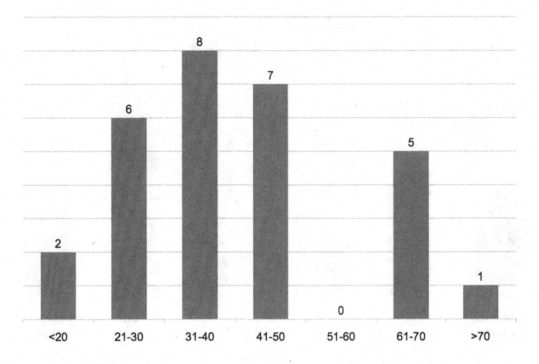

*Figure 9. Interest in Prague*

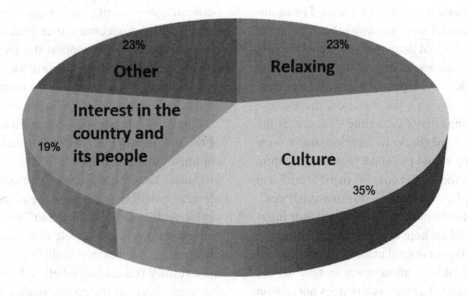

no problems using the guide, an additional 17% used the second best option on the 6-point Likert scale and only 7% expressed some problems. In other questions, 70% of test users were happy with the overall functionality of the guide and 93% found the MCG easy to use. To sum up, most of the users seemed to have operated the MCG without problems and liked the overall application.

Since acquiring ratings is one of the critical design issues in the collaborative recommender,

*Figure 10. Rating the items was annoying*

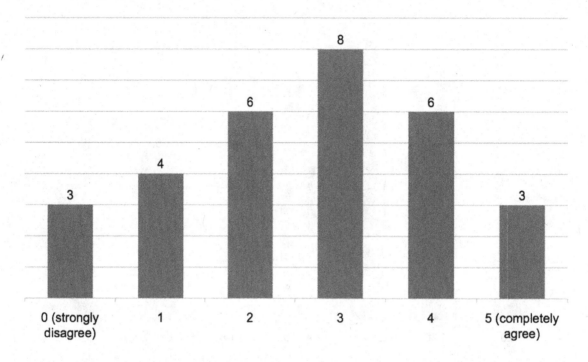

the evaluation contained questions on whether the explicit rating option (see Figure 5) was annoying to users. Figure 10 shows the distribution of answers. 10% of participants selected the best and worst grade each. Most of the answers were in the middle of the 6-point scale, with a tendency towards "annoying". So, the users did express mixed feelings about the rating dialogue at the end of audio playback. In addition, users were asked if they would prefer an "automatic" rating system over the current one shown in Figure 5 and a majority of users (almost 75%) answered "yes".

These results indicate that explicit user interaction should be kept at a minimum. However, giving a rating is optional in the MCG system and it is conceivable to allow users to turn off the rating dialogue. The framework does not rely on explicit rating. Instead, implicit ratings only could be used.

Furthermore, users were asked if they would provide additional information (e.g. more ratings or preferences) to get better recommendations. Figure 11 shows that most users would indeed

provide more information. Delicate trade-off between users wanting to get good recommendations and the necessary user interaction was noticed. These results support the idea to offer (collaborative) recommendations while keeping user interaction at a minimum by using implicit ratings based on media usage in the mobile guide.

Finally, the questionnaire asked if users were of the opinion that the MCG selected POIs according to their interests or preferences. Most of the participants choose the positive options (Figure 12). A similar question on the users' impression about the whole "guided tour" was included, with results very comparable to the ones shown in Figure 12. However, it is difficult to interpret these results. It is unclear whether the users liked the item selection, the corresponding audio file and other information, or the sight itself.

*Figure 11. " I would provide more information to get better recommendations"*

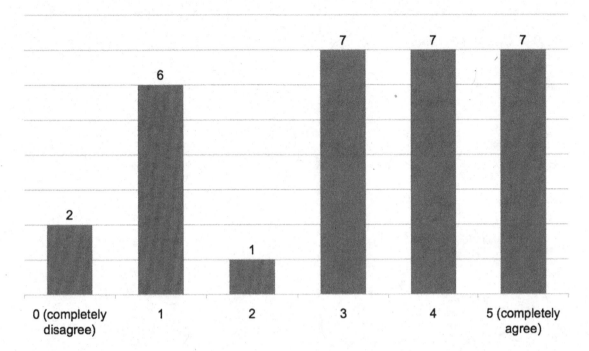

*Figure 12. "POIs were selected according to my interests"*

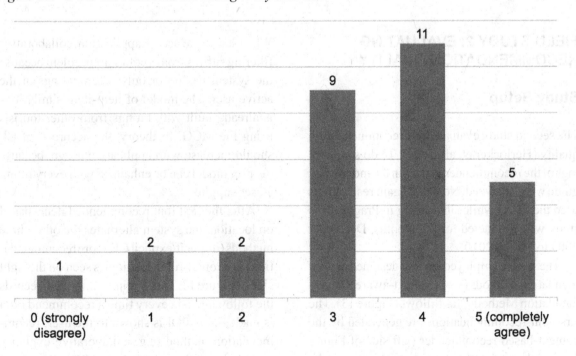

*Figure 13. Sequence of recommender methods*

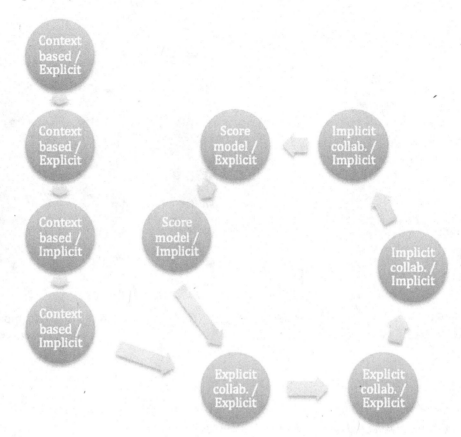

## FIELD STUDY 2: EVALUATING RECOMMENDATION QUALITY

### Study Setup

The second study evaluated the recommendation quality (Herlocker et. al., 2004). To do so, usage logs of the recommender system in the mobile city guide were analyzed. Note that again real tourists used the MCG while sightseeing in Prague. This study was conducted for 2.5 months, December 2009 to March 2010.

The system employed the four designed recommendation methods (see "Context-aware Recommendation Methods") as follows (Figure 13). The first four recommendations are generated by the context-based recommender (left side of Figure 13) and thus using the current user location only.

When starting to use the application, collaborative filtering suffers from a cold start problem because the system has no or only a few ratings of the active user. The model of item-item similarities is already built with ratings from other tourists using the MCG. In theory, the accuracy of CF should increase with application usage, because the user model can be enhanced with every rating a user supplies.

After the first four recommended items based on location, the system alternates the other three methods (implicit/explicit CF, score recommendation) in a round-robin fashion as seen on the right side of Figure 13. The logging component records the following data every time a recommendation is made, i.e. a POI is shown to the user: recommendation method (e.g. collaborative explicit), score value of the recommended POI, distance of

*Figure 14. Users per device*

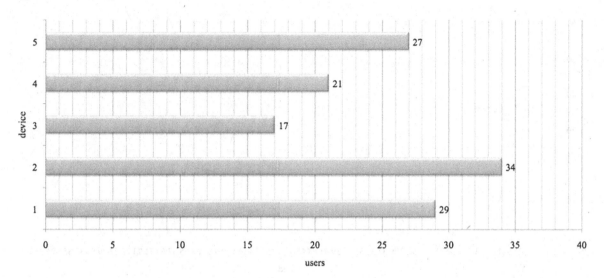

the recommended POI to the user, and (implicit or explicit) rating by the user for the item. This log allows us to compare the predicted item rating with the actual rating of the item by the user.

## Usage Data

During the field study, five devices with the extended recommendation system were distributed among the rental places in Prague. 128 tourists were using these devices. A detailed statistic of users per device is shown in Figure 14. The devices itself were identical, Figure 14 just provides some background information how many tourists used one device during the study period.

The 128 users generated a total of 1739 log entries about items. One log corresponds to one recommended item, i.e. one viewed POI. On average, 25.6 persons used one device generating 347.8 log entries. Figure 15 shows the count of views for each POI. The POIs are identified with a number from 1 to 93. Average view count is 18.63 per POI. The minimum value is 0, the maximum views for a POI is 50. So, not every POI was recommended in the study. The number of viewed POIs per user was 13.5 on average. The

maximum number is 34, therefore one tourist listened to 34 audio files during usage.

The MCG system also recorded the distance from the current user position to the POI location when a recommendation was made. Figure 16 summarizes the results. The "count" column shows the number of times a POI was selected by the four methods. Unsurprising, the items recommended by the context-based filter were nearest to the user since only location is used for selection. The distance value of these POIs was about 140 meters on average. The POIs recommended by the explicit and implicit collaborative filters were farther away, about 423 and 412 meters respectively. The distance value of the score recommender is between the context and CF scores, but nearer towards context. This result is in line with the design of the score recommendation method, because context is slightly favored over the CF score.

## Recommendation Quality and Discussion

This section discusses the results with regard to recommendation quality. The goal was to find out which recommendation methods lead to better

*Figure 15. Views per POI*

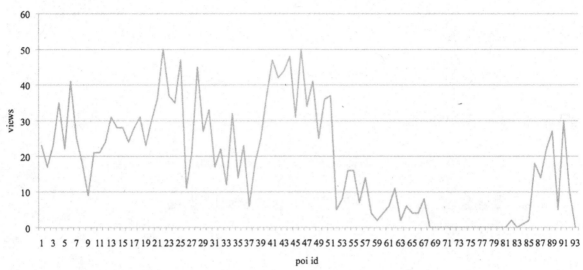

*Figure 16. Distance to POI*

|  | count | ∅ [m] | σ [m] |
|---|---|---|---|
| context based | 494 | 140,15 | 87,1588 |
| collaborative explicit | 412 | 432,74 | 215,8802 |
| collaborative implicit | 328 | 412,48 | 195,3476 |
| score recommendation | 497 | 201,14 | 136,6675 |

*Figure 17. Average user ratings*

|  | count | ∅ | var | σ |
|---|---|---|---|---|
| context based | 495 | 3,92 | 4,53 | 2,13 |
| collaborative | 741 | 4,02 | 4,55 | 2,13 |
| - explicit | 412 | 3,86 | 2,97 | 1,72 |
| - implicit | 329 | 4,22 | 3,05 | 1,75 |
| score rec. | 497 | 3,94 | 3,62 | 1,90 |

results. In other words, the ratings given by the user were compared to the POI recommended by the various methods. Figure 17 shows the outcome. The collaborative filter based on the model with implicit ratings performed best, with an average rating of 4.22. Items recommended by the collaborative filter based on the explicit model had an average rating of only 3.86. The context-based approach led to an average rating of 3.92, and the score recommendation slightly

higher (3.94). Overall, introducing collaborative filtering in the mobile city guide lead to better POI recommendations, especially with the implicit model, in comparison to the location-based method used before.

Figure 18 summarizes the result in a graphical way. Note that the evaluation analyzed the actual ratings of recommended items. In theory, an even better method is to directly compare the actual ratings with the predicted score values and com-

*Figure 18. Average user ratings*

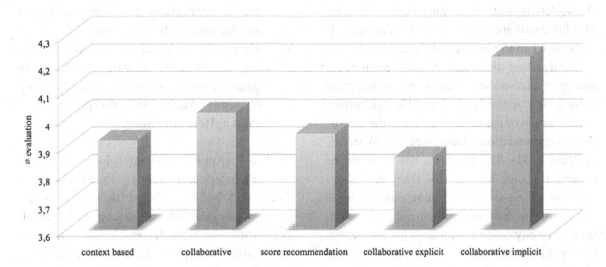

pute correlation between these numbers. In other words, to calculate the deviation of the predicted score with the actual one. However, in practice, the system recommends only the single best item at a time, and worse items are never shown to the user. Therefore, the predicted scores of recommended items are always very high, and the results of the correlation calculation were very similar to the analysis shown here.

## OTHER APPLICATION SCENARIOS FOR CONTEXT-AWARE RECOMMENDERS IN MOBILE SCENARIOS

The authors of this chapter have worked and are working on recommender systems in several other application domains related to the MCG presented (Woerndl, Brocco, & Eigner 2009). The overall goal is to identify and refine general guidelines for context-aware recommender systems and apply them in mobile scenarios. In addition to the following three subprojects, user interface issues for mobile recommenders and group recommendations are also investigated (Birnkammerer, Woerndl, & Groh, 2009). An application example

for the latter is a group of people intending to choose a movie in a cinema nearby.

## Proactive and Context-Aware Recommendations in an Automotive Environment

This subproject deals with a recommender system for in-car recommendation, e.g. suggesting a gas station based on the current fuel level and other context information that is available in a car. The question in this scenario is not only *which* item to recommend, but also *when*. This implies a proactive behavior of the recommender system (see "Other Characteristics of Mobile Recommender Systems").

One contribution is a new model for situation awareness tailored to proactive recommendations in automotive scenarios (Bader, Woerndl, & Prinz, 2010). Unlike other approaches, the model allows for incorporating past and future situations for assessing the relevance of a recommendation in the present. The model is based on fuzzy logic to cope with different sources of uncertainty. The implemented scenario is a proactive POI recommender for gas stations using situation descriptions for fuel level states and number of reachable gas stations. The evaluation shows that the model allows determining significant situation changes. This information can then be used to derive decisions on when to recommend a gas station.

For deciding which gas station to recommend, the approach consider item attributes such as price or location and also context data like the current time, location or gas level of the car at the time of the recommendation (Bader, Neufeld, Woerndl, & Prinz, 2011). The system is based on Multi-Criteria Decision Making (MCDM) methods to calculate scores on several dimensions. This is very similar to the score recommender applied in the mobile city guide as explained in this chapter. Utility functions modeling the importance of different route context elements were used. The functions were derived from a preliminary user study, among other information. In addition, the system performs contextual pre- and post-filtering to reduce the number of considered items. The evaluation showed that the approach produced reasonably good results in comparison with the assessment of users in a user study (Bader et. at., 2011).

## Recommending Mobile Applications

This recommender system is integrated in a framework supporting the development of mobile applications (Woerndl, Schueller, & Wojtech, 2007). Part of the framework is a deployment server where developers of mobile applications can register their services and end users can browse and search for relevant and interesting gadgets. One problem for users is to find interesting and – with regard to their current context – relevant applications on their mobile device. The (hybrid) recommender system developed in this scenario recommends mobile applications to users derived from what other users have installed and rated positively in a similar context (location, currently used type of device, etc.). Users can choose between several content-based and collaborative filtering components:

- **LocationAppRecommender**: recommends applications that were used in a similar location by other users
- **CFAppRecommender**: applies existing collaborative filtering algorithms to generate results using the "Taste" library (Lemire & Maclachlan, 2005)
- **PoiAppRecommender.**

The PoiAppRecommender does not recommend POIs but recommends mobile applications based on POIs in the vicinity of the user using triggers. An administrator can select among types of points-of-interests (such as restaurant, museum or train station) and specify the distance to an actual POI within which an application is recommended. When making a recommendation, the system retrieves the current user position, determines POIs in the vicinity and generates a recommendation based on this context information. For example, an administrator can specify that his/her mobile train table application shall be recommended when the user is near a train station. After applying the trigger rules, the approach uses collaborative filtering to rank found items according to user ratings of applications in a second step. User ratings are collected implicitly by automatically recording when a user installs an application within the framework. It is also optionally possible for users to explicitly rate applications after usage, similarly to rating items in the mobile city guide as explained above. The ratings are stored together with context information (time, location, used device etc.) to capture the situation when a rating was made. These context-based ratings can then be utilized by the CFAppRecommender to recommend applications.

## Mobile Semantic Personal Information Management

Personal Information Management (PIM) is intended to support the activities people perform to organize their daily lives through the acquisition,

*Figure 19. Finding current resources (left), displaying the relation path (middle) and displaying resources on a map (right)*

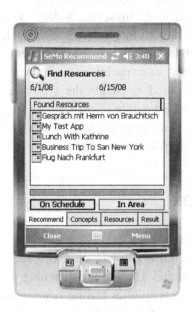  

maintenance, retrieval, and sharing of information (Teevan, Jones, & Bederson, 2006). One approach to deal with PIM is the "Semantic Desktop", aiming to integrate desktop applications and the data managed on desktop computers using semantic web technologies (Sauermann, Bernardi, & Dengel, 2005). While Semantic Desktop ideas to support PIM seem very useful in a mobile setting, most existing approaches are geared towards desktop computers.

The "SeMoDesk" system is an implementation of the Semantic Desktop idea for PDAs (Woerndl & Woehrl, 2008). Users can define and manage their personal ontology, and assign relations between resources. To do so, SeMoDesk assists the user as much as possible, for example calls or short messages on the mobile device are integrated automatically. It is possible to structure and browse the personal information space across different applications. In addition, SeMoDesk offers a context-aware recommendation function to improve information access and find relevant resources (Woerndl, Groh, & Hristov, 2009).

The recommender consists of the following two steps. In step one, users have the option to select resources either manually or by using a retrieval function based on date, time and location. As a result, the system displays a list of resources which are of current interest to the user (Figure 19, left). These elements mainly serve as possible starting nodes for the more advanced search in step two. The system traverses the graph of concepts and resources and analyzes every node according to an evaluation function using configurable heuristics. For example, if the recommender finds a path between the starting node and a particular project resource via "related" edges, this project is considered relevant in the current context. As a result, the system displays several resources and also information on why these results are suggested, i.e. the path from the starting node to the result entry (Figure 19, middle) (Woerndl, Groh, & Hristov, 2009).

To further improve location-awareness, the approach includes a location and sensor ontology as an extension to the personal ontology (Woerndl,

Schulze, & Yordanova, 2010). For indoor positioning, the application integrates an RFID infrastructure. For outdoor position, the system utilizes GPS capabilities of the mobile device. SeMoDesk allows for assigning resources to locations and displaying resources in the current vicinity on a map (Figure 19, right). To semi-automatically learn relevant user locations, a solution based on a time-based clustering algorithm has been designed and implemented. This method has been evaluated with good results in the considered scenario (Woerndl, Schulze, & Yordanova, 2010).

## FUTURE RESEARCH DIRECTIONS

Future work includes the integration of other media types. So far, the information about POI is mostly an image with optional text and a corresponding audio file. It is planned to integrate more information about the sights, e.g. videos, and also other POI types such as restaurants. The assumption is that with a more diverse item set, the improvement of the collaborative filter with regard to recommendation quality could turn out more significant. An additional goal is to refine the user interface of the mobile guide and improve the overall user experience when utilizing the system. With more POI types and POIs, users need better means to identify recommended items and select among them. Therefore it is important to investigate how to communicate to the user that some POI are more recommended than users without given up a simple map-based interface. Related to this question are explanations of recommendations (Tintarev & Masthoff, 2010).

Finally, an area of research is critique-based or conversational recommenders (McGinty & Smyth, 2006; McGinty & Reilly, 2010). In these systems, a user can express his/her preferences by criticizing items that the system recommended. The user is involved in a dialogue where the system's item suggestions interleave with the user's feedback to the recommended items (Ricci, 2010). Examples

in the considered scenario include the following options given to users: "Recommend a cheaper restaurant!", "No more museums today!" etc. One question is how to integrate user feedback on suggested POIs into the process. In general, the critique-based user preference elicitation procedure is assumed to enable the system to build a better understanding of the user's needs and preferences, and may ultimately improve the recommendations (Chen & Pu, 2009).

## CONCLUSION

This chapter summarizes experiences from integrating collaborative filtering in a real world mobile tourist guide. The guide is running on PDAs with GPS capabilities but no permanent Internet connection. This mobile city guide has been extended with options to rate items and get recommendations. The solution is based on item-based collaborative filtering. The system builds a model of item similarities based on explicit and implicit ratings and utilizes the model to generate recommendations. Predictions of CF ratings can be combined with a context score to take multiple dimensions into account. This idea has also been applied in other mobile recommendation scenarios, for example a context-aware in-car recommender for gas stations.

One of the most important results of the presented work is that the explained approach was able to introduce collaborative filtering within the requirements and restrictions of a real world mobile application. The approach did not obstruct with the overall user experience but enhanced the system with better recommendations for points-of-interests.

The approach has been evaluated in two separate field studies. In the first one, 30 international participants – real tourists visiting Prague – used the recommender function and were asked to fill out a questionnaire later. The results showed that the users were in general willing to provide information to get recommendations and felt that the proposed POIs fit their interests. The explicit rating process received mixed results. However, the system works with only implicit ratings as well. The implicit ratings were derived from audio playback duration. So one of the overall findings is that it may be preferable to skip explicit ratings and just rely on implicit ratings. A second field study evaluated the recommendation quality in more detail. Log data of 128 users and 1739 ratings of POIs were acquired. The results of the examination show an improvement of recommendations based on the collaborative filter in comparison to the location-based filter used before. The study also allowed for comparing recommendations based on explicit versus implicit user ratings. Recommendation based on implicit ratings outperformed the model based on explicit ratings.

## REFERENCES

Abowd, A., Atkeson, C., Hong, J., Long, S., & Pinkerton, M. (1997). Cyberguide: A mobile context-aware tour guide. *Wireless Networks*, *3*(5), 421–433. doi:10.1023/A:1019194325861

Adomavicius, G., Sankaranarayanan, R., Sen, S., & Tuzhilin, A. (2005). Incorporating contextual information in recommender systems using a multidimensional approach. *ACM Transactions on Information Systems*, *23*(1), 103–145. doi:10.1145/1055709.1055714

Adomavicius, G., & Tuzhilin, A. (2005). Towards the next generation of recommender systems: A survey of the state-of-the-art and possible extensions. *IEEE Transactions on Knowledge and Data Engineering*, *17*(6). doi:10.1109/TKDE.2005.99

Adomavicius, G., & Tuzhilin, A. (2010). Context-aware recommender systems. In Kantor, P., Ricci, F., Rokach, L., & Shapira, B. (Eds.), *Recommender systems handbook: A complete guide for research scientists and practitioners*. Berlin, Germany: Springer.

Anand, S. S., & Mobasher, B. (2003). *Intelligent techniques for Web personalization*. Revised Selected Papers of the 2nd Workshop on Intelligent Techniques in Web Personalization (ITWP 2003), Heidelberg, Germany: Springer.

Anegg, H., Kunczier, H., Michlmayr, E., Pospischil, G., & Umlauft, M. (2002). LoL@: Designing a location based UMTS application. *Elektrotechnik und Informationstechnik*, *119*(2), 48–51.

Bader, R., Neufeld, E., Woerndl, W., & Prinz, V. (2011). Context-aware POI recommendations in an automotive scenario using multi-criteria decision making methods. *Workshop on Context-aware Retrieval and Recommendation (CaRR 2011), Conference on Intelligent User Interfaces (IUI 2011)*. Palo Alto, CA, USA.

Bader, R., Woerndl, W., & Prinz, V. (2010). *Situation awareness for proactive in-car recommendations of points-of-interest (POI)*. Workshop Context Aware Intelligent Assistance (CAIA 2010), 33rd Annual German Conference on Artificial Intelligence (KI 2010). Karlsruhe, Germany.

Baldauf, M., Dustdar, S., & Rosenberg, F. (2007). A survey on context-aware systems. *International Journal of Ad Hoc and Ubiquitous Computing*, *2*(4), 263–277. doi:10.1504/IJAHUC.2007.014070

Berkovsky, S., Kuflik, T., & Ricci, F. (2007). *Distributed collaborative filtering with domain specialization*. ACM Conference on Recommender Systems. Minneapolis, MN, USA.

Birnkammerer, S., Woerndl, W., & Groh, G. (2009). *Recommending for groups in decentralized collaborative filtering*. Tech. Report TUM-I0927, Institut fuer Informatik, Technische Universitaet Muenchen, Germany.

Borchers, A., Herlocker, J., Konstan, J., & Riedl, J. (1998). Ganging up on information overload. *Computer, 31*(4), 106–108. doi:10.1109/2.666847

Brusilovsky, P., Kobsa, A., & Nejdl, W. (2007). *The adaptive Web*. Berlin, Germany: Springer. doi:10.1007/978-3-540-72079-9

Burke, R. (2007). Hybrid Web recommender systems. In Brusilovsky, P., Kobsa, A., & Nejdl, W. (Eds.), *The adaptive Web* (pp. 377–408). Berlin, Germany: Springer. doi:10.1007/978-3-540-72079-9_12

Chen, L., & Pu, P. (2009). Interaction design guidelines on critiquing-based recommender systems. *User Modeling and User-Adapted Interaction, 19*(3), 167–206. doi:10.1007/s11257-008-9057-x

Costa, R., Correia, A., & Moital, M. (2008). *The determinants of intention to adopt mobile electronic tourist guides*. First International Meeting for Tourism Management: The Public and Private Sector (EIGTUR 2008). Minas Gerais, Ouro Preto, Brazil.

Deshpande, M., & Karypis, G. (2004). Item-based top-n recommendation algorithms. *ACM Transactions on Information Systems, 22*(1), 143–177. doi:10.1145/963770.963776

Dey, A. K., Abowd, G. D., & Salber, D. (2001). A conceptual framework and a toolkit for supporting the rapid prototyping of context-aware-applications. *Human-Computer Interaction, 16*(2), 97–166. doi:10.1207/S15327051HCI16234_02

Ducheneaut, N., Partridge, K., Huang, Q., Price, R., Roberts, M., Chi, E. H., et al. (2009). *Collaborative filtering is not enough? Experiments with a mixed-model recommender for leisure activities*. 17th International Conference on User Modeling, Adaptation, and Personalization (UMAP). Trento, Italy.

Fuchs, M., Rasinger, J., & Höpken, W. (2007). Exploring information services for mobile tourist guides – Results from an expert survey. In Dimanche, F. (Ed.), *Tourism mobility & technology* (pp. 4-14). Travel and Tourism Research Association (TTRA), Nice, France.

Grun, C., Werthner, H., Proll, B., Retschitzegger, W., & Schwinger, W. (2008). *Assisting tourists on the move – An evaluation of mobile tourist guides*. In 7th International Conference on Mobile Business (ICMB '08). Barcelona, Spain.

Herlocker, J., Konstan, J. A., Borchers, A., & Riedl, J. (1999). *An algorithmic framework for performing collaborative filtering*. In 22nd Annual International ACM SIGIR Conference on Research and Development in Information Retrieval. Berkeley, CA.

Herlocker, J., Konstan, J. A., Terveen, L. G., & Riedl, J. T. (2004). Evaluating collaborative filtering recommender systems. *ACM Transactions on Information Systems, 22*, 5–53. doi:10.1145/963770.963772

Hinze, A., & Buchanan, G. (2006). The challenge of creating cooperating mobile services: Experiences and lessons learned. In *ACSC '06: Proceedings of the 29th Australasian Computer Science Conference* (pp. 207-215). Darlinghurst, Australia.

Jacobsson, M., Rost, M., & Holmquist, J. E. (2006). *When media gets wise: Collaborative filtering with mobile media agents*. In 11th International Conference on Intelligent User Interfaces (IUI '06). Sydney, Australia.

Jannach, D., Zanker, M., Felfernig, A., & Friedrich, G. (2010). *Recommender systems: An introduction*. Cambridge University Press.

Jones, M., & Marsden, G. (2005). *Mobile interaction design*. John Wiley & Sons.

Kantor, P., Ricci, F., Rokach, L., & Shapira, B. (2010). *Recommender systems handbook: A complete guide for research scientists and practitioners*. Berlin, Germany: Springer.

Kenteris, M., Gavalas, D., & Economou, D. (2008). Evaluation of mobile tourist guides. In *First World Summit on the Knowledge Society (WSKS 2008)* (pp. 603-610). Athens, Greece.

Kenteris, M., Gavalas, D., & Economou, D. (2010). Mytilene e-guide: A multiplatform mobile application tourist guide exemplar. *Multimedia tools and applications*. ISSN 1380-7501 (in press).

Kenteris, M., Gavalas, D., & Economou, D. (2011). Electronic mobile guides: A survey. *Personal and Ubiquitous Computing, 15*(1), 97–111. doi:10.1007/s00779-010-0295-7

Kobsa, A. (2007). Privacy-enhanced personalization. *Communications of the ACM, 50*(8), 24–33. doi:10.1145/1278201.1278202

Kray, C., & Baus, J. (2003). A survey of mobile guides. In *Proceedings of HCI in Mobile Guides, Fifth International Symposium on Human Computer Interaction with Mobile Devices and Services*, Udine, Italy.

Kray, C., Laakso, K., Elting, C., & Coors, V. (2003). Presenting route instructions on mobile devices. In *Proceedings of the 2003 International Conference on Intelligent User Interfaces (IUI'03)*. Miami, FL, USA.

Krüger, A., Baus, J., Heckmann, D., Kruppa, M., & Wasinger, R. (2007). Adaptive mobile guides. In Brusilovsky, P., Kobsa, A., & Nejdl, W. (Eds.), *The adaptive Web*. Berlin, Germany: Springer. doi:10.1007/978-3-540-72079-9_17

Lemire, D., & Maclachlan, A. (2005). *Slope one predictors for online rating-based collaborative filtering*. In SIAM Conference on Data Mining (SDM 2005). Newport Beach, USA.

Linden, G., Smith, B., & York, J. (2003). Amazon.com recommendations: Item-to-item collaborative filtering. *IEEE Internet Computing, 7*(1), 76–80. doi:10.1109/MIC.2003.1167344

McGinty, L., & Reilly, J. (2010). On the evolution of critiquing recommenders. In Kantor, P., Ricci, F., Rokach, L., & Shapira, B. (Eds.), *Recommender systems handbook: A complete guide for research scientists and practitioners*. Berlin, Germany: Springer.

McGinty, L., & Smyth, B. (2006). Adaptive selection: An analysis of critiquing and preference-based feedback in conversational recommender systems. *International Journal of Electronic Commerce, 11*(2), 35–57. doi:10.2753/JEC1086-4415110202

Miller, B., Konstan, J., & Riedl, J. (2004). PocketLens: Toward a personal recommender system. *ACM Transactions on Information Systems, 22*(3), 437–476. doi:10.1145/1010614.1010618

Poslad, S., Laamanen, H., Malaka, R., Nick, A., Buckle, P., & Zipl, A. (2001). *CRUMPET: Creation of user-friendly mobile services personalised for tourism*. In Second International Conference on 3G Mobile Communication Technologies. London, UK.

Pospischil, G., Umlauft, M., & Michlmayr, E. (2002). Designing LoL@, a mobile tourist guide for UMTS. In Paterno, F. (Ed.), *Mobile Human-Computer Interaction* (pp. 140–154). Berlin, Germany: Springer.

Rasinger, J., Fuchs, M., & Höpken, W. (2007). Information search with mobile tourist guides: A survey of usage intention. *Journal of Information Technology & Tourism, 9*(34).

Raubal, M., & Panov, I. (2009). A formal model for mobile map adaptation. *Location Based Services and Tele Cartography II - From Sensor Fusion to Context Models. Selected Papers from the 5th International Symposium on LBS & TeleCartography*. Salzburg, Austria.

Ricci, F. (2010). (in press). Mobile recommender systems. *International Journal of Information Technology and Tourism*.

Ricci, F., & Nguyen, Q. N. (2006). MobyRek: A conversational recommender system for on-the-move travelers. In Fesenmaier, D. R., Werthner, H., & Wober, K. W. (Eds.), *Destination recommendation systems: Behavioural foundations and applications* (pp. 281–294). CABI Publishing. doi:10.1079/9780851990231.0281

Ricci, F., & Nguyen, Q. N. (2007). Acquiring and revising preferences in a critique-based mobile recommender system. *IEEE Intelligent Systems, 22*(3), 22–29. doi:10.1109/MIS.2007.43

Sarwar, B., Karypis, G., Konstan, J., & Riedl, J. (2001). *Item-based collaborative filtering recommendation algorithms*. In 10th International Conference on World Wide Web (WWW10), Hong Kong, China.

Sauermann, L., Bernardi, A., & Dengel, A. (2005). *Overview and outlook on the semantic desktop*. 1st Workshop on the Semantic Desktop, ISWC 2005 Conference. Galway, Ireland.

Schafer, J., Frankowski, D., Herlocker, J., & Sen, S. (2007). Collaborative filtering recommender systems. In Brusilovsky, P., Kobsa, A., & Nejdl, W. (Eds.), *The adaptive Web* (pp. 377–408). Berlin, Germany: Springer. doi:10.1007/978-3-540-72079-9_9

Schneiderman, B. (1994). Dynamic queries for visual information seeking. *IEEE Software, 11*(6), 70–77. doi:10.1109/52.329404

Strang, T., & Linnhoff-Popien, C. (2004). A context modeling survey. *Workshop on Advanced Context Modelling, Reasoning and Management*. UbiComp Conference, Nottingham/England.

Stroobants, R. (2006). *Mobile tourist guides. Technical Report*. Belgium: Katholieke Universiteit Leuven.

Subramanya, S. R., & Yi, B. K. (2007). Enhancing the user experience in mobile phones. *Computer, 40*(12), 114–117. doi:10.1109/MC.2007.420

Teevan, J., Jones, W., & Bederson, B. B. (2006). Personal information management. *Communications of the ACM, 49*(1), 40–43. doi:10.1145/1107458.1107488

Tintarev, N., & Masthoff, J. (2010). Designing and evaluating explanations for recommender systems. In Kantor, P., Ricci, F., Rokach, L., & Shapira, B. (Eds.), *Recommender systems handbook: A complete guide for research scientists and practitioners.* Berlin, Germany: Springer. doi:10.1007/978-0-387-85820-3_15

Woerndl, W., Brocco, M., & Eigner, R. (2009). Context-aware recommender systems in mobile scenarios. *International Journal of Information Technology and Web Engineering, 4*(1), 67–86. doi:10.4018/jitwe.2009010105

Woerndl, W., Groh, G., & Hristov, A. (2009). Individual and social recommendations for mobile semantic personal information management. *International Journal on Advances in Internet Technology, 2*(2&3).

Woerndl, W., Muehe, H., & Prinz, V. (2009). *Decentral item-based collaborative filtering for recommending images on mobile devices*. Workshop on Mobile Media Retrieval (MMR'09), MDM 2009 Conference. Taipeh, Taiwan.

Woerndl, W., Muehe, H., Rothlehner, S., & Moegele, K. (2009). *Context-aware recommendations in decentralized, item-based collaborative filtering on mobile devices*. Workshop on Innovative Mobile User Interactivity, MobiCASE Conference. San Diego, USA.

Woerndl, W., & Schlichter, J. (2008). Contextualized recommender systems: Data model and recommendation process. In Pazos-Arias, J., Delgado Kloos, C., & Lopez Nores, M. (Eds.), *Personalization of interactive multimedia services: A research and development perspective*. Hauppauge, NY: Nova Publishers.

Woerndl, W., Schueller, C., & Wojtech, R. (2007). *A hybrid recommender system for context-aware recommendations of mobile applications*. In IEEE 3rd International Workshop on Web Personalisation, Recommender Systems and Intelligent User Interfaces (WPRSIUI'07). Istanbul, Turkey.

Woerndl, W., Schulze, F., & Yordanova, V. (2010). Modeling and learning relevant locations for a mobile semantic desktop application. *Journal of Multimedia Processing and Technologies, 1*(1).

Woerndl, W., & Woehrl, M. (2008). *SeMoDesk: Towards a mobile semantic desktop*. Personal Information Management (PIM) Workshop, CHI 2008 Conference, Florence, Italy.

## ADDITIONAL READING

Amatriain, X., Torrens, M., Resnick, P., & Zanker, M. (2010). In *Proceedings of the 2010 ACM Conference on Recommender Systems (RecSys 2010)*. Barcelona, Spain.

Anderson, M., Ball, M., Boley, H., Greene, S., Howse, N., Lemire, D., & McGrath, S. (2003). RACOFI: Rule-Applying Collaborative Filtering Systems. *IEEE/WIC COLA'03*. Halifax, Canada.

Balabanovic, M., & Shoham, Y. (1997). Content-Based, Collaborative Recommendation. *Communications of the ACM, 40*(3), 66–72. doi:10.1145/245108.245124

Bergman, L., Tuzhilin, A., Burke, R., Felfernig, A., & Schmidt-Thieme, L. (2009). In *Proceedings of the 2009 ACM Conference on Recommender Systems (RecSys 2009)*. New York, NY, USA.

Bridge, D., Göker, M., McGinty, L., & Smyth, B. (2006). Case-Based Recommender Systems. *The Knowledge Engineering Review, 20*(3), 315–320. doi:10.1017/S0269888906000567

Burke, R. (2002). Hybrid Recommender Systems: Survey and Experiments. *User Modeling and User-Adapted Interaction*, 1–29.

Carenini, G., Smith, J., & Poole, D. (2003). Towards More Conversational and Collaborative Recommender Systems. In *Proceedings of the 2003 International Conference on Intelligent User Interfaces*. Miami, FL, USA

Chen, A. (2005). Context-Aware Collaborative Filtering System: Predicting the User's Preferences in Ubiquitous Computing. In *Proceedings of the 1st International Workshop Location- and Context-Awareness (LoCA 2005)*. Oberpfaffenhofen, Germany

Church, K., Smyth, B., Cotter, P., & Bradley, K. (2007). Mobile Information Access: A Study of Emerging Search Behavior on the Mobile Internet. *ACM Trans. Web, 4 (1)*.

Goldberg, D., Nichols, D., Oki, B. M., & Terry, D. (1992). Using Collaborative Filtering to Weave an Information Tapestry. *Communications of the ACM, 35*(12), 61–70. doi:10.1145/138859.138867

Hong, J., Suh, E.-H., Kim, J., & Kim, S. (2009). Context-Aware System for Proactive Personalized Service Based on Context History. *Expert Systems with Applications, 36*, 7448–7457. doi:10.1016/j.eswa.2008.09.002

Horozov, T., Narasimhan, N., & Vasudevan, V. (2006). Using Location for Personalized POI Recommendations in Mobile Environments. In *Proceedings of the International Symposium on Applications on Internet (SAINT '06)*. Phoenix, Arizona, USA.

Horvitz, E., Koch, P., & Subramani, M. (2007). Mobile Opportunistic Planning: Methods and Models. In *Proceedings of the 11th international conference on User Modeling (UM '07)* (pp.228-237). Berlin/Heidelberg: Springer.

Jannach, D., Zanker, M., Felfernig, A., & Friedich, G. (2010). *Recommender Systems: An Introduction*. Cambridge University Press.

Kang, E.Y., Kim, H., & Cho, J. (2006). Personalization Method for Tourist Point of Interest (POI), Recommendation. *Knowledge-Based Intelligent Information and Engineering Systems*, 392–400.

Konstan, J., Riedl, J., & Smyth, B. (2007). *Proceedings of the 2007 ACM Conference on Recommender Systems (RecSys 2007)*. Minneapolis, MN, USA.

Lee, H., & Park, S. J. (2007). Moners: A News Recommender for the Mobile Web. *Expert Systems with Applications, 32*(1), 143–150. doi:10.1016/j.eswa.2005.11.010

Maimon, O., & Rokach, L. (2010). *Data Mining and Knowledge Discovery Handbook*. Berlin, Heidelberg: Springer. doi:10.1007/978-0-387-09823-4

Manouselis, N., & Costopoulou, C. (2007). Analysis and Classification of Multi-Criteria Recommender Systems. *World Wide Web (Bussum), 10*(4), 415–441. doi:10.1007/s11280-007-0019-8

Nakashima, H., Aghajan, H., & Augusto, J. C. (2009). *Handbook of Ambient Intelligence and Smart Environments*. Berlin, Heidelberg: Springer.

Opperman, R., & Specht, M. (2000). A Context-Sensitive Nomadic Exhibition Guide. In *Proceedings of the 2nd Symposium on Handheld and Ubiquitous Computing*. Brisol, UK.

Pazzani, M. J. (1999). A Framework for Collaborative, Content-Based and Demographic Filtering. *Artificial Intelligence Review, 13*, 393–408. doi:10.1023/A:1006544522159

Pu, P., Bridge, D., Mobasher, B., & Ricci, F. (2008). In *Proceedings of the 2008 ACM Conference on Recommender Systems (RecSys 2008)*. Lausanne, Switzerland.

Resnick, P., & Varian, H. R. (1997). Recommender Systems. *Communications of the ACM, 40*(3), 56–58. doi:10.1145/245108.245121

Ricci, F. (2002). Travel Recommender Systems. *IEEE Intelligent Systems, 17*(6), 55–57.

Ricci, F., Rokach, L., & Shapira, B. (2009). Introduction to Recommender Systems. Handbook. In Kantor, P., Ricci, F., Rokach, L., & Shapira, B. (Eds.), *Recommender Systems Handbook: A Complete Guide for Research Scientists and Practitioners*. Berlin, Heidelberg: Springer.

Shani, G., & Gunawardana, A. (2010). Evaluating Recommendation Systems. In Kantor, P., Ricci, F., Rokach, L., & Shapira, B. (Eds.), *Recommender Systems Handbook: A Complete Guide for Research Scientists and Practitioners*. Berlin, Heidelberg: Springer.

Sharda, N. (2009). *Tourism Informatics: Visual Travel Recommender Systems, Social Communities, and User Interface Design*. Idea Group Reference. doi:10.4018/978-1-60566-818-5

Su, X., & Khoshgoftaar, T. (2009). *A Survey of Collaborative Filtering Techniques*. Journal Advances in Artificial Intelligence.

Uchyigit, G., & Ma, M. Y. (2008). *Personalization Techniques And Recommender Systems*. World Scientific Publishing Company.

Wang, J., Pouwelse, J., Lagendijk, R. L., & Reinders, M. J. (2006). Distributed Collaborative Filtering for Peer-to-Peer File Sharing Systems. In *Pro,eedings of the 2006 ACM Symposium on Applied Computing (SAC 06)*. Dijon, France.

Woerndl, W., Manhardt, A., & Prinz, V. (2010). A Framework for Mobile User Activity Logging. In *Proceedings of Mining Ubiquitous and Social Environments (MUSE 2010) Workshop, 21st European Conf. on Machine Learning / 14th European Conf. on Principles and Practice of Knowledge Discovery in Databases (ECML/PKDD 2010)*. Barcelona, Spain.

Yuan, S., & Tsao, Y. (2003). A Recommendation Mechanism for Contextualized Mobile Advertising. *Expert Systems with Applications*, 24(4), 399–414. doi:10.1016/S0957-4174(02)00189-6

## KEY TERMS AND DEFINITIONS

**Collaborative Filtering:** Recommender systems that takes other users' opinion into account, e.g. utilizing other users' ratings for items.

**Context:** Any information that can be used to characterize the situation of entities (i.e. whether a person, place or subject) that are considered relevant to the interaction between a user and an application, including the user and the application themselves.

**Item:** Entity that may be recommended by a recommender system.

**Item-Item Matrix:** Model of an item-based collaborative filtering approach, contains item similarities.

**Mobile City Guide:** Application running on a mobile devices such as a smartphone and supporting visitors of a city.

**Recommender System:** System that recommends products such as books, restaurants and other items for an active user by computing the chance that a user likes an item.

**User Profile:** Contains information such as preferences, interests or knowledge, e.g. the user's favourite type of restaurant or a rating.

**User-Item Matrix:** Input data for collaborative filtering recommender system, contains ratings of users for items.

# Chapter 6
# Design and Implementation of Mobile–Based Technology in Strengthening Health Information System:
## Aligning mHealth Solutions to Infrastructures

**Saptarshi Purkayastha**
*Norwegian University of Science & Technology, Norway*

**ABSTRACT**

*In the context of developing countries, there is a mounting interest in the field of mHealth. This surge in interest can be traced to the evolution of several interrelated trends (VW Consulting, 2009). However, with numerous attempts to create mobile-based technology for health, too many experiments and projects have not been able to scale or sustain. How is it possible to design and implement scalable and sustainable mHealth applications in low resource settings and emerging markets?. This chapter provides lessons from case studies of two successful and large scale implementations of mHealth solutions and the choices that were made in the design and implementation of those solutions. The chapter uses Information Infrastructure Theory as a theoretical lens to discuss reasons why these projects have been able to successfully scale.*

DOI: 10.4018/978-1-4666-0080-5.ch006

## INTRODUCTION

India is the fastest growing mobile market in the world (ITU, 2010). Mobile phones are accessible in remote geographies and have become an integral part of the fabric of society. India has more number of mobile phones than landline phones. Thus, mobile phones are the most common medium of communication for long distances in India (TRAI, 2010). Such deep penetration in social structure and technological capabilities, make mobile phones relevant Information & Communication Technology for Development (ICT4D). Mobile technology has been identified as an important tool to strengthening of health information systems (Ganapathy & Ravindra, 2008). Provisioning of health services through the use of mobile technology is called mHealth. mHealth applications range from data collection for health services using mobile devices, delivery of health related information to medical practitioners or researchers or patients, monitoring of patient vital signs through mobile sensors or mobile networks and even direct interventions (telemedicine) through the use of mobile technology. Using Sweden as an example of developed economy, it can seen that mobile phone penetration in developing countries had reached the same penetration levels as that of Sweden within just 10 years (ITU, 2009); while for infant mortality, the rate in developing countries in 2007 was at the level where Sweden was 72 years earlier. This shows the irony between the progress made in mobile phone acceptance and health indicators.

The excitement around mHealth can be seen through the increased interest in mHealth applications as summarized by VW Consulting (2009). Recent studies (Pyramid Research, 2010) show that mHealth applications will "increase three-fold in the next two years by 2012". Thus, a whole network of actors which includes mobile operators, handset manufacturers, application developers, health providers, patients, researchers have large stake in the field of mHealth. With this increased interest in mHealth, we have also seen numerous attempts that have not been able to scale or meet the needs of the health sector. An analysis by Anderson and Perin (2009) of the VW Consulting (2009) report shows that only 7 out of 51 projects have been able to scale, while 36 out of 51 have been stuck in proposal or just small pilots that have not continued. Some of these pilots have stopped because the funding agencies stopped the project and were not taken up by the community or the government. Few others projects have been surpassed by better technology availability and highlights the fact that infrastructure in the field of mHealth changes quickly. The argument holds true for mHealth that any new field of research meets with initial failures. Experiments should be considered a commonplace in development of new science. But for how long can the investments continue without extending our pool of knowledge is a question that researchers and stakeholders ask of mHealth.

In the next section of the chapter the author present opportunities and challenges for mHealth, by looking at some failed examples of mHealth projects. The section presents mobile phones as the Information & Communication Technology (ICT) of choice compared to other technologies. In the later section, the case of scalable Indian mHealth solution, "SCDRT" and its technology choices is presented. Afterwards, author presents the case of using plain-text SMS in Kenya to scale health services. In Section 5, author look at his case studies through the lenses of information infrastructure theory and discuss the reasons why these two projects have been able to scale well. In the later section, open-source solutions and their advantages in emerging markets and ICT4D projects is discussed. In the last section, author provide avenues for future research and give concluding remarks. This chapter attempts to bring together successful cases of mHealth applications and theoretically explain why these projects have been able to successfully scale.

*Figure 1. Range of mHealth devices on increasing cost*

## OPPORTUNITIES AND CHALLENGES FOR mHEALTH IN EMERGING MARKETS

There are numerous examples of mHealth failures, which in itself would be enough to fill this entire chapter. Following is a concise list of solutions that were hailed as revolutionary, but have not been able to scale (Anderson & Perin, 2009):

- TeleDoc – Jiva Healthcare by The Soros Foundation and Jiva Institute in 15 villages in Haryana. Doctors receive diagnostics on Java-enabled mobile phones and medicines are delivered to the homes. Won award in 2003 World Summit at Geneva, but hasn't scaled.
- Tamilnadu Health Watch – by Voxiva was after 2004 Tsunami. Disease reporting by mobile calls, fixed-line and Internet. 300 PHCs trained, not in use now.
- Handhelds for health – IIM Bangalore & Encore Software for disease tracking & surveillance. Started with use of "Simputer" and now uses "Mobilis". These products are comparative to smart phones, but haven't scaled to be used across villages or state.
- AIIMS telemedicine project makes use of Windows Mobile based PDAs for capturing patient data and create an EMR (Electronic Medical Records). This project has been in development and test at Vallabhgarh in Haryana for couple of years, but has not scaled beyond its use in the village.

- AESSIMS (Acute Encephalitis Syndrome Surveillance Information Management System) by Voxiva in Andhra Pradesh to report cases of encephalitis has been piloted, but has not been adopted for scale.

Although mHealth encompasses all kinds of mobile computing devices - from wireless chip-based solutions to portable computers, we believe mobile phones are the most scalable, especially when considering emerging markets and low-resource contexts. Other than mobile phones, even PDAs, laptops, specialized telemedicine equipment or video conferencing devices rely on mobile phone networks in rural contexts. Thus, the underlying wireless network of mobile phones is crucial to all kinds of mHealth solutions (Figure 1).

Another important consideration in terms of emerging markets and low-resource contexts is that there is a big problem in the capacity of the health staff. The number of doctors, nurses or health facilities available in comparison to the population is very low. This means that for a health beneficiary (anyone who receives medical services) to reach a location where medical services are delivered is problematic. Many times the health beneficiary has to travel long distances to avail proper primary care and mHealth provides an opportunity to bridge the distances between the patient and health providers or between skilled medical practitioners and less skilled ones. In emergency situations, a low-skilled health worker can easily contact a medical officer and deal with the situation appropriately by using mobile phones. Where travelling distances for a skilled medical practitioner might be tedious as well as costly, mHealth may prove to be a useful solution that can be both efficient and low cost.

From a health information system perspective, mHealth provides a unique opportunity to get data quickly from the field-level health workers. Getting data on a timely basis is currently a big problem with health systems from emerging markets. From examples of mHealth initiatives (Mukherjee & Purkayastha, 2010), we see that mHealth solutions also help improve the data coverage rate because data comes from lowest levels of the health system. There is also a hypothesis presented in the above mentioned paper that since data comes from the lowest levels, quality and accuracy of data is better than if they came only from the higher levels. Although there is not enough evidence to support this hypothesis, it provides good research direction and opportunities for research in the field of health information systems for practitioners.

Braa and Purkayastha (2010) show that mobile phones have some important characteristics that make them suited to large-scale deployment and used in low resource settings.

- **Greater Market Penetration:** Mobile phones are easily available and people are less intimidated by the technology of mobile phones. This market penetration is also called the "installed base" in infrastructure theory. The installed base provides a form of inertia to the technology artifact and hence has greater chance of success in the system

- **Small Learning Curve:** There are only limited ways in which a mobile phone can work compared to a Computer/Smartphone where it can perform many operations at a time. Mobile phones also only have a keypad for input instead of keyboard, mouse etc. available in computers. This makes the learning curve for using mobile phones much smaller.

- **Low Power Consumption:** Mobile phones consume less power compared to other kinds of mobile computing devices. This is extremely critical in low resource contexts where stable and continuous power is a rarity. Approximately 1.6 billion people worldwide live without access to electricity, of which 25% live in India.

- **Wide-coverage Area**: Mobile phone networks work in remotest of places in these emerging markets. In emerging markets, the governments also have an inclusive agenda where they want people from all places to get access to mobile services.

- **Low cost of Device:** Mobile phones ought to be between Rs.2000 ($40) to Rs.5000 ($100) to be purchased in mass quantities. This makes applications on mobile phones to have a wider audience compared to other mobile computing devices.

But along with these opportunities, there are also challenges for using mobile phones as the devices for mHealth applications. Mobile phones have some inherent limitations that need to be highlighted compared to computers or other mobile devices. These are embedded in the artifact of mobile phones, but with time could also possibly change. There is a clear trend that we see mobile devices are getting more and more powerful. But in today's context, the following limitations of mobile phones need to be taken into consideration when designing mHealth solution (Braa & Purkayastha, 2010):

- **Limited Processing Power:** The cost of the device is generally directly proportional to the processing power of the device. So, cheaper the phones are their processing capabilities also keep on decreasing. Most of the processors in mobile phones do not support multi-tasking and cannot allow processing large amounts of data.

*Figure 2. Infrastructure considerations for mHealth applications*

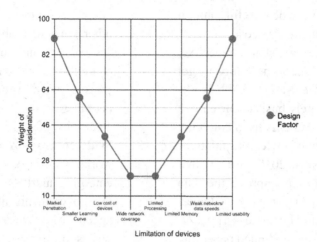

- **Small Screen Sizes:** The normal display sizes for cheap mobile phones are less than 3 inches with resolutions less than 220dpi. Although small screen sizes help improve the battery life and make the devices low-powered, they are a limitation on the usability of the devices. We understand that most of the health workers in India are middle-aged women and hence it is extremely important to keep the font-size large, but still not waste too much of screen area.
- **Limited Visualization:** As people with low technological skills are the target audience, it would be great to display as much visualization as possible. Sadly, mobile phones are limited by the visualizations that they can show. Thus, the designer of mHealth application needs to intelligently place graphics and icons on the screen, so as to make the applications attractive and more usable, but considering how clear and easy to understand the images can be.
- **Limited Memory:** Compared to computers or other devices, mobile phones cannot store a lot of data. This makes storing all data entered by the health worker for many months or years pretty much impossible. Even complete patient records for a large number of patients cannot be stored on the phone. Thus, mHealth application designers need to create applications that can sync records with online servers through mobile phone networks
- **Weak Mobile Networks:** At many of these contexts, the mobile phone networks are unstable or have weak signal strength. This means that mHealth solutions need to be robust enough to be able to deal with situations where no wireless communication is possible.

These pros and cons need to be carefully balanced by designers of mHealth applications and governments or implementing organizations need to realize the advantages of mobile phones as devices for mHealth.

In Figure 2, the graph is an interpretive understanding from the author of the chapter that can assist new mHealth designers in balancing the pros and cons of mobile phones as a device in mHealth. The graph shows the infrastructure factors that a designer of mHealth application needs to consider. From left to right, we see increasing limitations to designing mHealth applications and from bottom to top we see the increasing weight of factors that need to be considered. For e.g. although market penetration is an advantage of huge importance, equally important limitation is the limited intuitiveness/usability of mobile phones. Similarly, although the low-cost of mobile phones is an advantage over smartphones, equally important disadvantage is that mobile phones have limited memory. This is not an exhaustive list of factors that need to be considered. But it is an indication of the factors that were considered by the designers of the cases mentioned later in this chapter. Thus, all these factors mentioned in the graph need to be considered when designing mHealth applications.

To build a mobile health information infrastructure, a balance between the complexities of the context and technology needs to be analyzed. As mentioned by Hanseth and Ciborra (2007) simplicity of ICT can help reduce risk. It is important that mHealth solutions be simple and manageable in their approach. As a first step to designing large-scale information systems, the chapter prescribes that simple technological solutions be tried first and then more complex solutions be developed and deployed as the designers as well as users become more and more used to the system. In the next section the importance of looking at mHealth applications as part of the infrastructure is discussed.

## RESEARCH METHODOLOGY

The *Indian mHealth case* described below in the chapter has been developed through Scandinavian action research tradition in IS development, such as user participation, evolutionary approaches and prototyping (Sandberg, 1985; Bjerknes, Ehn, Kyng, & Nygaard, 1987; Greenbaum & Kyng, 1991). The research is part of a global network of action researchers and aims to generate knowledge by taking part in the full cycle of design, development, implementation, use and analysis. The above mentioned steps are done together with all the involved parties (Government of India/State of Punjab, mobile phone operators, handset distributors and health workers) before the interventions are adjusted accordingly, and the next cycle begins again (Susman & Evered, 1978). The chapter is a result of the participation in this action research network for the last 3 years and provides insights from the central role in design of the system and participatory role in implementation and maintenance of the system after implementation. The research has been done within the framework of interpretive research (Walsham, 1995). Data was collected through group discussions, requirement meetings, developer discussions, feedback reports along with one-to-one interviews in health workers, health officers, doctors and ministers.

The first phase of the research for this project was started as part of the pilot and gathering requirements by involving the state health department and their officers through interviews and group discussions. After the requirements gathering phase, prototype demonstration were made to the national and state ministries of health working under the National Rural Health Mission (NRHM) programme. The feedbacks from these demos have been recorded in the meeting notes. These meeting notes have also helped in improving the software and served as further requirements for the next iterations. The next phases involved development of the software, which have been done using Agile Methodology of software development. Regular iterations and emails to the developer mailing lists have served as useful data to interpret the development of the software and project as a whole. The next phase of the research is the involvement in user training and after the user training, recording feedback on a set of questions from the health workers. The data collected has been analyzed and quantitatively represented in the report given to the ministries of each state, using which they have been able to participate in monitoring the progress and usefulness of the application. The research covers involvement in customization and training of the application and the data interpretation involves documents from implementations, health system manuals and being part of meetings. The research for Kenyan case has been done over a period of 6 months, by looking at the documents produced by the designers of the system and communicating with the implementation field workers for the project. The data interpretation for this case is through ex-post-facto observations of the implementation and documents made available by the implementers.

## THE SCALABLE INDIAN mHEALTH SOLUTION

As governments seek to make health services more "patient centric", there is increasing demand for the health service provider (called Auxiliary Nurse Midwife (ANM) in India) to be more mobile and be able to cover the catchment population in her jurisdiction (typically 5-7 villages or population of about 5000) for providing health services. Thus, the health worker's activities are mobile in nature and hence the tool required to assist her in service delivery should be mobile as well. mHealth solutions in this context are only going to be used by health workers when these solutions can provide support and assistance to the health workers in their day-to-day activities. If mHealth solutions do not do so, they are considered to be an additional burden for the health workers who are already covering a large population base and results in large scale protests and shutdown of the mHealth application.

Implementing software solutions at the lower levels of the Indian health system is a huge undertaking due to its enormous scale in terms of the vast number of installations, system maintenance and training activities. So when HISP India (a decade old NGO with global footprint) was approached in January 2009 by the National Health Systems Resource Center with an idea to empower health workers and get information from sub-centers, use of mobile phones became the obvious choice. The author of this chapter being the Director of

R&D of the project closely followed the project through design, development and implementation. The project was initiated as pilot in 5 states to represent varied geographies, varied user profiles, varied infrastructure and user base. The project was successfully evaluated a year later and has currently been implemented in full-scale in state of Punjab to over 5000 mobile phones and is under discussion for implementation in two other states. This is probably an example of one of the largest implementation of any mHealth solution around the world and reasons for that are simplicity, installed base and designing the solution by aligning to the available infrastructure.

The application in the pilot was called Sub Centre Data Registration and Transmission (SC-DRT) application and had the following aims and outcomes:

1.  To develop a very simple paper form-like data collection tool on mobile phones
2.  Efficient transmission through SMS of the data from the sub-centre to the higher levels
3.  Establish a basis for improved data quality and validation
4.  Explore the potential of other value added mobile phone based applications, such as:
    a.  Providing feedback to the ANM on activity scheduling.
    b.  Strengthening processes of communication of the ANM with other functionaries (such as the medical doctor).
    c.  As a training tool, such as to help orient the ANM on new data elements.
5.  Integrating the SCDRT information with the mainstream district based health information systems. In this case, it was the software application called District Health Information System (DHIS2 - a standard aggregate health information system used in many states in India) used in the pilot sites.

The details of the implementation of the pilot and its evaluation (Mukherjee & Purkayastha,

*Figure 3. SCDRT infrastructure and flow*

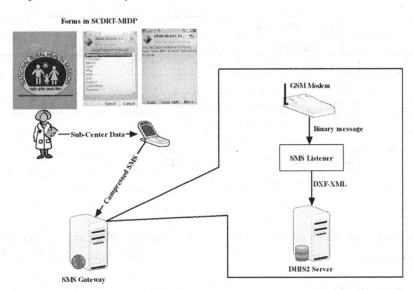

2010), highlights the different steps required for successful implementation of mHealth projects and also the challenges faced in the process of implementation. The paper highlights that there are enough challenges in the implementation of such a widely used mHealth solution and is a good example for implementers and policy makers. This chapter on the other hand is geared towards mHealth application designers and what they need to consider when designing scalable mHealth solutions.

SCDRT enables the ANM to send monthly report via an SMS through the mobile phone to the next level(s). SCDRT was designed not as a stand-alone application for sending the report, but integrated with the mainstream district health information system for national level reporting. By design SCDRT was envisaged to be similar to the paper formats to maintain the familiarity and identification of the ANMs to the application. Figure 3 describes the SCDRT infrastructure and flow.

The flow is summarized:

1. ANM fills the Sub-Centre Monthly Dataset (monthly reporting form) on the mobile phones.
2. After completing the form, the report is sent as a text message (SMS) to desired phone numbers, which are located at the PHC/ Block/District.
3. The SMS is compressed as a binary message. The compression is about 70%, resulting in more data being reported with lesser number of SMS messages.
4. These messages are received at the PHC/ Block/District into the GSM Gateway Cards using a software application called SMSListener, and messages are imported into the state HMIS application (called DHIS2).
5. SMSListener is a utility application developed and configured to listen/receive SMS on a stand-alone system. When SMSListener receives the compressed message, it first decompresses it and writes complete information on XML files which then is easily imported into the DHIS2 application
6. The staff at PHC/Block/ District performs data quality checks using the built-in features

*Table 1. Mobile phone specification*

| GENERAL | GSM 900 / 1800 / 1900 |
|---|---|
| KEYPAD | 5-way navigation key |
| DISPLAY | 128 x 160 pixels (or more) |
| JAVA: | MIDP 2.0 (JSR 118) |
| | Connected, Limited Device Configuration (CLDC) 1.1 (JSR 139) |
| | Wireless Messaging API 2.0 (JSR 205) |
| | |
| | Scalable 2D Vector Graphics API (JSR 226) * |
| | FileConnection and PIM API (JSR 75) * |

called validation rules which are part of the DHIS2 application.

## Managing Complexities in Design and Development

The SCDRT application and its evolution from the pilot can be described through the perspective of managing complexity in design, development and implementation. The SCDRT application uses very simple technology and depends in the background on the existing system of DHIS2 to store data on the server-end.

The application is developed using basic Mobile Information Device Profile (MIDP) 2.0 components that are part of all Java-enabled mobile phones and thus meant that the application could be run on basic Java phones. The specification required to the run the application was the basic Java-enabled handset shown in Table 1.

## SCDRT Technology Requirements

1. **Mobile Phones:** Any Java-enabled phone; one for every ANM at the sub-centre.
2. **GSM Modems:** Each PHC/Block/District installation will receive SMS using these into DHIS2.
3. **Software:** Free & Open-source applications Mobile-SCDRT, SMSListener and DHIS2.
4. **SIM cards:** Every mobile phone and GSM modem requires a SIM card with a phone number.

Other than the handset at each health facility/health worker, there is not much infrastructure required to run the project. A GSM modem needed to be installed at the DHIS2 server. Any SIM card could be installed in the phones and GSM modem and they need not be the same. But to encourage communication and make the services cheaper, it is useful to establish a Closed-User Group (CUG) connection between all the mobile phones and GSM modems. The CUG connection from the operator makes it free to make calls within the health system and about 200 SMS are free in a month. It was seen that a CUG connection helps foster communication between health workers and also between the data managers and health officers. Besides the data coming and contributing to improving the health information system, the communication and improved efficiency at which the health system now communicates through mobile phone is a very important consequence. Most evaluations of this project suggests that health workers became more social with each other and the community in which they provide health services with the introduction of mobile phones. The CUG is thus a recommendation for any mHealth solution and is an important infrastructural consideration with the mobile phone operators.

The SCDRT-MIDP is a JavaME MIDP 2.0 application which uses plain JavaME components. It consists of Form, TextField, Choice, Command and built-in MIDP 2.0 layouts and CommandActions. A total of 9 screens are shown to the health worker based on the category of data reported by the sub-center. After the data collection is complete, a screen is displayed to notify the health worker of the fields that have not been filled. If acceptable, the "Send SMS" button is displayed and the data collected is sent as SMS to any (max 3) phone numbers, which are actually numbers of the SMS gateway. The data from the mobile devices is sent through SMS and the choice of SMS was to manage the complexity and align to the existing infrastructure in rural India where data speeds are slow and latency of the connections is huge. The SMS is created through a simple formatted string. Data from each TextField or Choice is collected and this is added as a pipe separated ("|") string. The month/week/date of reporting is separated from the rest of the string with a ("$") dollar character. After the full string is created, it is sent to the Compressor class. The Compressor creates a byte[] array as output containing the data to be sent as an SMS (sent as EMS binary message in 8-bit format).

The Compressor is a Java class which compresses Java String to byte[] and decompresses a byte[] to a Java String. It uses range encoding for the compression and decompression. Range encoding is a simple algorithm for compression and does not require huge amounts of memory. This was an important technical decision when considering that the application was made for low-end mobile phones and should be usable for a large number of devices. The range encoding algorithm is also an open-standard and can be used by anyone easily. The compression ratios provided by range encoding are not superlative, but these are exactly the technical choices that the designer needs to make, when making solutions

that can scalable and still be useful. The SMS sent is compressed through range encoding algorithm, which is simple yet elegant and gives an average compression rate of 67%. It does not include any salt and hence it would be incorrect to call it encryption. But when the data is sent, it is sent as a binary message and it cannot be understood if received on a normal phone, since it is binary data. On phones which send binary messages as only Nokia Picture Format (not as EMS), the decompression has some problems on the server side. On such phones, we have to disable compression before deploying. Most of these phones are outdated and not manufactured any more, but still it is one of the limitations of sending binary messages through the phones.

The SMS sent by the health worker is received by the GSM gateway, which according to the implementation model is located at 3 places (PHC, Block and District). The SMS Listener is kept running on the computer which acts as the GSM gateway. Whenever the SMS is received, the SMSListener first de-compresses the SMS and then converts it into an XML in a simplified DXF format. This XML is saved to a mi/pending folder in the DHIS2_HOME and is available to DHIS2 for importing. The "Mobile Importing" module in DHIS2 includes a start mobile importing button. When the process of importing to DHIS2 starts, it runs through the XML's located in the pending folder and imports the data in each of them. If the importing is successful the XML file is moved to the mi/completed folder. If some error occurs in the importing process, then the user is shown a message and the XML file is sent to the mi/bounced folder. The mobile importing module is an additional module that has to be included when building DHIS2 web application. It adds a "Mobile Importing" section to the Services menu of DHIS2. Each mobile phone is registered to a single sub-center in the organization unit of DHIS2. The XML contains the phone number from which the SMS was received and when importing the data, the phone number acts as the information through

*Figure 4. Punjabi mHealth application user interface*

which data is associated to a sub-center. Data after importing can be seen in the Data Entry screen within DHIS2. Thus, the mobile phone sends the same data that a person would do data-entry into DHIS2 through the keyboard.

On the mobile phone interface, all elements can be localized. In the pilot states, the application was localized to Gujarati, while in other states it was in English. In the state-wide implementation in Punjab, the application was localized to Punjabi using Gurmukhi text. The design of the application is simple and all localizations can be done through a single messages properties text file. This text file is external to the application and can easily be modified by through a word processor without involving any change in code. This allows flexibility and ease to translate the application to multiple languages very easily. These localizations are lightweight and do not require a lot of processing power on the phones. The designers of SCDRT solution took into consideration that any localization required can be easily done by even non-developers and can be done with simple tools without requiring any programming knowledge.

The user-interface of the application deployed state-wide in Punjab is also extremely simplistic and easy for the user to understand. It has been designed to look similar to the paper forms and the pages are arranged similar to the paper forms (see Figure 4).

The entire project including the DHIS2 server components have been developed through open-source collaboration. The projects have been publically hosted and contributions have been made by developers from at least 4 different countries.

At the start of development of the project, other alternative open-source software solutions were evaluated. JavaRosa is a popular open-source framework for developing mobile applications. JavaRosa is the Java implementation of the OpenRosa standard created by the OpenRosa consortium (Klungsøyr, Wakholi, Macleod, Escudero-Pascual, & Lesh, 2008). JavaRosa renders Xforms on any mobile device that supports execution of Java on phones. Xforms are XML-based forms which separate the data being collected from the markup of the controls collecting the individual values (Boyer, 2007). The OpenRosa mobile consortium has defined some standard tags for Xforms on mobile devices, so that these can be commonly represented across devices. Thus, JavaRosa was an implementation of the OpenRosa standard on Java-based mobile phones. The JavaRosa platform is very popular for mHealth projects and has been used by more than 15 different mHealth applications around the world (Klungsøyr et al., 2008). So ideally a solution based on JavaRosa would require the developer to create an Xform for whatever data one wants to collect on a server. The JavaRosa J2ME application would then send a request to the server and download the Xform to the mobile device. The user would then be able to see the form for data collection on the mobile phone and fill the data. After the data is filled in, a new Xform with data contents is submitted to the server and the server understand the data and extracts it out for processing. There were 3 major issues that cropped up when using JavaRosa:

- **Data Services/XML data:** One of the characteristics of low resource settings was the weak networks. These places have very slow connectivity for data services from mobile operators. Even the data services are not very robust in such settings. JavaRosa internally tried to solve this with fail through checks and resends, but that was an overhead for low resource settings. Additionally the cost of data services is quite expensive compared to SMS, especially when we are dealing with sending of monthly reports which consume only 1-2 SMS per report for 1 month.
- **Required a Powerful Phone:** Since the JavaRosa client was trying to read XML and render the forms on the fly, it required good amount of memory and processing capabilities on the mobile phone. This made the application display form elements slowly and would not work on the low-end phones in the earlier mentioned price range.
- **Lack of server-side integration:** OpenRosa Xforms were not linked or integrated with the DHIS in any way. This meant that additional development effort would have to be made on DHIS to be able to understand and parse Xforms sent through JavaRosa client. Thus, JavaRosa required much more efforts and polishing so that it can become an end-to-end solu-

tion for an already established health information system based on DHIS2.

Thus, to manage the complexities of GPRS, XForms, separate libraries for UI elements etc. a simplified application platform was chosen instead of JavaRosa. Another project which we looked at was Kiwanja's FrontlineSMS. It is a powerful community supported project which was designed for the purposes of low resource communities. It had proved to be successful in 2007 during the monitoring of the Nigerian elections (Banks, 2007). FrontlineSMS was a simple, yet powerful solution because it used Short Messaging Service (SMS) for sending information from mobile phones to the server. FrontlineSMS could be setup with only a GSM modem/phone connected to a computer and receive SMS. The SMS could then be put into a database or mapped to create an XML. Although FrontlineSMS seemed like an appropriate solution, the usability of the solution was in question. A similar approach to FrontlineSMS is used by the case presented below in Kenya, but the same solution was not considered appropriate for the Indian context. The Indian context wanted people to see the forms similar to what they would see in the paper forms and not type messages from the phone's SMS system. It was considered too tedious for middle aged health workers from India to be able to type messages correctly with the correct keywords and message formats

The software development for SCDRT did not take more than a couple of weeks and simplicity was the basic design goal of the project. After the software had been developed for the pilot, continuous feedback was received from the health workers through refresher trainings. From these training it was observed that the health workers needed better feedback and wanted to store more information on the mobile phones for their own use.

Thus, development was done in an iterative fashion with regular updates of the applications being installed on the phones. The advantage of the software being open-source was that other developers who were working on similar projects contributed to the applications and improved the usability and flexibility of the application over a period of time. There were projects in other countries in Africa that used the same framework, customized the forms and deployed the application to the health workers. The regular updates to the application on the deployed handsets were done and immediate improvements in the user experience could be seen.

All these factors of managing risk and complexity ensured that the mHealth application could be scaled and implemented successfully to over 5000 phones and is continues to be scaled.

## USING SMS IN KENYA TO FIGHT MALNUTRITION AND MALARIA

The Millennium Villages Project (MVP) in Sauri, Kenya has worked on a pilot project to fight malnutrition and malaria in children under the age of 5 years through the use of SMS and the most basic mobile phones. This mHealth solution is called ChildCount and has been developed by the Earth Institute at Columbia University. Community Health Workers (CHW) are empowered with mobile phones and are trained to create SMS to report malnutrition cases. The project used a popular open-source platform known as RapidSMS as the back-end system and data is sent as formatted SMS from the phones of health workers.

Unlike the Indian mHealth application (SC-DRT), this case makes the CHW to use the phone's SMS functionality and does not provide any form-like user interface. This makes the learning curve for creating the SMS longer, but highlights the issue of aligning solutions to infrastructures. The infrastructure of the health system in Kenya did not allow for buying new handsets and distributing it to health workers. Thus, basic mobile phones which did not have Java in them had to be made usable. Even the infrastructure for data services is unavailable and costly in Kenya. Thus, SMS was the most obvious infrastructural choice for such a mHealth solution.

The system provides the following functionality:

- Ability to the CHWs to register themselves by sending an SMS and identifying themselves with a phone number for sending reports.
- Ability of the CHW to register new cases of children with malnutrition and generate a unique ID number for each child.
- Ability to report different observations and measures of a child's case through SMS
- Automated notification and alerts to health workers for follow-ups and treatment of children whom they have registered
- A web interface through which data for individuals can be monitored as well as an aggregate view of indicators generated from individual records.

The following is an example of a formatted SMS that is sent to register a child into the system:

NEW LAST FIRST GENDER(M/F) Date Of Birth(DDMMYY) Parent

new sumi john m 010609 Joanna

After this SMS is received into the RapidSMS servers, there is a routine run to see if the child is already registered in the system. If the child is not registered in the system, a new patient record linking the child to the system is generated and sent back to the health worker as an SMS.

PATIENT REGISTERED> 121 SUMI, John. M/13M

Mother: Joanna, Kangaba Village.

There also other features like **user-to-user messaging**, as shown in Figure 5.

*Figure 5. User-to-user messaging*

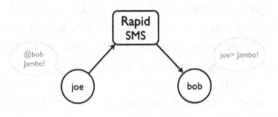

For example:

@bob Jambo!

Would result in @yndour receiving:

joe> Jambo!

Group Messaging is shown in Figure 6.

## Reminders and Alerts

Core to ChildCount is an automated reminder system to help ensure that no patient falls between the cracks. When certain events occur, like when a patient enters into a home-based Supplemental Feeding treatment program with Community based Management of Acute Malnutrition (CMAM), a follow-up alert can be assigned to the caregiver assigned to that patient. When the time for a follow-up visit nears, in this case seven-

days, a message will go out to the patient's health care provider requesting a follow-up malnutrition monitoring report. After one day, if ChildCount does not receive an update for that patient, the alert status for the patient will be elevated to urgent. At this point a reminder will be sent out on a daily basis not only to the health care to their manager and team mates for follow-up.

As experienced by the health staff and data managers, one of the unintended benefits of the system was the ability for the managers to provide better feedback to the health workers. The data coming in allows the health workers to work more optimally than before. Prior to ChildCount implementation, the health managers had a tough time to monitor the activities of the health workers. While the system is not encouraged to be used a monitoring system, there have been reports created to compare the performance of CHWs.

The project from the start looked at the proposed solution through an infrastructure perspective. The technology choices of using RapidSMS, formatted SMS, SMS as a delivery medium were all because of the existing infrastructure available. Also partnerships were established between different players of the ecosystem. Ericsson as a partner helped provide handsets of Sony Ericsson to each health worker, but to scale the system, existing handsets owned by health workers could

*Figure 6. Group user messaging*

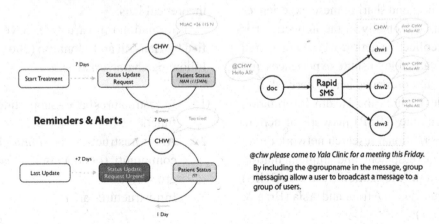

also be used. Zain, the local GSM mobile operator provided a toll-free SMS number that allows CHWs to send messages to the system for free. The health facilities and the health officers in the pilot area were involved in training and consulting during the implementation of the project.

As seen in the report (Berg, Wariero, & Modi, 2009), there have been over 20000 reports that have been sent by CHWs. There have been over 9500 child registrations and 7,646 nutrition screening reports, 839 Rapid Diagnostics Tests results and registrations of 7803 measles vaccinations. Along with the malnutrition and malaria, the solution is also being adapted for other programs such as immunization, PMTCT (Preventing Mother To Child Transmission of HIV). It is also being scaled at multiple MVP sites in different countries and also in a lot of countries in Africa. The fast scale at which the solution has grown and evolved highlights the importance of consider and aligning to the existing infrastructure when designing and developing mHealth solutions.

## AN INFRASTRUCTURE PERSPECTIVE

The research approach is action-oriented and interpretative and characterized as a 'network of action' methodology. The network of action approach is based on the principle of creating learning and innovation through multiple sites of action and use, and sharing these experiences vertically and horizontally in the network. It is premised on collective action where connected research units are able to share experiences and learning. The case presented here is derived from units (or nodes) within the Health Information System Programme (HISP) network of action. HISP is an international research network doing open source development and implementation of District Health Information Systems (DHIS) in over 10 countries in Africa and Asia (Braa & Humberto, 2007).

Along with being simple and manageable, designers of mHealth solutions need to think of the infrastructure in emerging markets. As emphasized by Monteiro and Hanseth (1995), there exists the social shaping of infrastructure. We need to think of infrastructure as a socio-technical network and consider the technical aspects of design, but also the social capacities of individuals who are going to be part of the mHealth infrastructure. Mobile phones, unlike smartphones, PDAs, portable computers are already socially relevant. They signify empowerment of health workers who are majority women and the use of mobile phones as a communication tool is equally significant as much as a data collection and reporting tool. The ease of use of mobile phones and an existing installed base weighs a lot more than technological prowess of other ICT tools for the scalability and success of mobile phones

It is important not to look at the mobile phone as a standalone device, but with a systems perspective including various other kinds of infrastructure – such as the paper registers at the sub-centers, the computers at the district levels, the mobile phone networks, internet access, the servers at the state level, and also the basic infrastructure required to support the mobile phone use (charging facilities, support centers from mobile operators and handset manufacturers, network coverage etc). To understand all of the factors of a mHealth infrastructure, the chapter suggests looking at Information Infrastructure theory as a theoretical lens.

Information Infrastructures have been identified by Hanseth and Monteiro (2001) to have the following characteristics:

1.  Infrastructures have a supporting or *enabling* function.
2.  An infrastructure is *shared* by a larger community (or collection of users and user groups).
3.  Infrastructures are *open.*

4.  Infrastructures are more than "pure" technology; they are rather *socio-technical networks*.
5.  Infrastructures are connected and interrelated, constituting *ecologies of networks*.
6.  Infrastructures develop through extending and improving the *installed base*.

From the above cases of mHealth applications, we look at the following aspects of Information Infrastructure:

**Enabling:** A mHealth infrastructure should be designed to support a wide range of activities, not especially tailored to one. It is enabling in the sense that it is a technology intended to open up a field of new activities, not just improving or automating something existing. Rather than just capturing data from health workers, mHealth applications should become part of the work culture of the health workers and assist them in their day-to-day activities. In the above cases, we see this as one of the establishing principle. Although both the applications were designed for data collection and reporting, the infrastructure created due to the mHealth applications enabled other forms of use. These include peer-to-peer communication, communication between community and the health workers, requests for leave and information exchange about the new developments in the community. Health workers in the Indian case were given mobile phones which had a digital camera in them. This made it possible for health workers to capture images of mosquito pits or skin rash in patients, so that further action can be taken on these. When these images were shown during monthly meetings at the district-level, officers responded in a better way and health workers felt more empowered. Thus, the mHealth infrastructure created by these applications acted as enablers for new ways of working in the health system. This enabler of change may sometimes be expected by the designers of the mHealth applications or sometimes there may be unintended consequences.

**Shared:** The mHealth infrastructure should be able to bring in more partners and users. Mobile phones assist in communication between the medical officers, state health & data officers and the field-level health workers. The infrastructure allows shared information between members of the health structure. In the above cases, we see that the information received through mobile phones is shared between different health staff and officers. In the Kenyan case we see that different programs have been integrated with the advent of this mHealth application. Since the same health worker provides data for different health programs like HIV/AIDS, PMTCT, Malnutrition and Immunization, it makes sense that the same application can report for all this programs. Thus, it avoids duplication of work for the health workers and increases the efficiency and interest of health workers.

**Open:** The mHealth infrastructure should be made open in the sense that there are no limits for number of user, stakeholders, vendors involved, nodes in the network and other technological components, application areas or network operators. There are already different standards that allow open-access to technology in mobile networks. GSM, CDMA, SMS are standards that allow communication between different mobile handsets. There are also standards like XForms that allow standard communication of data and data collection forms on multiple devices. When you look at the Kenyan case, use of plain-text SMS allows different kinds of keywords that can be created by the administrators of the system and change the way in which they report data. This openness in the infrastructure allowed the system to be spread to other programs and health workers. The Kenyan mHealth application also allows new health workers to register themselves and start reporting data. This allows inclusion of many more people into the system and training for the usage of the application is done by one health worker to the other through the word of mouth. In the Indian case, we see that the back-end system of DHIS2 allows creating many flexible datasets and this can be used with mobile phones for reporting. In both the cases, we see that the solutions are open for use with any mobile operator, any mobile handset and open standards.

**Socio-Technical Networks:** mHealth is not just a technology piece in the health system, but rather involves the socio-technical network. Although ICT and mobile phones are part of the infrastructure, they are to be symmetrically seen with the people and processes involved in the system. Infrastructures are heterogeneous concerning the qualities of their constituencies. They encompass technological components, humans, organizations, and institutions. We see from our cases that the designers of the mHealth solutions have given importance to how their users are going to use the system and made quick changes based on the user's feedback. The human aspects of the system like user capacity, health system hierarchy, processes of reporting were considered in design of the systems. The designers of the system did not look at the mHealth solutions as a technology tool in standalone, but as an assemblage of the health worker and the mobile phone + application. Also the assemblage is between this entity and different elements of the health system that needs to be considered by the designers of the mHealth application. The use of CUG in the above cases is an example of how social networks were tapped and better use of the infrastructure was done to foster communication and the humanness in all of this.

**Ecologies:** Infrastructures are layered upon each other just as software components are layered upon each other in all kinds of information systems. This is an important aspect of infrastructures, but one that is easily grasped as it is so well known. In mHealth, there is an ecological dependence between mobile operators, handset manufacturers, health workers, health information systems. All these players act together to make a mHealth system work and finally result in better health services delivery. From the Indian case, we see

that the implementers first negotiated with the handset manufacturers and mobile phone operators for best deals for the solution. The low price and Java-enabled compatibility of the handsets were important components of the ecology that enabled the solution to scale well. The mobile operators providing CUG connection at low prices was also important factor of the ecology. The mHealth solution in both the cases provided a win-win situation for all the parties in the mobile ecosystem.

**Installed base:** There is already an installed base of mobile phones, mobile networks, health data capture forms and health information systems. There is inertia within an infrastructure and designers of mHealth application should realize this inertia. For e.g. some areas have better mobile networks from certain mobile operator. People from some areas are better used to certain languages and certain handsets. This inertia of the installed base is hard to change and mHealth applications need to align themselves to the installed base. In the Indian case, we see that the DHIS2 system was an already existing system which is widely used in different states for management of the health system. It was important for the mHealth application to make use of this installed base and is one of the important factors for its success. Similar installed base of a large number of Java-enabled mobile phones being available in the market improves the success rate of this solution.

## USE OF OPEN-SOURCE COLLABORATION

In both the cases we see a common use of open-source software and collaboration between different developers and players. We consider this to be an important reason for success of these mHealth applications. Open-source software allowed the designer and developers of the solution to quickly improve the software through the help of other programmers who had the knowledge of similar projects. The same mHealth applications could be easily adapted to other contexts due to the open-source nature of the solutions. These mHealth applications are also being implemented at new locations and customized according to those contexts much more quickly due to the open-source nature of these applications.

Raymond (2001) discusses that open-source became a phenomenon through an accidental revolution. But when designing mHealth applications, we need to consider open-source not by accident, but by looking at the pros and cons that open-source software development brings to the table. Today we see a billion dollar industry for free and open-source software with headline products such as Linux and Apache (Feller & Fitzgerald, 2002), the Free and Open-Source Software (FOSS) community and business of FOSS in health care or mHealth is yet to emerge. The same rules of success or failures of FOSS that have been researched in technology artifacts like operating system and web servers can be applied to mHealth applications is still an open question. We still need more evidence to see if mHealth applications can use the open-source model and be successful and prove to be as wide-spread as the other mentioned technology artifacts. But one underlying concept of "Freedom" is something that FOSS in mHealth provides is beyond doubt. Let us discuss what freedoms we observe in our cases of mHealth solutions.

**Freedom of pricing:** The cases of mHealth applications that are presented in this chapter were created by a number of developers through open-source collaboration. Although most of the developers are involved in other activities, their contributions have enabled in creation of these applications. The overall cost of development of the mHealth applications is low because these were primarily contributions of a group of global developers. In the case of the Indian mHealth solution, the collaboration and the number of developers participating initially were less, but later after the pilot finished and lessons were learnt, a larger group of developers found interest and

took up the development to forefront. Due to the open-source development model, any new implementation of this mHealth solution can directly use the application "as is" or make changes to the application freely. This makes the cost of entry and development for the new implementation very less compared to if they would have to start developing their own solution from scratch. It is also evident that after the pilot, the state of Punjab did not have to pay anything in development cost for the mHealth solution. The software was quickly translated by the state's own developers within a day and the dataset for reporting was also easily customized by looking at the existing source code of the pilot applications. Similarly the ChildCount application from Kenya is being used at many other sites. Some of these are independent of the support of MVP or the original developers of the software. Other developers and implementers are able to pick the software and start using and customizing it without any paying any cost to the original software developers. The original developers are not getting paid and that could be seen as their loss, but since they created the software initially for themselves only and now it is being used at other places, proves to be beneficial for the community of mHealth initiatives on the whole. Also the original developers of the tool have the opportunity of being requested by other implementers for customization and support, which can be a useful business model, as seen in other FOSS software.

**Freedom of choice:** Along with freedom of pricing and cost, the health system has the freedom to choose which organization to use for supporting the development or maintenance of the mHealth solution. The state where the solution is implemented also gets the freedom to modify or change the solution without having to pay or ask for permission to the original authors of the system. Sen (1999) refers to such freedom of choice as a social opportunity where markets provide people with liberty, helps foster efficiency and build democratic institutions. With open-source mHealth applications, there can be democratic processes for development of solutions. It also gives local development opportunity of international and global solutions for customization to local contexts. This freedom of choice is important for mHealth applications because that allows for continuous improvements and change. It also allows software to be flexible and change according to the changes in the health system. This makes the ICT tool to be more in sync with the actual processes of the health system.

**Freedom of interoperability:** Integration has been considered as the holy-grail of Management Information Systems (Kumar & van Hillegersberg, 2000). People would like to see data from different sources for comparison and management of the available resources. This is more important for the health system in developing countries where resources are limited and these have to be effectively managed. Thus, when we talk about integrated view of data for management, it is important to have inter-operability between different software systems that are used in the health system. Open-source software as a principle allows better interoperability. Since the different algorithms for storing data, formats for data exchange and ways to process and manipulate them is open and accessible, these systems are easy to interoperate with. The cases of mHealth applications in this chapter allow interoperability through very simple technology. The Indian case of SCDRT, allows any application that can send data through compressed SMS can communicate with the DHIS2 server. The compression algorithm is open and so is the XML format for data exchange. Any other mobile application can be used with the server end of the system and use to send data to DHIS2. Similarly, the Kenyan case uses HL7 as a standard for data exchange after data has been received by the server. The different keywords that are setup in the server side for reporting of data can be exported to other systems. The underlying platform of RapidSMS has a open-format for configuring SMS structure and this makes it easy

for other people to build upon. Anyone who wants to create another application can interface with the RapidSMS through its API and easily create new solutions similar to ChildCount.

**Freedom to collaborate:** The community around these mHealth solutions encourages the process of sharing knowledge. At many times during implementer or developer debates, the teams would put the question to the community at large and the community members who faced similar challenges would respond back. Similarly, other organizations with similar requirements for such solutions share use-cases which helps build a more generic application. Since the community is already sharing code, it also encourages people to share stories of success, failures, challenges and politics. Thus, other than just code being shared, open-source enables sharing of ideas and helps in the times of challenges.

As with proprietary and closed system, exchanging data with these systems or changing them involves many unfreedoms. The unfreedoms are inherently part of the business model of closed-source or proprietary software. Some of these are not purely technical choices, but also organizational or business choices.

- Intellectual Property (IP) challenges related to proprietary ownership - When a designer or a developer of a mHealth solution does not want their software to be changed by others or wants to share the revenue of modification open-source isn't a good choice. Proprietary software is more useful when we want to establish an Intellectual Property on the mHealth solution and want to give this technology to someone at a price tag.
- Restricted access to way in which information is stored and managed. Restrict the flexibility to exchange information - Proprietary software generally provides these restrictions. These restrictions hamper the inter-operability of the software

with other systems. Software developers would be able to charge extra fees for modifying the software for inter-operability with other products.

- Restrictions to modification of existing hardware/software platforms - Some people do not want their solutions to be modified or there can be a business logic to charge for such modifications. Either ways, open-source software is not good if you want to imply such restrictions. Proprietary software creates unfreedoms such as these and implies restrictions on hardware/software changes to mHealth solutions.

- Cost of initial ownership as well as recurring cost of modifications to the system - as mentioned earlier and in the example of our cases, due to open-source software the cost of ownership or entry for a new implementer is very low. Also people can modify and customize to their free will. This is not possible with proprietary software or people have to pay more or get special access to make such customizations.

- Non-compliance to standards of data exchange - Although you can have proprietary software systems to be complaint with standards, it is much harder in proprietary systems to understand the way these store and manipulate data. Thus it becomes much harder for people to inter-operate with closed-source / proprietary software systems.

With Free and Open-Source Software (FOSS) there is choice to be able to replace these unfreedoms. Sen (1999) concludes through his observations and reasoning that increasing freedom to individuals or a social group leads to development. Not only that, but removing unfreedoms is an important criteria to development. With FOSS initiatives we suggest that stakeholders get freedoms and this in turn would lead to better chances in developmental projects. Thus if

mHealth initiatives are being implemented with a social development focus, then there is a strong suggestion to use free and open-source software.

An important prescription of this chapter is to show how can mobile applications be sensitively designed and introduced in order to support the development of an integrated mobile based health information infrastructure. Information infrastructure is used in the broader sense, meaning the technological and human components, networks, systems, and processes that contribute to the functioning of the health information system. Strengthening and reinforcing the social network of health workers by creating communication possibilities across hierarchy and among peers, as well as creating feedback channels for the lowest level will be crucial. Existing research into the topic of how sustainable mobile health information can be effectively deployed and scaled is limited, and hence this topic lies in the frontiers of health information systems research (Donner, 2008).

## FUTURE RESEARCH DIRECTIONS

Today with the mounting interesting in mobile applications in health (mHealth) in emerging markets it is an interesting field of research for a number of reasons:

**Evolving Infrastructures**: Mobile phones represent the fastest growing ICT in emerging markets. Mobile phones reach lower-levels of the system and show promise to bridge the digital divide (Warschauer, 2004). As time goes by, we see more and more computing power being built into mobile devices and the prices of mobile devices coming down greatly. Smartphones with more powerful operating systems and those that enable hosting more powerful applications are more commonplace and these devices become cheaper and cheaper (Raento et al., 2009). Most emerging markets are also deploying 3G services that enable high-speed data and internet on mobile phones. Thus, with time data over IP packets would make more sense as data transmission medium instead of SMS. People also will gain more skills over time with the use of mobile phones. Thus, mHealth applications should be adapted to this improved user skills.

As the infrastructure improves, how these scaled and established base of mHealth applications evolve will be an interesting challenge for the implementers of these solutions. There is also a need for research to be able to highlight a typology of applications that suit a given infrastructure. Given that the mHealth applications are customized and are being used in different contexts, it is interesting to study the evolution and changes in the solutions and how they are being used in the different contexts. How technology from one place can be directly adapted for use in another context is a challenge and has potential for further research on these mHealth solutions.

**Changing Mobile Standards**: Mobile standards are also changing rapidly. Recent rulings on mobile net neutrality (FCC, 2010) show that mobile telephony could slowly be moving towards more operator driven access to internet from an open-access internet. Mobile computing represents the future of access to the internet, but a change in the open nature of this infrastructure could result in a different evolution of the mobile world. The change of power to the mobile operators could result in different network configurations. There is a continuous change in the power struggle between mobile operators, handset manufacturers and government regulators. As can be seen in the 3G auctions in India, suddenly the cost of operations in India for mobile operators have increased to 10-folds. If the same pricing for mobile services are available after this change is an interesting trend to study. Any change in these factors need to be considered by developers of mHealth applications. Also with 3G, the opportunity to do a lot of data transmission increases and the solutions that can use these fast speed of data exchange will be interesting to research.

**Quality Studies**: As shown by Akter et. al. (2010), there is a need to study the quality of

mHealth applications and how quality affects the growth of the field of mHealth. Along with perceptions of the people who use these systems, it is also important to see the improvements in the health service delivery after the continuous use of mHealth applications. Call for quality studies should result in research in fields of CSCW (Computer Supported Co-operative Work), HCI (Human Computer Interaction) and Anthropology to study the behavior of health workers with mobiles. There are some very limited conclusions that have been made by studies on quality of the mHealth applications by independent researchers and organizations. What methods to use, what kind of studies will be required for evaluating the quality of mHealth applications is an interesting and emerging area of research.

## CONCLUSION

In this chapter, it is evident that considering the available infrastructure is critical to the success of implementation of mHealth solutions. Through the cases of two large-scale, successful mHealth applications, it can be seen that from the planning and design phases of mHealth projects the socio-technical networks needs to be studied as the infrastructure. The inertia of existing work practices and systems need to be dealt with and integration with existing systems is crucial for scaling mHealth solutions. We also see that minimizing risks and managing complexity correctly is another step in building scalable mHealth solutions.

## REFERENCES

Akter, S., D'Ambra, J., & Ray, P. (2011). Trustworthiness in mHealth information services: An assessment of a hierarchical model with mediating and moderating effects using Partial Least Squares (PLS). *Journal of the American Society for Information Science and Technology, 62*(1). doi:10.1002/asi.21442

Anderson & Perin. (2009). *Case studies from the Vital Wave mHealth report.* Retrieved November 10, 2010, from http://www.cs.washington.edu/homes/anderson/docs/2009/ mHealthAnalysis_v1.pdf

Banks, K. (2007). Then came the Nigerian elections: The story of frontline SMS. *SAUTI: The Stanford Journal of African Studies, (Spring/Fall)*, 1–4.

Berg, M., Wariero, J., & Modi, V. (2009). *Every child counts - The use of SMS in Kenya to support community based management of acute malnutrition and malaria in children under five.* Retrieved December 25, 2010, from http://www.childcount.org/reports/ ChildCount_Kenya_InitialReport.pdf

Bjerknes, G., Ehn, P., Kyng, M., & Nygaard, K. (1987). *Computers and democracy: A Scandinavian challenge.* Gower Pub Co.

Boyer, J. (2007) *XForms 1.0* (3rd ed.). W3C. Retrieved December 25, 2010, from http://www.w3.org/TR/xforms/

Braa, J., & Humberto, M. (2007). *Building collaborative networks in Africa on Health Information Systems and open source software development– Experiences from the HISP/BEANISH Network.* BEANISH Network.

Braa, K., & Purkayastha, S. (2010). Sustainable mobile information infrastructures in low resource settings. *Studies in Health Technology and Informatics, 157*, 127.

Consulting, V. W. (2009). *mHealth for development: The opportunity of mobile technology for healthcare in the developing world.* Washington, DC: UN Foundation-Vodafone Foundation Partnership.

Donner, J. (2008). Research approaches to mobile use in the developing world: A review of the literature. *The Information Society, 24*(3), 140–159. doi:10.1080/01972240802019970

Federal Communications Commission (FCC). (2010). *Report and order on open Internet rules* Federal Communications Commission, December 2010, Washington, DC. Retrieved from http://www.fcc.gov/Daily_Releases/Daily_Business/2010/db1223/FCC-10-201A1.pdf

Feller, J., & Fitzgerald, B. (2002). *Understanding open source software development.* Boston, MA: Addison-Wesley Longman Publishing Co., Inc.

Ganapathy, K., & Ravindra, A. (2008). *mHealth: A potential tool for health care delivery in India.* Rockefeller Foundation.

Greenbaum, J. M., & Kyng, M. (1991). *Design at work: Cooperative design of computer systems.* CRC.

Hanseth, O., & Ciborra, C. (2007). *Risk, complexity and ICT.* Edward Elgar Publishing.

Hanseth, O., & Monteiro, E. (2001). *Understanding information infrastructure.* Retrieved from http://www.ifi.uio.no/*oleha/Publications/bok.html

ITU (International Telecommunication Union). (2009). *World telecommunication/ICT indicators database 2009.* Geneva, Switzerland: International Telecommunication Union (ITU). Retrieved from www.itu.int/ITU-D/ict

ITU (International Telecommunication Union). (2010). *World telecommunication/ICT indicators database 2010.* Geneva, Switzerland: International Telecommunication Union (ITU). Retrieved from www.itu.int/ITU-D/ict

Klungsøyr, J., Wakholi, P., Macleod, B., Escudero-Pascual, A., & Lesh, N. (2008). *Open-ROSA, JavaROSA, GloballyMobile - Collaborations around open standards for mobile applications.* International Conference on M4D Mobile Communication Technology for Development, Karlstad University, Sweden.

Kumar, K., & van Hillegersberg, J. (2000). ERP experiences and evolution. *Communications of the ACM, 43*(4), 22–26. doi:10.1145/332051.332063

Monteiro, E., & Hanseth, O. (1995). Social shaping of information infrastructure: On being specific about the technology. *Information Technology and Changes in Organizational work. Proceedings of the IFIP WG8. 2 Working Conference on Information Technology and Changes in Organizational work, December 1995* (pp. 325–343).

Mukherjee, A., & Purkayastha, S. (2010). Exploring the potential and challenges of using mobile based technology in strengthening health information systems: Experiences from a pilot study. *AMCIS 2010 Proceedings*, (p. 263).

Pyramid Research. (2010). *Health check: Key players in mobile healthcare*. Pyramid Research. Retrieved from http://www.pyramidresearch.com/store/RPMHEALTH.htm

Raento, M., Oulasvirta, A., & Eagle, N. (2009). Smartphones. *Sociological Methods & Research, 37*(3), 426. doi:10.1177/0049124108330005

Raymond, E. S. (2001). *The cathedral and the Bazaar: Musings on Linux and open source by an accidental revolutionary*. Sebastopol, CA: O'Reilly & Associates, Inc.

Sandberg, A. (1985). *Socio-technical design, trade union strategies and action research. Research Methods in Information Systems* (pp. 79–92). Amsterdam, The Netherlands: North-Holland.

Sen, A. (1999). *Development as freedom*. Oxford University Press.

Susman, G. I., & Evered, R. D. (1978). An assessment of the scientific merits of action research. *Administrative Science Quarterly, 23*(4), 582–603. doi:10.2307/2392581

TRAI (Telecom Regulatory Authority of India). (2010). *Annual telecom report*. India: Ministry of Telecom.

Walsham, G. (1995). Interpretive case studies in IS research: Nature and method. *European Journal of Information Systems, 4*(2), 74–81. doi:10.1057/ejis.1995.9

Warschauer, M. (2004). *Technology and social inclusion: Rethinking the digital divide*. The MIT Press.

## ADDITIONAL READING

Asangansi, I., & Braa, K. (2010). The Emergence of mobile-supported National Health Information Systems in developing countries. *Studies in Health Technology and Informatics, 160*, 540.

Braa, J., Hanseth, O., Heywood, A., Mohammed, W., & Shaw, V. (2007). Developing Health Information Systems in developing countries: The flexible standards strategy. *Management Information Systems Quarterly, 31*(2), 381–402.

Braa, J., Kanter, A. S., Lesh, N., Crichton, R., Jolliffe, B., Sæbø, J., Kossi, E., et al. (2010). Comprehensive Yet Scalable Health Information Systems for Low Resource Settings: A Collaborative Effort in Sierra Leone, *2010*, 372-376.

Braa, J., Monteiro, E., & Sahay, S. (2004). Networks of Action: Sustainable Health Information Systems across Developing Countries. *Management Information Systems Quarterly, 28*(3), 337–362.

Bults, R., Wac, K., Van Halteren, A., Nicola, V., & Konstantas, D. (2005). Goodput Analysis of 3G Wireless Networks supporting m-Health services. *8th International Conference on Telecommunications (ConTEL05)*.

Ciborra, C. U., & Hanseth, O. (1998). From tool to: Agendas for managing the information infrastructure. *Information Technology & People, 11*(4), 305–327. doi:10.1108/09593849810246129

Dick, M. H. (2010). Weaving the "Mobile Web" in the Context of ICT4D: A Preliminary Exploration of the State of the Art. In. *Proceedings of the American Society for Information Science and Technology, 47*(1), 1–7. doi:10.1002/meet.14504701097

Donner, J. (2004). Innovations in mobile-based public health information systems in the developing world: An example from Rwanda. *Retrieved-November18*, 2008.

Donner, J., Verclas, K., & Toyama, K. (2008). Reflections on MobileActive08 and the M4D Landscape. In *Proceedings of the First International Conference on M4D* (pp. 73–83).

Fraser, H. S. F., & Blaya, J. (2010). Implementing Medical Information Systems in developing countries, what works and what doesn't. *AMIA... Annual Symposium Proceedings / AMIA Symposium. AMIA Symposium, 2010*, 232.

Hanseth, O., & Aanestad, M. (2003). Bootstrapping Networks, Communities and Infrastructures. On the evolution of ICT solutions in Health Care. *Methods of Information in Medicine, 42*(4), 385–391.

Hanseth, O., & Monteiro, E. (1997). Inscribing behaviour in Information Infrastructure Standards. *Accounting Management and Information Technologies, 7*, 183–212. doi:10.1016/S0959-8022(97)00008-8

Hanseth, O., Monteiro, E., & Hatling, M. (1996). Developing Information Infrastructure: The tension between standardization and flexibility. *Science, Technology & Human Values, 21*(4), 407. doi:10.1177/016224399602100402

Haux, R. (2006). Health Information Systems-past, present, future. *International Journal of Medical Informatics, 75*(3-4), 268–281. doi:10.1016/j.ijmedinf.2005.08.002

Heeks, R. (2002). Information systems and developing countries: Failure, Success, and Local Improvisations. *The Information Society, 18*(2), 101–112. doi:10.1080/01972240290075039

Hoe, N. S. (2006). *Breaking barriers: The potential of Free and Open source software for Sustainable Human Development; A compilation of case studies from across the world.* New Delhi, IN: Elsevier.

Istepanian, R. S. H., Jovanov, E., & Zhang, Y. T. (2004). Guest editorial introduction to the special section on m-health: Beyond seamless mobility and global wireless health-care connectivity. *Information Technology in Biomedicine. IEEE Transactions on, 8*(4), 405–414.

Istepanian, R. S. H., & Pattichis, C. S. (2006). *M-health: Emerging Mobile Health Systems.* Springer-Verlag New York Inc.

Ling, R. S. (2004). *The Mobile Connection: The cell phone's impact on society.* Morgan Kaufmann Pub.

Mansell, R. (2002). From digital divides to digital entitlements in knowledge societies. *Current Sociology, 50*(3), 407. doi:10.1177/0011392102050003007

Mishra, S., & Singh, I. P. (2008). mHealth: A developing country perspective. *Making the eHealth connection.* Bellagio, Italy.

Monteiro, E. (2000). Actor-network theory and Information Infrastructure. *From Control to Drift*, 71–83.

Sahay, S., & Walsham, G. (2006). Scaling of Health Information Systems in India: Challenges and Approaches. *Information Technology for Development, 12*(3), 185–200. doi:10.1002/itdj.20041

Sarker, S., & Wells, J. D. (2003). Understanding mobile handheld device use and adoption. *Communications of the ACM, 46*(12), 35–40. doi:10.1145/953460.953484

Tongia, R., & Subrahmanian, E. (2007). Information and Communications Technology for Development (ICT4D)-A Design Challenge? *Information and Communication Technologies and Development, 2006. ICTD, 06,* 243–255.

## KEY TERMS AND DEFINITIONS

**Auxilliary Nurse Midwife (ANM):** The peripheral health worker who provides care to childbearing women during pregnancy, labour and birth, and during the postpartum period. They also care for the newborn and assist the mother with breastfeeding as well as provide family planning knowledge and counseling.

**Complexity:** It is the measure of the number of linkages to properties in an object. This collection of properties is known also known as state.

**Free & Open-Source Software:** Software that is liberally licensed to grant the right of users to use, study, change, and improve its design through the availability of its source code.

**Health Information Systems:** Information systems that capture data related to health of an individual, community or an entire nation and allow meaningful use as information.

**Information Infrastructure:** A structure of the people, processes, procedures, tools, facilities, and technology which supports the creation, use, transport, storage, and destruction of information.

**mHealth (or m-Health):** Provisioning of health services through the use of mobile technology is called mHealth. mHealth has also been referred to as Mobile Health Information Systems.

**Scalability:** The ability of a system, network, or process, to handle growing amounts of work in a graceful manner or its ability to be enlarged to accommodate that growth.

**SMS:** Short Messaging System is the text communication standard in phone or mobile systems allowing non-voice or non-data communication between different users of telephone networks.

# Chapter 7
# Weather Nowcasting Using Environmental Sensors Integrated to the Mobile

**Srinivasa K G**
*M S Ramaiah Institute of Technology, India*

**Arhatha B**
*M S Ramaiah Institute of Technology, India*

**Harsha R**
*M S Ramaiah Institute of Technology, India*

**Abhishek S C**
*R V College of Engineering, India*

**Sunil Kumar N**
*M S Ramaiah Institute of Technology, India*

**Harish Raddi C S**
*M S Ramaiah Institute of Technology, India*

**Anil Kumar M**
*M S Ramaiah Institute of Technology, India*

## ABSTRACT

*With the vagaries of nature being unpredictable, it's now more important to have access to weather forecast for short periods of time. Many businesses, including those in agriculture and the fishing industry, depend on an hourly update of the weather. The access to such weather nowcasting data has, until now, been through traditional media like the television, radio, et cetera, while new media of communication such as mobile devices, have been largely unexplored. The advancement in MEMS (Micro Electrical Mechanical System) technology has now brought forth various sensors that are miniaturized and can be integrated or embedded into various other systems used presently. Environmental sensors that measure weather parameters are miniaturized to fit the size of a mobile. The system aims to integrate these sensors along with a mobile device so as to provide the capability of data measurement to the vast population that use mobile devices and thus create regional grid networks. The system aims to use the mobile for updating weather parameters as well to be the focal point of communication of the weather nowcasting information. As a result, the mobile device would provide targeted distribution of the weather information, which is more advantageous than the traditional means of mass distribution of information; also, as mobile technology acts as a focal point of gathering weather related parameters, it provides a twofold advantage for setting up a low cost, region specific weather monitoring system.*

DOI: 10.4018/978-1-4666-0080-5.ch007

## INTRODUCTION

Weather forecasting is a technology used to predict the atmospheric state for a future time for a given location. Prior to nineteenth century informal weather prediction methods were common which were succeeded by formal methodologies. Weather forecasts are made by collecting quantitative data about the current state of the atmosphere and using scientific understanding of atmospheric processes to project how the atmosphere will evolve.

Nowcasting comprises the detailed description of the current weather along with forecasts obtained by extrapolation for a period of 0 to 6 hours ahead. In this time range it is possible to forecast small features such as individual storms with reasonable accuracy. A forecaster using the latest radar, satellite and observational data is able to make analysis of the small-scale features present in a small area such as a city and make an accurate forecast for the following few hours. It is, therefore, a powerful tool in warning the public of hazardous, high-impact weather including tropical cyclones, thunderstorms and tornadoes which cause flash floods, lightning strikes and destructive winds. In broad terms, nowcasting contributes to the:

1.  reduction of fatalities and injuries due to weather hazards
2.  reduction of private, public, and industrial, property damage
3.  Improved efficiency and savings for industry, transportation and agriculture.

Over the years there have been several methods of data aggregation and corresponding models to forecast weather. Most of them rely heavily on manual observatories and also satellite imagery (Nehrkorn et al., 1993). In a country like India where investment on manual or automatic weather observatories is a costly investment option there is a need for an alternate method of aggregation of weather parameters.

The most widely used weather prediction model is the computer simulations of the atmosphere. They take the recorded data as the starting point and evolve the state of the atmosphere forward in time using physics and fluid dynamics. The complicated equations which govern how the state of a fluid changes with time require supercomputers to solve them. The output from the model provides the basis of the weather forecast. Most end users of this forecasting are the general public. Knowing that forecasting is of great use to sectors like agriculture where in farmers rely on weather forecasts to decide what work to do on any particular day. The major setback faced by the farmers is that they do not have easy access to the forecast at any point of time. Also the present forecasting is more on a general scale and does not provide the farmers with an accurate localized weather report. With this idea system mainly aims at addressing these issues faced by the agriculture sector.

There is a distinct need for better data aggregation of the weather parameters without having the cost burden to improve the weather forecast in India. Nowcasting is essential in developing countries like India, where the harvest season depends on the vagaries of nature and a sudden torrential downpour destroys crops worth millions of rupees. Nowcasting provides weather information for the short period of 0-6 hours which is essential as a general weather forecast never provides granular weather information which is region specific. The infrastructure for such an exercise is missing in the Indian subcontinent and the cost of setting it up is not economical for any governing agency to bear it. Hence there is a need for setting up an alternative infrastructure where the people themselves participate to support weather data aggregation and are beneficiaries of the weather nowcasting information which is provided to them. This sustainable loop is essentially created in proposed system which will be described in the following sections. The mobile being a ubiquitous device is an excellent option which provides the distribution of weather related sensors and the ease of aggregation of data as the ease of communication is now increasing with the ever expanding capability of the telecommunication network. The mobile would also provide targeted distribution of the weather information which is advantageous than the traditional means of mass distribution of information.

This chapter is organized as follows. Background section deals with various nowcasting and weather forecasting approaches such as Radar-based nowcasting, Image sequence extrapolation and Spatio-temporal autoregressive technique testing on extensive data set of satellite image sequences covering different meteorological conditions. M2M-based machine and process control, building and facility automation and RFID-based traceability systems are also supported with relevant research work. Micro Electrical Mechanical System (MEMS) technology is enabling the development of inexpensive, autonomous wireless sensor nodes. The System overview includes various modules of the system such as Data aggregation sensors which includes MEMS components which serve the purposes of sensors, the activated MEMS sensors would measure the corresponding weather parameter and the output would be amplified In role of mobile (2G) description is provided on how it is used to collect the data, package it and transmit it via GPRS to a local server which acts as a node in the distributed computing network. In Centralized computational unit one of the main strategies of grid computing is to use middleware to divide and apportion pieces of a program. The data collected by the phone is transferred to a centralized computing unit which behaves as a node in the distributed computing environment. Data is stored in a database for efficient lookup. As there are no well defined algorithms for nowcasting, a heuristic learning based algorithm is used, which runs in two modes. In the Working details, explanation about the processing of the data collected by the MEMS device on the central computing. The

atom processor would also update the central supercomputer at frequent intervals of time is explained in the working state diagram. In the Specification section, the hardware and software requirements of the system have been described. Several real world Applications are provided in the Applications section. Future Research Directions is given next followed by Conclusion.

## BACKGROUND

In the over a few years, nowcasting has played an increasingly important role in the field of Meteorology. Significance of nowcasting weather events can be seen in matters that range from routine daily planning (Purdom, 1976) to pivotal situations such as hurricane preparedness (Goerss, Velden, & Hawkins, 1998). Radar-based nowcasting quickly developed to the use of real-time reflectivity pattern recognition techniques (Bellon & Austin, 1978). Methodologies such as image sequence extrapolation which is demonstrated, particularly for the purpose of near-real-time weather nowcasting during satellite launches. The highlight of this model is its ability to produce a sequence of simulated satellite images extended in time scale, which is very important for forecasting the evolution of a meteorological system. For this, three different models based on Spatio-Temporal Autoregressive Technique, Discrete Fourier Transform, and Hybrid Approach are developed and tested on an extensive data set of satellite image sequences covering different meteorological conditions (Shukla, Pal, & Joshi, 2010).

Geostationary Operational Environmental Satellites (GOES) sounder performance has mostly been meeting or exceeding expectations with regard to signal to noise, calibration, and navigation; noise in the longwave bands is greater than specified. The GOES sounder data and products are produced on a higher space and time scale than conventional observations. The overall quality of the GOES sounder measurements and the added information from the products has been verified by comparison to independent observations. Sounder products from both GOES-8 (at 75 W) and GOES-9 (at 135 W) are being used operationally by the National Weather Service. GOES soundings are available hourly to help assist the NWS produce accurate forecasts/nowcasts of severe weather (Paul et al., 1998).

Sensor networks applied in agriculture and food production for environmental monitoring, precision agriculture, M2M-based machine and process control, building and facility automation and RFID-based traceability systems are given in Ning and Naiqian (2006) and Huang, Tseng and Lo (2004). The application of MEMS technology is enabling the development of inexpensive, autonomous wireless sensor nodes with volumes ranging from cubic mm to several cubic cm.(Howard, Mataric, & Sukhatme, 2002) These tiny sensor nodes can form rapidly deployed, massive distributed networks to allow unobtrusive, spatially dense, sensing and communication. MEMS enable these devices by reducing both the volume and energy consumption of various components (Warneke & Pister, 2002).

Radar-meteorology is another popular approach in collection of data for weather nowcasting for predicting based on the cloud, data collection is based on a complex physical process of interaction between cloud droplets, snowflakes, rain drops and electromagnetic waves (Austin & Bellon, 1982). Presently satellites allow the macroscopic evaluation of many atmospheric parameters, the meteorological radar provides detailed information about the composition of clouds and precipitation. As a result the meteorological radar is a fundamental instrument for the operational very short term nowcasting. Other popular approach is Meteorology from satellite, using geostationary satellites, such as Meteosat 8, have the unique possibility to observe the atmosphere and its cloud coverage at every 15 minutes (Shukla, Pal, & Joshi, 2010; Benjamin et al., 1998). Therefore, the data collected from

their sensors constitute the essential tool for the nowcasting observations and meteorological for short period forecasts.

## SYSTEM OVERVIEW

The system uses the current supporting infrastructure of mobile phones to aggregate and transfer the measured parameters to a centralized computational server (Figure 1 and Figure 2).

MEMS system is present as an USB interfaced accessory for each mobile handset and used to measure the weather parameters. The MEMS system would measure the following parameters.

- Air temperature
- Relative humidity
- Barometric pressure
- Wind direction
- Wind speed
- Sunlight intensity

The measured parameters would then be transferred to the centralized computational unit via the WAP through the customer network operator. This unit would receive and update the data from multiple sources and simultaneously process the received data to update the weather forecasts. A request via the network operator as a Value Added Service (VAS) would then send a request to the centralized unit for the latest weather reports. The received information is transferred to the customer through the network operator thus enabling the farmer and fishermen to access latest weather reports using a mobile.

The computing framework facilitates collection of data, aggregation, forecast model execution and publishing the results. Data normalization is done with respect to geographical location determined using existing referential coordinate systems. As there are no weather forecasting models that have been built that can predict short term weather accurately, the heuristic feedback based

*Figure 1. The overview of system*

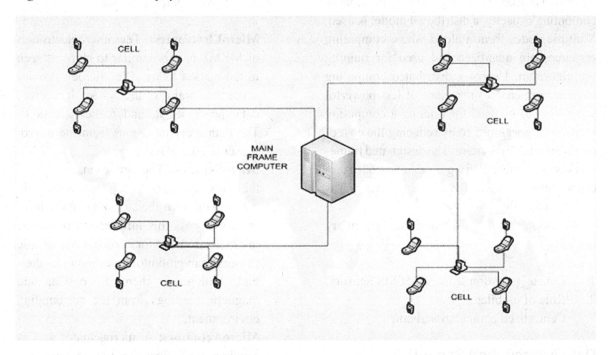

*Figure 2. The cell view of system*

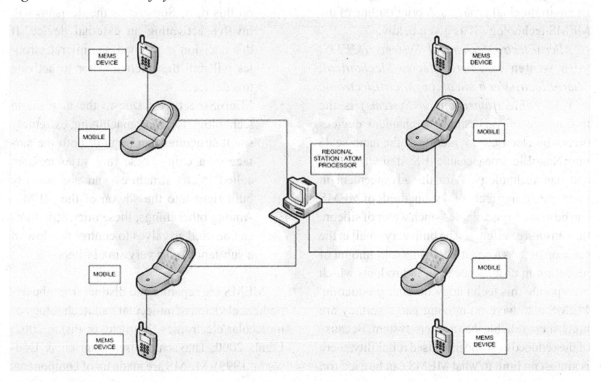

algorithm is used. As this algorithm requires large computing capacity, a distributed model is used. Multiple nodes form a cloud where computing resources are quantized and used for running the algorithm. Use of a distributed computing environment enables the usage of less powerful processors to be used for running a computing intensive algorithm thereby reducing the cost of deployment of the system. The designated framework is implemented using a modular strategy, in order to ensure an easy functionality extension as well as scalability.

The system can be divided into 3 primary modules.

1. Data aggregation sensors (MEMS Sensors)
2. Role of mobiles
3. Centralized computational unit

## Data Aggregation Sensors

The data aggregation sensors are built by Crossbow company which uses MEMS technology to build its environmental products. A brief outline of the MEMS technology is as given below:

*MicroElectroMechanical Systems (MEMS)* (also written as *Micro-Electro-Mechanical, MicroElectroMechanical* or *MicroElectronic and MicroElectromechanical Systems*) is the technology of very small mechanical devices driven by electricity; it merges at the nano-scale into NanoElectroMechanical Systems (NEMS) and nanotechnology. With the advancement in micromachining technology, hundreds of MEMS can be made from a single 8-inch wafer of silicon, thus an entire system can be built very small in the scale of $10^{-6}$. As a result a considerable amount of reduction in cost can be seen in products which incorporate this technology on bulk production. MEMS also have no moving parts, so they are much more reliable than a macro system. Because of the reduced cost and increased reliability, there is almost no limit to what MEMS can be used for.

Basic MEMS components are:

- **MicroElectronics:** The microelectronics of MEMS is very similar to chips as seen in the market today. The microelectronics act as the "brain" of the system. It receives data, processes it, and makes decisions. The data received comes from the microsensors in the MEMS.

- **MicroSensors**: The microsensors act as the arms, eyes, nose, etc. They constantly gather data from the surrounding environment and pass this information on to the microelectronics for processing. These sensors can monitor mechanical, thermal, biological, chemical, optical and magnetic readings from the surrounding environment.

- **MicroActuators**: A microactuator acts as a switch or a trigger to activate an external device. As the microelectronics is processing the data received from the microsensors makes decisions on what to do based on this data. Sometimes the decision will involve activating an external device. If this decision is reached, the micrelectronics will tell the microactuator to activate this device.

- **MicroStructures**: Due to the increase in technology for micromachining, extremely small structures can be built onto the surface of a chip. These tiny structures are called micro structures and are actually built right into the silicon of the MEMS. Among other things, these microstructures can be used as valves to control the flow of a substance or as very small filters.

MEMS are separate and distinct from the hypothetical vision of molecular nanotechnology or molecular electronics (Warneke & Pister, 2002; Frank, 2000; Thaysen, Boisen, Hansen, & Bouwstra, 1999). MEMS are made up of components between 1 to 100 micrometres in size (i.e., 0.001 to 0.1mm) and MEMS devices generally range in size from 20 micrometres (20 millionths of a metre) to a millimetre. They usually consist of a central unit that processes data, the microprocessor and several components that interact with the peripheral devices such as microsensors. At these size scales, the standard constructs of classical physics are not always useful. Because of the large surface area to volume ratio of MEMS, surface effects such as electrostatics and wetting dominate volume effects such as inertia or thermal mass. The potential of very small machines was appreciated before the technology existed that could make them - for example, in the Richard Feynman's famous 1959 lecture on *There's Plenty of Room at the Bottom*. MEMS became practical once they could be fabricated using modified semiconductor device fabrication technologies, normally used to make electronics. These include molding and plating, wet etching and dry etching, Electro Discharge Machining (EDM), and other technologies capable of manufacturing small devices. An early example of a MEMS device is the resonistor – an electromechanical monolithic resonator.

The basic techniques for producing all silicon based MEMS devices are deposition of material layers, patterning of these layers by photolithography and then etching to produce the required shapes. MEMS devices can also be made from polymers by processes such as injection molding, embossing or stereolithography (Rotting, Ropke, Becker, & Gartner, 2001).

## MEMS Module

The output of the MEMS sensors to measure the different weather parameters are amplified and fed to the Multiplexer. The multiplexer output is fed to a Band-pass filter whose band-pass frequency is controlled by the 8 bit microcontroller. The output of the band-pass filter would be fed to the Analog-to-Digital Converter (ADC) which would

send the digital output to the microcontroller for storage and further retrieval. The microcontroller controls the operation of the MEMS module. Each of the MEMS sensors are activated serially for a certain period of time. The activated MEMS sensors would measure the corresponding weather parameter and the output would be amplified. The microcontroller would determine the center-pass or the band-pass frequency for each of the MEMS sensor thus eliminating noise from the output of the MEMS sensor. The microcontroller would store the output data parameters and after the full cycle of operation is complete, would send a signal to the mobile to retrieve the data parameters. The retrieval of the data parameters would be through a USB data cable (Figure 3).

## Role of the Mobile

2G is Second Generation (2G) telephone technology based on GSM or in other words Global System for Mobile communication. Second generation 2G cellular telecom networks were commercially launched on the GSM standard in Finland by Radiolinja (now part of Elisa Oyj) in 1991. Three primary benefits of 2G networks over their predecessors were that phone conversations were digitally encrypted; 2G systems were significantly more efficient on the spectrum allowing for far greater mobile phone penetration levels; and 2G introduced data services for mobile, starting with SMS text messages (Kar & Banerjee, 2003).

International Mobile Telecommunications - 2000 (IMT - 2000), better known as 3G or 3rd generation, is a generation of standards for mobile phones and mobile telecommunications services fulfilling specifications by the International Telecommunication Union. Application services include wide-area wireless voice telephone, mobile Internet access, video calls and mobile TV, all in a mobile environment. Compared to the older 2G and 2.5G standards, a 3G system must provide peak data rates of at least 200kbit/s according to the IMT-2000 specification. Recent 3G releases

*Figure 3. MEMS sensor system layout*

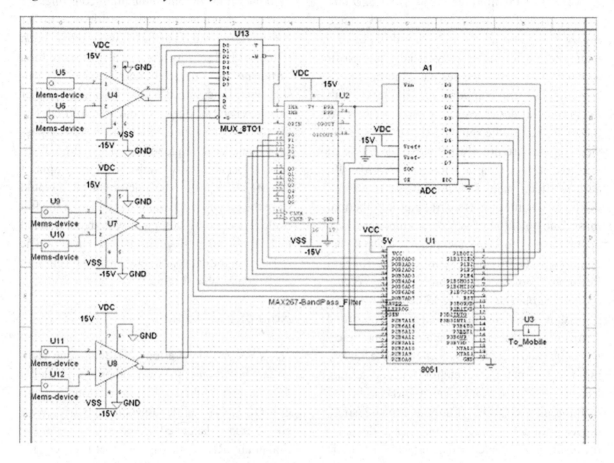

often denoted as 3.5G and 3.75G also provide mobile broadband access of several Mbit/s to laptop computers and Smartphones.

The role of a 2G or 3G enabled mobile phone is twofold. Firstly it is used to collect the data, package it and transmit it via GPRS to a local server which acts as a node in the distributed computing network (Michelini, Hijazi, Nassar, & Zhiqiang, 2003). The MEMS device is connected to the mobile phone as an add-on accessory, from which data is periodically collected and aggregated by a JME based application that runs as a daemon process (Adrian, 2001). The geographical location can be identified using either the GPS (Global Positioning System) module of the phone, or identifying the first hop tower. In phones that don't have support for GPRS, the data is normalized and packaged

unto a SMS and transmitted. Secondly, another JME application can be installed on the phone that can retrieve the results of the forecasting model from the centralized computing unit. This can also be provided as a Value added service by network operators to those subscribers who do not contribute to data collection (Figure 4).

Data collected via MEMS devices is collected by mobile devices; the collected data includes temperature details, wind details. In mobile end of the system the processing takes place in the following steps: The status of the device is checked for hardware and software based errors, such errors are prompted via a proper fail safe mechanism to get out risk. In case of hardware based error the user is prompted as to which component in the device is not working or failed. In case of

*Figure 4. State diagram of mobile end*

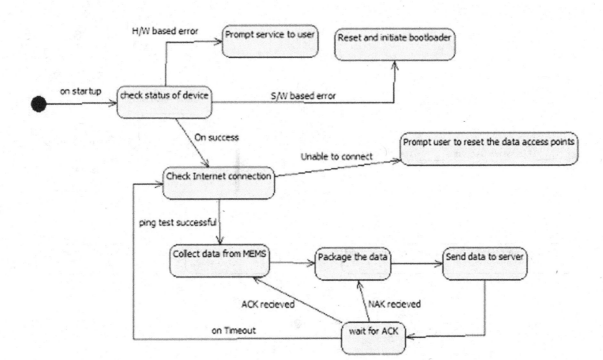

software based error the device is reset to initiate a bootloader. On success the status of internet connection is checked, here again the user is prompted if there are any connection errors, so that the user can check status of access points. On successful ping, the collected data from the above mentioned MEMS device is packaged and sent to the server. This mechanism is coupled with appropriate acknowledgement and no acknowledgement signals, if the above process reaches a time out, the entire process is restarted.

## Centralized Computing Unit

The objective of proposed system is to create a grid computing network ultimately which would replicate the behavior of a supercomputer by use of processors used in mobile. The processors recently developed by Qualcomm and other notable OEM of mobile processors is powerful enough to be comparable to desktop computers.

Their usage has been mainly on the graphics part which enables applications to run high definition and video content on mobiles. This architecture in recent processors can be used to run applications which would enable grid computing to be a reality. A note on grid computing is as below.

Grid computing is a term referring to the combination of computer resources from multiple administrative domains to reach a common goal. The grid can be thought of as a distributed system with non-interactive workloads that involve a large number of files. What distinguishes grid computing from conventional high performance computing systems such as cluster computing is that grids tend to be more loosely coupled, heterogeneous, and geographically dispersed. Although a grid can be dedicated to a specialized application, it is more common that a single grid will be used for a variety of different purposes. Grids are often constructed with the aid of general-purpose grid software libraries known as middleware.

*Figure 5. State diagram of the server end*

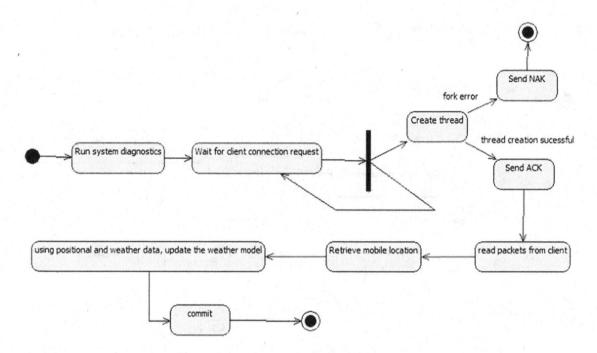

Grid size can vary by a considerable amount. Grids are a form of distributed computing whereby a *super virtual computer* is composed of many networked loosely coupled computers acting together to perform very large tasks. Furthermore, *distributed* or *grid* computing, in general, is a special type of parallel computing that relies on complete computers (with onboard CPUs, storage, power supplies, network interfaces, etc.) connected to a network (Private, Public or the Internet) by a conventional network interface, such as Ethernet. This is in contrast to the traditional notion of a supercomputer, which has many processors connected by a local high-speed computer bus.

Grid computing combines computers from multiple administrative domains to reach a common goal to solve a single task, and may then disappear just as quickly. One of the main strategies of grid computing is to use middleware to divide and apportion pieces of a program among several computers, sometimes up to many thousands. Grid computing involves computation in a distributed fashion, which may also involve the aggregation of large-scale cluster computing based systems as shown in Figure 5.

The size of a grid may vary from small - confined to a network of computer workstations within a corporation, for example - to large, public collaborations across many companies and networks. *The notion of a confined grid may also be known as intra-nodes cooperation whilst the notion of a larger, wider grid may thus refer to inter-nodes cooperation.* Grids are a form of distributed computing whereby a "super virtual computer" is composed of many networked loosely coupled computers acting together to perform very large tasks. This technology has been applied to computationally intensive scientific, mathematical, and academic problems through volunteer computing, and it is used in commercial enterprises for such diverse applications as drug discovery, economic forecasting, seismic analysis, and back office data processing in support for e-commerce and web services. The details of the application that is developed is as given below.

The data collected by the phone is transferred to a centralized computing unit which behaves as a node in the distributed computing environment. Data is stored in a database for efficient lookup. As there are no well defined algorithms for nowcasting, a heuristic learning based algorithm is used, which runs in 2 modes. In the learning mode, it forms numerous patterns from the data collected. As it has been proven that weather patterns are fairly accurate, the system intends to logically map weather patterns. Weight-ages are assigned to each path in the pattern depending on the occurrence of the pattern. The pattern is stored in the form of graphs with nodes defining the various weather states. The measured weather parameters are quantized to ensure that the system has a finite set of state.

Weather parameters received at a finer geographical level as well as those calculated for larger granular regions are considered to define each state. As wind direction is highly influential in the change in weather conditions, the direction of the air currents are used to determine the impact of neighboring geographical regions. As air currents change gradually, it can be used as the change in weather parameters (deltas) are used to define the paths. Before the system is deployed, data from previous months is fed, however the geographical locations are granular and hence this is used as a global initialization graph for all levels. On deployment the system utilizes this set of patterns which are then modified at a finer geographical level, as weather patterns are not highly accurate for large geographical regions. As data is received, the patterns are updated. If the system receives a request for weather forecast then these patterns are traversed. Depending on the current weather condition at that geographical location, a weather state closest to it is identified and the path it will traverse is then identified using standard graph traversal algorithms. The path with the highest weight-age is considered. The weather parameters

are then inferred from the destination's weather state. This useful data is then returned through a SMS or GPRS services to the application that requested it.

On the server end the system diagnostics tests are run which is followed by the acknowledgement from the client for successful connection setup. Thread based processing is done on the server end where the number of threads correspond to the clients in the connection. Each thread reads the packets from the client. The data from the mobile is processed to obtain the location of the mobile device, using the positional and weather data the weather model is updated and committed.

## WORKING

Once the MEMS module is initiated the weather conditions are measured by the MEMS devices present in the module. These parameters are then stored in the microcontroller. After the operation is complete the mobile would automatically retrieve these data parameters from the microcontroller and transfer via the mobile web. The data would be sent to a particular IP address which also serves as the address of the regional atom processor. Each of the atom processor would be assigned a unique id which would be based on the geographical location. The atom processor would receive the multiple sources of data serially and update the data aggregation table which is quantized on an hourly basis. After every hour the parameters are averaged and this average value is stored in the updated data aggregation table. A copy of the updated table would be sent to the central supercomputer. The atom processors would also request for other weather parameters apart from those measured by the MEMS module for further processing of data. The data communication would be through the internet. This could be either through LAN or through the mobile web. The request would be segregated on the type of request and on priority basis to avoid overwhelming of the system. The

*Figure 6. System state diagram*

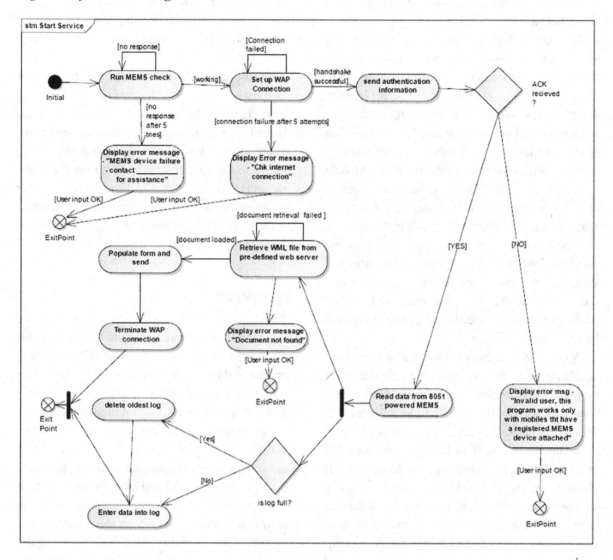

atom processor would also update the central supercomputer at frequent intervals of time. The whole system state diagram is shown in Figure 6.

The mobile would initially retrieve the weather data parameters and would then initiate the WAP connection. If the connection is established with the web server then the request for the WML document containing the weather data parameters would be requested. Upon connection failure the operation to setup the connection is tried for five more times after which the entire operation is suspended for a certain period of time. On suc-

cessful establishment of the connection the atom processor would then retrieve this WML document from the predefined web server and update its database. The WAP connection is then terminated after the operation is successfully performed.

## Data Transfer to Mobiles with WAP Technology

WAP, Wireless Application Protocol is a technology for delivering information to wireless devices (primarily Mobile phones, but also Pagers and

Personal Digital Assistants). WAP is a collection of protocols and specifications that work together to give mobile phones access to Internet-like information. One part of the WAP standard is the Wireless Markup Language (WML), which is used to create wireless applications just like HTML that is used to create web pages. WML is an application of XML, meaning that it is defined in a document type definition.

The reason for using WAP is not to port existing web sites onto mobile phones, as the small limited bandwidth and screen size make it impractical. Rather, it is a system for creating special applications for the handheld devices. WAP is good for delivering short, tiny bits of data, such as temperature details, humidity details, intensity of sunlight, wind speed and so on. It is not very useful for the complex data with visual layouts that have made the World Wide Web what it is today. A WAP based application is formatted using WMLScript and WML, the scripting language for adding interactivity to wireless applications. WAP-enabled devices communicate with the web server through a WAP gateway. Internet and the mobile network are supported by an intermediate WAP gateway, converting WAP requests into HTTP requests. A wireless device (such as a mobile phone) requests information via the airwaves to the gateway. The WML document from the web server is received by the gateway through HTTP Request.

The document received by the gateway, gets compiled into WAP Binary XML (WBXML) before it is sent to the phone. This compilation process reduces the file size by replacing tags with specific single-character codes, and removing comments and extra white space. Once the document has been compiled, it is returned via the airwaves to the mobile phone. WAP gateways are normally owned by wireless service providers, so you don't need your own gateway to provide WAP applications.

## Sample use Case Scenario

Consider a farmer or a fisherman X. X is provided with a MEMS system which is basically a device which contains a series of MEMS modules measuring each of the aforementioned parameters. The system is interfaced with the mobile as an USB accessory. These devices would be monitored by a microcontroller which would multiplex these devices to function serially once activated. It would then store the measured parameters in its cache memory. After all the parameters are measured the system would automatically connect to the internet via the network operator. With the onset of 3G technology data transfer from a mobile can be achieved at much higher speeds. The measured parameters are then transferred to the central computing unit through the internet. Since every mobile transfers data through established network towers it is possible to approximately determine the location and time from where the data has been sent.

The central computing unit receives the data from X, updates its log and processes the data to update the weather forecasts. Considering 5000 farmers from a region of 20sqkm to send the data, the central computing unit would receive the measured weather parameters from multiple sources located within a specific region. Thus by averaging of the received data for a certain parameter there is less room for error and accuracy is not compromised. The central processor would provide the atom processor with general weather parameters measured and the atom processor would be aided with the received local parameters to make region specific weather forecasts. The atom processor would also send the received local weather data to the central processor. Hence, multiple atom processors would update the central processor to give a comprehensive picture of the weather patterns across the country. Now, farmer X would request the weather forecast as a VAS through the network operator. The network operator would then send a request to the specific atom processor for furnishing the information of the same. The region specific weather forecasts along with the general weather pattern predictions across the country would then be provided to the farmer via the network operator thus completing the cycle.

## SYSTEM SPECIFICATIONS

### MEMS Module

Hardware requirements:

- MEMS sensors for measuring the required parameters
- Pre-amplifiers - lm221/lm321
- Max-267 universal and band pass filter
- Intel 8-bit control oriented microcontroller
- 8-bit up compatible analog-to-digital converters with 8-channel
- Multiplexer: adc0808 from national semiconductor
- 8:1 multiplexer
- USB cable

The MEMS sensors would be procured for measuring the weather parameters as listed initially in the abstract. The MEMS output is usually in the range of micro volts. The output voltage would be initially passed through a pre-amplifier and these would be sent to a 8:1 multiplexer which would be then connected to a band-pass filter. This is to ensure that the noise is eliminated from the output signal. The output is then converted to a digital form using an ADC. The microcontroller would store the data parameter for further retrieval from the mobile. The microcontroller would also control the sequence of operation of the MEMS sensors and the band-pass frequency to eliminate the noise. The data parameters would be stored in the microcontroller after each of the MEMS sensors collect data from the surrounding environment. The microcontroller would then initiate the mobile to retrieve the data. C language would be used to program the microcontroller to perform the above mentioned operations.

## HARDWARE REQUIREMENTS

- Mobile with 3G technology enabled.
- Web server powered by the Atom Processor N270.
- Other peripheral devices as required.

The preferred mobile for implementing the system would be 3G enabled as this helps in faster data transfer. GPRS enabled phones would help in aggregation and transfer of other crucial weather parameters like the vertical profile of the atmosphere at a particular place. This would help in the weather forecast of thunderstorms. The mobile would be JavaME enabled to help run the application for automatic retrieval and transfer of data parameters.

### JavaME Platform for Converged Services

The JavaME platform covers everything from small limited devices with intermittent network connection to capable on-line mobile devices. The platform's design enables it to flexibly and efficiently support the need for services covering all mobility channels. Services are easily portable between different configurations and profiles, and the same service can be delivered via different channels.

### Java Device Test Suite (JDTS)

The Java Device Test Suite simplifies quality assurance and reduces time-to-market for JavaME software implementations by providing comprehensive tests and a robust test manager. These enable suite users to evaluate, validate, and verify the quality of implementations of the Connected Limited Device Configuration (CLDC) and the

Mobile Information Device Profile (MIDP) on a particular device. The Java Device Test Suite helps device manufacturers and service providers ensure their reputation for quality, while building customer satisfaction and loyalty.

## SOFTWARE REQUIREMENTS

The client device must be a Java and GPRS/UMTS enabled device that supports the following specifications:

- Mobile Information Device Profile (MIDP) 2.0: MIDP is a specification published for the use of Java on embedded devices such as cell phones and PDAs. MIDP is part of the JavaME framework. MIDP 2.0 was developed under the Java community process as JSR-118.
- Either Connected Limited Device Configuration (CLDC) 1.1 or Connected Device Configuration (CDC) 1.1.2 CLDC is a specification of a framework for JavaME applications targeted at devices with very limited resources, such as pagers and mobile phones. CLDC 1.1 was developed under the Java community process as JSR-139.
- Scalable 2D vector graphics API for J2ME 1.1: JSR 226: The scalable 2D vector graphics API specification defines an API for rendering 2D graphics in the World Wide Web Consortium (W3C) Scalable Vector Graphics (SVG) tiny format. SVG makes it possible for developers to create interactive graphical content, with the ability to zoom and resize on displays with different resolutions and aspect ratios. JSR 226 also defines a subset of the Micro Document Object Model ($\mu$DOM) API to allow user interaction and dynamic manipulation of SVG content.

## INTEL ATOM AND ITS INTEGRATION

The Intel atom is a multithreaded processor with an in order pipeline. It supports two-way hyper-threading technology, with the key components highlighted, including the following four core clusters:

- FEC: front-end cluster (with l1 instruction cache)
- FPC: floating-point cluster
- IEC: integer execution cluster
- MEC: memory execution cluster (with l1 data cache)

In the atom core, the FEC is responsible for performing instruction fetch. Instructions flow from the 8-way set associative 32KB L1 i-cache into a set of fetch buffers, and from there into the decode unit. The processor's decode unit features two hardware decoders and one microcode decoder, and can decode up to two instructions per cycle. The FEC also includes the microcode ROM, which implements the more complex x86 instructions, e.g., Operations like REPNZ, repeat while not zero. The IEC and the FPC are responsible for integer and floating-point execution, respectively. On the data side, the address generation units in the MEC can perform two address calculations per cycle, and memory accesses are supported by the 6-way set-associative 24KB l1 D-Cache.

While the atom core consists of the four clusters listed above, all the remaining structures in the processor form what is referred to as the Uncore (The Uncore is a term used by Intel to describe the part of a microprocessor that is not the core). The uncore includes the l2 Cache, Fuses, Pads, and the Bus Interface Unit (BIU), which translates memory requests from the internal MEC format to the appropriate Front Side Bus (FSB) protocol. The BIU Cluster contains a 512KB on-die l2 Cache with in-line error correction, and a FSB interface that can perform 400mt/s or 533mt/s.

Out of a total of 47.2 million transistors, the four clusters in the atom core account for 13.8 million transistors. In comparison, the Pentium processor design used in has about 3.1 million transistors. The Intel atom processor is used to power the web server. The web server would be assigned a unique IP address which would receive the weather data parameters from different sources (using the mems module). The atom processor would perform all the operations as indicated above. It would handle mobile operator requests, receive and allocate the weather data parameters, retrieve and process these data parameters using a weather model to make weather forecasts. It would retrieve essential data parameters from the central supercomputer and update the same at regular intervals of time.

## APPLICATIONS

There are a variety of end uses to weather nowcasts.

- **Public uses:** The general public constitute the major section of people being benefited by nowcasting. A major part of modern weather forecasting is the weather alerts and advisories which the national weather services issue in case of severe or hazardous weather in order to protect life and property.
- **Air traffic:** Because the aviation industry is especially sensitive to the weather, accurate weather forecasting is essential. Fog or exceptionally low ceilings can prevent many aircraft from landing and taking off.
- **Marine:** Commercial and recreational use of waterways can be limited significantly by wind direction and speed, wave periodicity and heights, tides, and precipitation. These factors can each influence the safety of marine transit.

- **Agriculture:** Farmers rely on weather forecasts to decide what work to do on any particular day.
- **Forestry:** Weather forecasting of wind, precipitations and humidity is essential for preventing and controlling wildfires. Different indices, like the forest fire weather index and the Haines index, have been developed to predict the areas more at risk to experience fire from natural or human causes.
- **Utility companies:** Electricity and gas companies rely on weather forecasts to anticipate demand which can be strongly affected by the weather. They use the quantity termed the degree day to determine how strong of a use there will be for heating (heating degree day) or cooling (cooling degree day). By anticipating a surge in demand, utility companies can purchase additional supplies of power or natural gas before the price increases, or in some circumstances, supplies are restricted through the use of brownouts and blackouts.
- **Private sector:** Increasingly, private companies pay for weather forecasts tailored to their needs so that they can increase their profits or avoid large losses. For example, supermarket chains may change the stocks on their shelves in anticipation of different consumer spending habits in different weather conditions.
- **Military application:** Military weather forecasters present weather conditions to the war fighter community. Military weather forecasters provide pre-flight and in-flight weather briefs to pilots and provide real time resource protection services for military installations.

Apart from the direct advantages of weather forecasts as shown above, this system would be greatly helpful in making more accurate regional weather forecasts. An accurate weather forecast is very helpful for the farmers and this could lead to a better revenue generation for the mobile operators who offer the service of the system to farmers as a VAS. A better weather mapping of the whole nation would be possible as this system would be updating the central supercomputer and this would help in a more accurate and a large database for the central supercomputer to update the weather forecasts.

## FUTURE RESEARCH DIRECTIONS

Further research work would be along multiple levels in this system. Presently there is no numerical method which pertains to nowcasting. The numerical methods which have been developed rely heavily on radar system data. The system is developed due to the fact that radar systems cannot be installed everywhere and hence complex numerical systems are required to make the analysis.

- With the proposed system, the aggregation of the essential data is now simplified and hence the numerical method need not have to heavily rely on radar or the satellite images for its numerical calculations. These data could augment the system for better predictions and hence much more efficient and a more lean programming model could be developed in the future.
- Integrating the sensors directly into the mobile would be another avenue of research. Right now, System focuses on integrating the sensors with the mobile for data communication between the two. As the adaptability to the system increases, there would be a distinct need to integrate this system into a single product/package. As the mobile industry is now adopting various sensors such as the accelerometers with increasing acceptance from the end users, it is predicted there would be a similar demand which would lead the research

work on the adopt ability of the environmental sensors directly into the mobile.
- Another area of research would include the integration of a laser into the mobile which could collect data on the vertical profile of the atmosphere. The atom processor would have the necessary processing power for computing the measured parameters to determine the vertical profile of the atmosphere using a laser.

The point of distribution of weather nowcasting was solved by the use of mobile as the targeted end user is the direct beneficiary of the same. The communication protocol needs to be essentially in the native language or through universally identifiable symbols which the vast illiterate rural population of a country like India could easily recognize and assimilate. The system can be developed further with respect to this needs to account for the native language spoken in the region of implementation and could be initiated by incorporating the major native languages spoken in a particular state. The use of Google translation tool is highly recommended as a starting point for providing such a service. The use of universally accepted symbols for communication of the weather related information could be supported for better adoption of the service.

## CONCLUSION

With the increasing expenditure of launching satellites or setting up a radar enabled weather station, it is necessary to explore other avenues which would provide the necessary information at a lower cost. The mobile being a ubiquitous device is an excellent option which provides the distribution of weather related sensors and the ease of aggregation of data as the ease of communication is now increasing with the ever expanding capability of the telecommunication network. The mobile would also provide targeted distribution of the weather information which is advantageous than the traditional means of mass distribution of information. As the mobile also acts as a focal point of gathering weather related parameters, it provides a twofold advantage for setting up a low cost region specific weather monitoring system.

# REFERENCES

Adrian, P. (2001). Wireless integrated micro-systems will enhance information gathering and transmission. *Sensor Business Digest* (September issue).

Austin, G. L., & Bellon, A. (1982). Very-short-range forecasting of precipitation by the objective extrapolation of radar and satellite data. In Browning, K. A. (Ed.), *Nowcasting* (pp. 177–190). Academic Press.

Bellon, A., & Austin, G. L. (1978). The evaluation of two years of real-time operation of a short-term precipitation Forecasting Procedure (Sharp). *Journal of Applied Meteorology, 17*, 1778–1787. doi:10.1175/1520-0450(1978)017<1778:TEOTYO>2.0.CO;2

Benjamin, S. G., Brown, J. M., Brundage, K. J., Devenyi, D., Schwartz, B. E., & Smirnova, T. G. … Dimego, G. L. (1998). The operational RUC-2. In *Sixteenth Conference on Weather Analysis and Forecasting*, (pp. 249-252).

Frank, R. (2000). *Understanding smart sensors* (2nd ed.). Norwood, MA: Artech House.

Goerss, J. S., Velden, C. S., & Hawkins, J. D. (1998). The impact of multispectral GOES-8 wind information on Atlantic tropical cyclone track forecasts in 1995. *Monthly Weather Review, part II. Nogaps Forecasts, 126*, 1219–1227.

Howard, A., Mataric, M. J., & Sukhatme, G. J. K. (2002). An incremental self deployment algorithm for mobile sensor networks. *Autonomous Robots, 13*(2), 113–126. doi:10.1023/A:1019625207705

Huang, C., Tseng, Y., & Lo, L. (2004). *The coverage problem in three-dimensional wireless sensor networks*. In Global Telecommunications Conference, GLOBECOM '04.

Kar, K., & Banerjee, S. (2003). Node placement for connected coverage in sensor networks. In *Proceedings of the Workshop on Modeling and optimization in Mobile, Ad hoc and Wireless Networks (WIPOT'03)*. Sophia Antipolis, France, 2003.

Michelini, M., Hijazi, S., Nassar, C. R., & Zhiqiang, W. (2003). Spectral sharing across 2G-3G systems. Signals, Systems and Computers, 2003. In *Conference Record of the Thirty-Seventh ASILOMAR Conference*.

Nehrkorn, T. H., Thomas, C., & Lawrence, W. (1993). *Nowcasting methods for satellite imagery*. In Atmospheric and Environmental Research Inc. Cambridge.

Ning, W., & Naiqian, Z. (2006). Wireless sensors in agriculture and food industry - Recent development and future perspective. *Computers and Electronics in Agriculture, 50*(1).

Paul, M. W., Frances, C. H., Timothy, J., Schmit, R., & Aune, M. (1998). Application of GOES-8/9 soundings to weather forecasting and nowcasting. *Bulletin of the American Meteorological Society.*

Purdom, J. F. W. (1976). Some uses of high resolution goes imagery in the Mesoscale forecasting of convection and its behavior. *Monthly Weather Review, 104*, 1474–1483. doi:10.1175/1520-0493(1976)104<1474:SUOHRG>2.0.CO;2

Rötting, O., Röpke, W., Becker, H., & Gärtner, C. (2001). Polymer microfabrication technologies. *Springer Link, 8*, 32–36.

Shukla, B. P., Pal, P. K., & Joshi, P. C. (2010). Extrapolation of sequence of geostationary satellite images for Weather nowcasting. *Geoscience and Remote Sensing Letters, 8*.

Thaysen, J., Boisen, A., Hansen, O., & Bouwstra, S. (1999). Digest of technical papers, transducers 99. In *10th International Conference on Solid-State Sensors and Actuators* (pp. 1852-1855). Sendai, Japan.

Warneke, B. A., & Pister, K. S. J. (2002). *MEMS for distributed wireless sensor networks*. In 9th IEEE International Conference on Electronics, Circuits and Systems (ICECS).

## ADDITIONAL READING

Cheng, M., Ruan, L., & Wu, W. (2005). Achieving minimum coverage breach under bandwidth constraints in Wireless Sensor Networks. *IEEE Info-Com 2005, March.*

De Pondeca, M. S. F. V., & Zou, X. (2001). Moisture retrievals from simulated zenith delay observations and their impact on short-range precipitation forecasts. *Tellus, 33A*, 192–214.

Dorling, S. R., & Gardner, M. W. (1998). Artificial Neural Networks (The multilayer perceptron) - A review of applications in the atmospheric sciences. *Atmospheric Environment, 32*, 2627–2636. doi:10.1016/S1352-2310(97)00447-0

Duato, J. (1996). A necessary and sufficient condition for deadlock-free routing in cut-through and store-and-forward networks. *IEEE Transactions on Parallel and Distributed Systems, 7*(8), 841–854. doi:10.1109/71.532115

Enke, W., & Spekat, A. (1997). Downscaling climate model outputs into local and regional weather elements by Classification and Regression. *Climate Research, 8*, 195–207. doi:10.3354/cr008195

Intergovernmental Panel on Climate Change (IPCC). (2008). Climate Change (2007). The Physical Science Basis. *World Meteorological Organization (WMO) and UN Environment Programme (UNEP),* 2007. *Sensors (Basel, Switzerland), 8,* 167.

Janjic, Z. I. (2002). A Nonhydrostatic Model Based on a New Approach. In *Meteorology and Atmospheric Physics.*

Larson, J. W., Jacob, R. L., Foster, I., & Guo, J. (2001). The Model Coupling Toolkit. In *Proceedings. 2001 International Conference on Computational Science.*

Madou, M. (1997). *Fundamentals of Microfabrication.* Boca Raton: CRC Press.

Meesookho, C., Narayanan, S., & Raghavendra, C. (2002). Collaborative classification applications in Sensor Networks. In *Proceedings of Second IEEE Multichannel and Sensor Array Signal Processing Workshop.* Arlington, VA; Megerian, S., Koushanfar, F., Potkonjak, M., & Srivastava, M. (2005). Worst and best-case coverage in Sensor Networks. *IEEE Transactions on Mobile Computing, 4*(1), 84–92.

Michalakes, J. (2000). RSL: A parallel runtime system library for regional atmospheric models with nesting, In *Structured Adaptive Mesh Refinement (SAMR) Grid Methods*, IMA Volumes in Mathematics and Its Applications (117), 59-74. Springer, New York.

Michalakes, J., Bettencourt, M., Schaffer, D., Klemp, J., Jacob, R., & Wegiel, J. (2003; 2004). Infrastructure Development for Regional Coupled Modeling Environments. *Parts I and II, Final Project Reports to Program Environment and Training Program in the U.S.* Dept. of High Performance Computing Modernization Office, Contract No. N62306-01-D-7110/CLIN4.

Michalakes, J., Chen, S., Dudhia, J., Hart, L., Klemp, J., Middlecoff, J., & Skamarock, W. (2001). Development of a next generation regional weather research forecast model. In *Developments in Teracomputing: Proceedings of the Ninth EC-MWF Workshop on the use of high performance computing in meteorology. In W. Zwieflhofer & N. Kreitz, (Eds.), World Scientific* (pp.269–276).

Michalakes, J., Dudhia, J., Gill, D., Klemp, J., & Skamarock, W. (1999). Design of a next-generation regional weather research and forecast model. In *Towards Teracomputing* (pp. 117–124). River Edge, New Jersey: World Scientific.

NASA. (2002). *Report from the NASA Earth Science Enterprise Computational Technology Requirements Workshop*. April 30 – May 1, 2002. [Available on line at http://esto.gsfc.nasa.gov/].

Nguyen, C. T. (1999). Frequency-selective mems for miniaturized low-power communication devices. *IEEE Transactions on Microwave Theory and Techniques, 47*, 486–1503. doi:10.1109/22.780400

Seinfeld, J. H., & Pandis, S. N. (1998). *Atmospheric Chemistry and Physics – From Air Pollution to Climate Change*. New York: John Wiley and Sons Inc.

Smith, T. K., Elmore, K., & Dulin, S. A. (2004). A damaging downburst prediction and detection algorithm for the WSR-88D. *Weather and Forecasting, 19*, 240–250. doi:10.1175/1520-0434(2004)019<0240:ADDPAD>2.0.CO;2

So, A., & Ye, Y. (2005). On solving coverage problems in a Wireless Sensor Network using Voronoi diagrams. In *Proceedings on Workshop on Internet and Network Economics (WINE'05)*. Hong Kong.

Stumpf, G. J., Smith, T. M., Manross, K. L., Lakshmanan, V., & Hondi, K. D. (2003). Severe weather warning and application development at NSSL using multiple radars and multiple sensors. Preprints, *31st Conference on Radar Meteorology* (pp. 555-558). Seattle, WA, Amer. Meteor. Soc.

Tapia, A., Smith, J., & Dixon, M. (1998). Estimation of convective rainfall from lightning observations. *Journal of Applied Meteorology, 37*(11), 1497–1509. doi:10.1175/1520-0450(1998)037<1497:EOCRFL>2.0.CO;2

Wang, Y., Hu, C., & Tseng, Y. (2005). Efficient deployment algorithms for ensuring coverage and connectivity of Wireless Sensor Networks. In *WICON Proceedings of the First International Conference on Wireless Internet* (pp. 114 – 121).

Xue, M., Wang, D. H., Gao, J. D., Brewster, D., & Droegemeier, K. K. (2003). The Advanced Regional Prediction System (ARPS), Storm-scale Numerical Weather prediction and Data Assimilation. *Meteorology and Atmospheric Physics, 82*, 139–170. doi:10.1007/s00703-001-0595-6

## KEY TERMS AND DEFINITIONS

**2G:** 2G is the second-generation wireless telephone based on digital technologies. It is basically for voice communications only, except SMS messaging is also available as a form of data transmission for some standards.

**Environmental sensors:** Device that produces a measurable response to a change in a physical condition, such as temperature or thermal conductivity, or to a change in chemical concentration.

**Grid:** Combination of computer resources from multiple administrative domains for a common goal.

**Nowcasting:** Form of very short-range weather forecasting, covering only a very specific geographic area.

**MEMS:** A technology that embeds mechanical devices such as fluid sensors, mirrors, actuators, pressure and temperature sensors, vibration sensors and valves in semiconductor chips.

**WAP:** An open international standard for application-layer network communications in a wireless-communication environment.

# Section 3
# Protocols and Technologies

# Chapter 8

# Network Mobility Management in the ITS Context:
## Protocols for Managing Vehicle–to–Infrastructure Communications

**Nerea Toledo**
*University of the Basque Country, Spain*

**Marivi Higuero**
*University of the Basque Country, Spain*

## ABSTRACT

*This chapter presents the most significant approaches developed so far for addressing vehicle-to-infrastructure communications by means of NEtwork MObility (NEMO[i]) management, including the most outstanding solution (NEMO Basic Support, NEMO BS), and highlights their most interesting features. The demanded key features to the NEMO protocols to be applied in the ITS context are also defined, and an analysis of the fulfillment of these key features by the NEMO protocols is provided.*

*The proposals suggested so far can be can be classified in two different categories: on the one hand, NEMO solutions that consider MIPv6 as the base host mobility management protocol, and on the other hand, solutions that consider alternative base host mobility management protocols like SIP, LIN6, or HIP. Besides, a taxonomy on MIPv6 based NEMO protocols classifying them by considering which characteristics they aim to enhance is provided. It is important to point out that the selection of the base host mobility management protocol is fundamental to have as many demanded key features satisfied as possible by the NEMO protocol to be applied in the Intelligent Transportation System (ITS) context.*

DOI: 10.4018/978-1-4666-0080-5.ch008

*Figure 1. Network mobility management scenario*

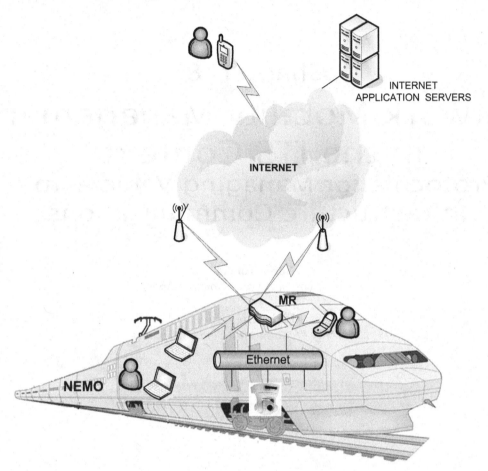

## INTRODUCTION

The provision of internet services while traveling is a flourishing market that should be covered by means of emerging mobile computing and communication techniques. This scenario presents specific challenges that current communication architectures do not cover. Indeed, Intelligent Transportation System (ITS) standardization bodies are currently defining a dedicated communication architecture that satisfies the scenario necessities and the increasing demands.

ITS communication standards are being designed to support multiple classes of applications including those that assist in vehicle operation and internet-based applications (European Telecommunications Standardization Institute [ETSI], 2010). In fact, internet-based applications are considered to be beneficial for safety, and fundamental for non-safety purposes (Baldessari, Festag, & Lenardi, 2007b). Furthermore, the ITS communication architecture should not only focus on the vehicle communication needs, but also in the demands of users onboard. On the other hand, it is assumed that an ITS-compliant vehicle is a moving network, where a set of nodes should obtain anytime-anywhere connectivity to the internet through one or more Mobile Routers (MR). Figure 1 shows the network mobility scenario.

*Figure 2. Approach for managing the mobility of entire networks: Individually or as a whole*

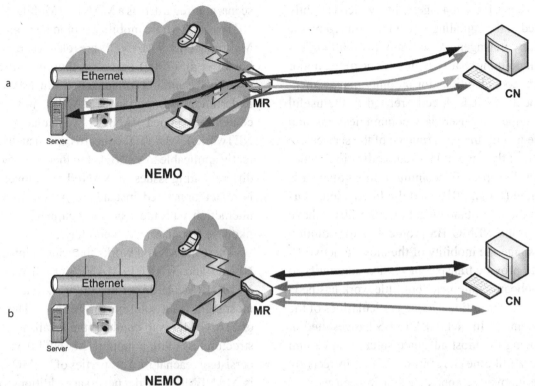

With the aim of obtaining anytime-anywhere connectivity to the internet, the mobility of the vehicle has to be managed. Two approaches for managing the mobility in this context can be distinguished:

- *Individual management or host mobility management.* The mobility of each node located in the vehicle is managed individually. Therefore, this approach requires managing one by one the handovers of each node located inside the mobile network in order to be them reachable from the internet. In this approach the Mobile Routers has basically routing functions. The most outstanding protocol to manage host mobility are MIPv6 (Johnson, Perkins, & Arkko, 2004) and its improvements, HMIPv6 (Soliman, Castellucia, ElMalki, & Bellier, 2008), FMIPv6 (Koodli, 2009) or PMIPv6

(Gundavelli, Devarapalli, Chowdhury, & Patial, 2008).

- *Management as a whole or network mobility management.* The mobility of the entire moving network is managed jointly through a single entity, the MR. This approach has the advantage of reducing the signaling overhead in the wireless access link and consequently, less power is consumed. In addition, when the MR performs a handover all the nodes inside the mobile network can access the outside network immediately. NEtwork MObility (NEMO) Basic Support (Devarapalli, Wakikawa, Petrescu, & Thubert, 2005) which is an extension of MIPv6 is the *de facto* protocol for managing network mobility.

Figure 2 shows both network mobility management approaches.

The key difference between the aforementioned approaches is the manageability, which is tightly related to the signaling exchange required by the mobility management solution. In addition, it is worth pointing that signaling is one of the major design considerations for mobility management protocols, as it is directly related to the useful throughput for user data communications and consequently the performance of the services.

Recently, several ITS standardization bodies like the European Telecommunications Standards Institute (ETSI, 2010) and the International Organization for Standardization (ISO, 2010), have chosen the NEMO BS protocol as the solution to manage the mobility of the moving networks and to provide transparent session continuity to the onboard equipment, but little work has been carried out adapting it to the particularities of the ITS context. In fact, NEMO BS has resulted in not being the most adequate solution in certain scenarios like the ITS context. This is why recently more innovative approaches have been worked out improving different aspects of NEMO BS.

The internet community is aware of the need to adapt NEMO BS to the ITS context and consequently works related to the application of NEMO solutions in specific ITS scenarios have been carried out. Most relevant ITS scenarios are the vehicular context and the aeronautics scenario, where not only the IETF (Internet Engineering Task Force) is proposing solutions to provide communication services but also industry and the international research community are working ahead in these fields. In addition, the railway context is also a relevant scenario where NEMO solutions can be applied, but currently, no specific work related to this scenario has been carried out in the IETF, but industry and the research community are focused on this context.

Apart from Network Mobility management by means of NEMO protocols, which is mainly focused on the hop between the moving networks and the infrastructure (i.e. Internet), there are other concerns in the vehicle-to-vehicle communica-

tions that have to be solved. The vehicle-to-vehicle scenario is regarded as a MANET (Mobile A-hoc NETworks), whose mobility should be managed. Main concerns of MANETs are related to routing, but Li et al. (2009) present a thorough discussion on the mobility management of MANETs underlining its importance. MANETs in the ITS context are named VANETs (Vehicular Ad-hoc NETworks). MANET concerns are usually directly applicable to VANETs but there are several differentiating issues: in VANET the movement is rather organized instead of random and the interaction with the roadside equipment can be likewise characterized accurately.

There are several works that aim to integrate NEMO and MANET. MANEMO (MANET + NEMO) (McCarthy et al., 2009) is one of the most significant solutions in this area. The goal of MANEMO is to combine the localized infrastructureless routing support (MANETs) with the persistent reachability properties of NEMO. That is, MANEMO provides necessary additions to the existing protocols (Neighbor Discovery, IPv6, NEMO) to find the most suitable route towards the infrastructure in nested moving networks or MANETs. Notice that in the NEMO perspective internet reachability is the higher requirement while the applicability of MANET solutions is for optimizing the route towards internet. Further description on the problem statement and requirements for MANEMO can be found in the IETF draft defined by Wakikawa et al. (2007).

Because of the field of the study the most significant advances in the integration of NEMO and VANET are enumerated herein. Bernardos and his colleagues (2007) present a route optimization solution for enhancing car-to-car communications, integrating it with NEMO, and perform simulation studies. Baldessari et al. (2007a) analyze the deployability approaches for NEMO in a VANET scenario. Jemma et al. (2010) implement an architecture that combines geographic-based routing for NEMO with VANET. In the NEMO and MANET/VANET integration solutions, vehicle-to-infrastructure communications are managed by means of the NEMO protocol while vehicle-to-vehicle communications are managed through MANET/VANET solutions, so, integration solutions aim at solving a wide heterogeneous scenario. However, each communication type (vehicle-to-infrastructure and vehicle-to-vehicle) is perfectly differentiated, with specific requirements and necessities. This is why manifold approaches have been proposed so far not only in addressing each field separately but also in integrating them.

In this chapter vehicle-to-infrastructure communications are considered, i.e. internet reachability, so, Network Mobility Management approaches are thoroughly studied herein. First an overview of the NEMO BS protocol is provided, which is the most outstanding solution for managing the mobility of networks and the solution adopted by the standardization bodies, highlighting its main shortcomings. Because of this, several approaches have been proposed so far for enhancing the NEMO BS protocol. In this chapter most relevant proposals are described and an analysis of their characteristics outlining their advantages and disadvantages is given. This chapter fills out the gap found in the literature analyzing the network mobility solution space. That is, solutions supporting other characteristics to route optimization like security, or route optimization schemes also supporting security, are studied herein. In addition, the demanded key features by NEMO protocols to be applied in the ITS context are underlined and map these key features with the aspects of the analyzed proposals. Based on the state of the art in NEMO protocols and the trends of the ITS context, further research directions are also outlined.

## AN OVERVIEW OF NETWORK MOBILITY BASIC SUPPORT

As mentioned before, the NEMO BS protocol is the *de facto* protocol for managing the mobility of a moving network. Furthermore, it has been adopted by standardization bodies to manage the mobility of entire moving networks. In fact, the NEMO concept has been successfully validated in a real scenario in the CVIS (Cooperative Vehicle-to-Infrastructure Systems) project CVIS, 2006). In addition, the NEMO BS protocol has settled the basis of network mobility management. Therefore because of its importance, the NEMO BS protocol describing its underlying architecture and its main procedures is introduced. An overview of the security provisioning is also provided, which is a major concern in several scenarios like the ITS context, and other performance issues like routing and multihoming support.

NEMO BS has been defined by the IETF NEMO Working Group to provide a solution for moving networks. It provides session continuity to every node located inside the moving network as the network moves, also allowing every node in the moving network to be reachable. The NEMO BS protocol is based on MIPv6 and runs on the network layer; thus it inherits its basic procedures. Before detailing the protocol operation, the elements that make up the NEMO architecture are described next[ii]:

- *Mobile Network (NEMO):* An entire network that moves as a unit, changing dynamically its point of attachment to the internet, and thus, its reachability and the topology. The mobile network can be composed by one or more IP subnets and it is connected to the outside network by means of one or more Mobile Routers.
- *Mobile Router (MR):* The MR is the only entity which communicates directly with the internet, and thus, every single communication between the mobile network and

the internet goes through the MR. The MR is defined as an extension of the MIPv6 Mobile Node; with routing capabilities between its point of attachment and the subnet that moves with the MR. The operation of a MR is a combination of functions of a Mobile Host and a Mobile Router. The difference between both behaviors remains in the fact that if the MR acts as a Mobile Host, it does not maintain any prefix information related to the MR's HoA, but only MR's HoA is stored and it neither sends prefix related information to its HA. On the other hand, when the MR acts as a Mobile Router, it manages prefix information.

- *Mobile Network Node (MNN):* The nodes inside the moving network can be fixed, Local Fixed Node (LFN), mobile, Local Mobile Node (LMN), or visiting nodes, Visiting Mobile node (VMN). These nodes can only be reached through the MR by means of a bidirectional tunnel between the MR and its Home Agent (HA). Meanwhile, network mobility has to be hidden to the operation of these nodes.

- *Correspondent Node (CN):* The correspondent node refers to any node located in the outside network, the internet, and is communicating with one or more MNNs. The CN can be either fixed or mobile.

It is assumed that the NEMO has a Home Network, where it resides when it is not moving. Since the NEMO is part of the Home Network, its addressing structure is part of the address block assigned by the Home Network. In addition, the MR has a unique Home Address (HoA) configured from a prefix advertised by its HA and an address block is assigned to the NEMO. In NEMO BS the MNNs should own globally routable addresses; therefore, and in order to be reachable, the prefix of the IP addresses of these MNNs should be the same as the MR's currently in use IP address, Care of Address (CoA).

When the MR changes its point of attachment, a new prefix is advertised by the MR in the NEMO as well as the HA is notified about the new prefix that the MR will be using. The NEMO prefix is included in the Binding Update (BU) message. The BU message exchange between the MR and the HA includes additional information related to network mobility management to the BU messages defined in MIPv6. More specifically a Mobile Router Flag is included to inform the HA that the BU is from a MR. A Mobile Network Prefix Option is also introduced in order to notify the HA the new network prefixes that should be forwarded to the NEMO, which is placed in a cache together with MR's HoA and CoA. Being the MR in charge of notifying the new prefix to the HA, the manageability level is enhanced compared to the individual mobility management case.

In NEMO, not only traffic destined to the MR is forwarded to the HA but also the traffic destined to any mobile node of the NEMO. In fact a bidirectional tunnel is established between the MR and its HA. IP-in-IP encapsulation is used for tunneling packets. This tunnel is set up when the MR sends a successful BU to its HA informing about its current point of attachment. Having established the bidirectional tunnel when the HA receives a packet destined to the mobile network, it tries to forward the packet to the next hop, MR's HoA and it looks for a binding cache for the HoA. So, the HA obtains the MR's CoA and tunnels the packet to this address. In the same way, if the MR receives a packet with a source address belonging to the mobile network prefix, it reverse tunnels the packet to the HA. Figure 3 shows the scenario.

This routing through the HA is commonly known as *dog-leg routing*. This non-optimal routing, named *suboptimal routing* is directly translated into increased delays, infrastructure load, packet overhead, processing delay and possible bottlenecks in the HA. In addition, the inefficient routing that traffic tunneling implies involves adding 40 bytes to each data packet. Thus, for

*Figure 3. NEMO BS suboptimal routing*

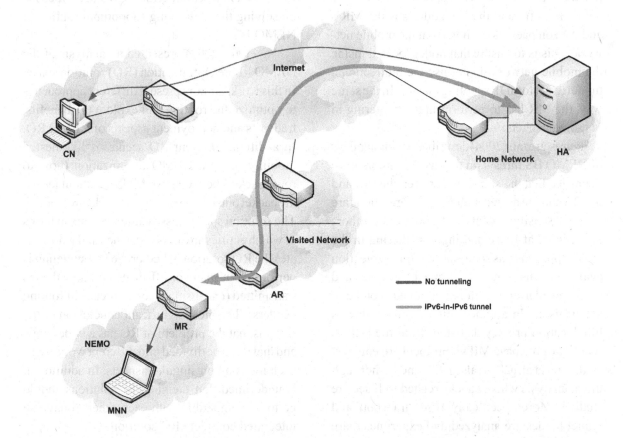

applications generating small data packages like VoIP, the overhead introduced is considered significant. In addition congestion at Home Network could even lead to packet loss, causing degradation to applications. Additionally, with the nesting of mobile routers (a mobile network that attaches another mobile network and obtains connectivity to the outside network by means of the root MR), data packets may go though different HAs and consequently several levels of encapsulation will be required. Therefore, with nested mobile networks, as several tunnels have to be established suboptimal routing gets worse, aggravating aforementioned problems.

## Security Support in NEMO BS

Regarding security provisioning, NEMO BS does not introduce improvements to the security support of MIPv6. Consequently, MIPv6 alike, IPsec is adopted to protect the signaling message exchange between the MR and the HA, which Arkko et al. (2004) describe in detail. IPsec provides integrity and authentication of the messages and it can optionally be used for encrypting packets in the MR-HA tunnel. More specifically, IPsec can be particularized for NEMO BS distinguishing two cases: IPsec transport mode for exchanging signaling packets between the MR and the HA and IPsec tunnel mode for data protection.

Being the HA the anchor point of the communications between the mobile network and the internet, it has to perform ingress filtering (check that the source address corresponds to the mobile network prefix and that the address is not MR's address) on packets coming from the mobile network. This is to ensure that nodes located inside the mobile network do not use the bidirectional tunnel to perform IP spoofing attacks. In the same way, the MR has to perform ingress filtering to outgoing packets.

Petrescu et al. (2004) described a wide analysis of NEMO BS threats. In this work, authors have identified that the signaling between the MR and the HA and the nested mobility configurations are the most sensitive points of the protocol. Authors underline that IPsec and ingress filtering in the HA are presented as solutions to afford protection against threats, but several attacks are outlined even considering the aforementioned protection mechanisms. In addition, other security threats like location privacy that stem from the lack of security of the base MIPv6 protocol are enumerated. Jung et al. (2004) also performed a thorough threat analysis where attacks related to IPsec are studied. More specifically IPsec transport and tunnel modes are analyzed and experiments are described where some attacks were confirmed. Authors in this work point out that the problem does not reside in the IPsec protocol itself but in the interaction with other functional components like packet forwarding or ingress filtering required in NEMO.

## Route Optimization Support in NEMO BS

As mentioned before, the inherent suboptimal routing of NEMO BS implies performance related problems, hence, it is seen that one of the most important enhancements to NEMO BS will be to optimize the routing that will result in improving the performance. The IETF worked on analyzing the sub-optimality of the *dog-leg routing* and published a draft document where the motivation behind RO is presented (Ng, Zhao, Watari, & Thubert, 2007). Following the described motivation

by the IETF, research efforts have been focused on solving the underlying suboptimal routing of NEMO BS.

Ng et al. (2007) presented an analysis of the NEMO Route Optimization (RO) solution space. In this work, authors present different approaches to optimize the routing and explore the benefits, tradeoffs and deployment aspects of NEMO RO. In addition, different RO scenarios like nested mobility or intra-NEMO optimization (how to route packets between two MNNs without letting the packet outside the mobile network) are studied. The solution space classification is done in terms of which entities are involved, when and who initiates the RO solution, whether the RO signaling is performed in-plane or off-plane and how data is transmitted (encapsulated or by means of routing headers). The most significant conclusion of the study is that the problem of RO has a wide scope and that it can be divided into different work areas, each focusing on singular aspects. In addition, it is underlined that the singular solutions should be to be compatible with each other to have an integrated complete RO solution.

Shahriar, Atiquzzaman and Ivanvic (2010) go further in the NEMO RO study providing a deep survey of the RO solutions. This work underlines the challenges of RO and main issues are given to classify and compare the different approaches suggested through the literature in the last five years. The classification is done embracing delegation, hierarchical, source routing or BGP-assisted RO schemes. In the comparison between the described RO approaches, characteristics like signaling, memory requirement, deployability or intra-RO provisioning are considered. Authors also point out future research directions indicating that performance evaluation of RO schemes or security threat analysis for RO scenarios are hot topics to study. However, other important issues like the lack of multihoming support are not mentioned.

## Multihoming Support in NEMO BS

The term multihoming has no formal definition in the IETF. In general, it is used to indicate a scenario where a node is available through several locators. From this definition, a multihomed node can be understood as any entity with capacity to connect to the internet using different IP addresses. Multihoming aims at ensuring continuous connectivity to the internet, anytime-anywhere, even while on the move. It is worth pointing that in a NEMO context, multihoming not only refers to having more than one egress interface, but also the case where more than one MR is considered. However, it is considered that this latter scenario is more complex and that it has its own particularities like MR operation switching or MR context sharing, and that it presents its own challenges that multihoming solutions do not necessarily solve.

The work carried out by Ernst and Charbon (2004) is the most significant regarding NEMO multihoming support. They analyze the behavior of NEMO BS when the mobile network is multihomed. Authors in this work analyze different multihoming configurations classifying them in terms of number of MRs, number of HAs and number of mobile network prefixes, simplified as *(n,n,n)*. The studied configurations are supported by NEMO BS but additional mechanisms to provide redundancy, load sharing and policy routing are required. More specifically, authors determine that a mechanism to identify which CoA is updated by the NEMO BS BU is required, a priority option to provide load sharing is needed and policy information in the NEMO BS BU is demanded for policy management. The conclusions stemming from this work were proposed to the IETF to standardize the NEMO multihoming support, but up to now, the IETF has not adopted any approach and hence, there is no standardized solution for solving the multihoming support in NEMO scenarios.

In conclusion, it can be outlined that although NEMO BS solves elegantly the manageability aspect of a moving network introducing a new entity and modifying the BU messages, flaws regarding security, route optimization or multihoming support are present in its specification.

## DEMANDED KEY FEATURES OF NEMO MANAGEMENT PROTOCOLS FOR THE ITS CONTEXT

Ernst (2001) first specified the requirements that a general network mobility solution should or must support in his Ph.D. Thesis work. These requirements are at the same time a refinement and an extension of the requirements discussed for host mobility support by Myles and Skellern (1993) or Bhagwat et al. (1996). Next, the network mobility support design requirements are highlighted.

- *Wide-Area Mobility.* A mobile network should be able to attach any point in the internet topology. That is, in practice a mobile network must be able to roam between administratively distinct access networks and via any available access technology (IEEE802.11, EPS, satellite, etc.).
- *Transparency.* Mobility management has to be transparent to MNNs and maintain continuous access to the internet. Indeed, these nodes should not be aware of the mobility of the entire network. However, the MNNs will perceive transmission delays of packet loss as the network moves, but these effects should be minimized to have *seamless* mobility. Furthermore, the topological location change must not have impact in layers above the IP layer. Therefore, in practice node identifiers used in the upper layer should be independent form the IP addresses used at the network layer for routing.
- *Optimal routing.* As the non-optimal routing increases the transmission delays, and the amount of traffic intended for and com-

ing from the mobile network is more significant than in the case of the single mobile node, the non-optimal routing is even a more important requirement in the mobile network support.

- *Minimum signaling overload.* Control traffic is usually required to efficiently route packets from the CN to a node inside the mobile network. In addition, the amount of signaling overload has always been a concern for host mobility management. In the case of mobile networks, as a result of a potentially large number of CNs and MNNs, the signaling overload becomes a more stringent issue.
- *Scalability.* Mobility support has to assume an increasing number of mobile nodes and assume that a major part of nodes composing the internet will be mobile. Therefore, network mobility support should scale with regards to mobile networks, CNs, and MNNs comprising the mobile network.
- *Nested Mobility.* Network mobility support must allow a mobile network attaching another mobile network.
- *Mobile CN.* Network mobility support should perform efficiently in the case where the CN is mobile or is located in a mobile network.
- *Backward compatibility.* The introduction of network mobility support must not prevent MNNs from operating any standardized protocol.
- *Minimum impact on existing protocols and infrastructure.* To have a quickly deployable solution, the network mobility support should introduce minimum changes in existing protocols and infrastructure.
- *Security.* Unlike fixed nodes, MNNs are more exposed to security threats, so, network mobility support should at least offer the same security level as for fixed nodes, and shouldn't introduce vulnerabilities.

Having described the basic requirements that a network mobility solution should support, next the specific key features of network mobility management protocols considering the particularities of the ITS context are specified.

- *Security.* Safety is the rule in the ITS scenario. Consequently, every control message involved in the network mobility management must be exchanged securely, ensuring node authentication, reliability, data integrity, confidentiality, non-repudiation and privacy.
- *Efficiency.* Network mobility support is demanded to be "efficient". That is, low latency, little data loss, minimum delay, minimum signaling and minimum bandwidth consumption are required when managing mobility. In fact, the appropriate performance level has to be guaranteed in the message exchange between the moving network and the outside network. The efficiency in NEMO protocols is often achieved by RO solutions that avoid traversing the Home Agent (HA) like in the NEMO BS protocol. Being the RO a fundamental characteristic of an efficient NEMO protocol, specific requirements have been defined for NEMO RO schemes to be applied in the vehicular context (Eddy, Ivanvic, & Davis, 2009) and Baldessari et al. (2009) did the same for the aeronautics context.
- *Global reachability of nodes inside the Mobile Network.* Nodes inside the mobile network have to be reachable from any point of the outside network while moving as well as internet access should be guaranteed for them anytime-anywhere. These nodes can be end-users travelling, or nodes in charge of the control of the vehicle operation that have to be reachable from the control centre.

- *Mobile Router (MR) multihoming.* Several wireless communication technologies will coexist in the ITS context, consequently and with the goal of enhancing the performance, the MR should be a multihomed node. Thus, managing several locators has to be supported by the NEMO protocol and issues like the availability of different addressing schemes inside the mobile network have to be considered.

- *Mobility support inside the Mobile Network.* As nodes inside the moving network would have the capacity to make use of wireless communications, this will provide the basis to allow nodes to move inside the NEMO while still being connected, which is an interesting feature to offer in certain scenarios like in large ships. Therefore, micromobility solutions to support mobility inside the moving network are necessary to be considered by the NEMO protocol.

- *Multiple MR support.* A solution that considers multiple MRs provides reliability to communications. Having reliable communications is fundamental in the ITS context, because of the stringent safety requirements. If the goal of the additional MR is to provide redundancy and protection against failures, MR configuration concerning ingress and egress interfaces, addressing scheme and migration of functionalities have to be solved. If the additional MR is for load sharing purposes, challenges of simultaneous operation of the MRs should be solved by the NEMO protocol.

- *Nested Mobility.* The ITS scenario is prone to have nested mobility situations as the case of a mobile network visiting another mobile network will be common, i.e. a WPAN (Wireless Personal Area Network) (Prasad & Muñoz, 2003) attaching the mobile network of a vehicle.

- *Backward compatibility and adaptability:* The NEMO protocol has to be easily deployed in the current internet architecture. In addition, new transport protocols, IP options or even post-IP solutions should be able to be introduced with no or minimum impact to the NEMO solution. The adaptability feature is requested because of the lack of penetration of communication solutions in the ITS context. Because of this, the application protocols, user data characteristics, transport protocols etc. that will be used in the transition process or in the final operational deployment are not defined yet, and may change with the research and development experience.

Aspects like *Backward compatibility and adaptability* or *Minimum impact on existing protocols and infrastructure* are also maintained in the demanded key features for network mobility support in the ITS scenario. However, it should be pointed out that these requirements are relaxed. This is because the introduction of IT solutions in the ITS context is slowing down mainly due to its conservative nature, the number of involved agents and resultant complexity. Furthermore, the useful life of cars is not tuned with the evolution of the technologies, so, long term and steady technologies are required to be introduced in the ITS scenario.

Table 1 shows the mapping between the basic design requirements and demanded key features for network mobility support for the ITS scenario. Notice that the mapping merges some basic aspects into wider concepts, extracts relevant characteristics like *Mobility support inside the mobile network* to the basic requirements and introduces additional features not considered in the basic specification like *Multiple MR support*.

It is worth pointing that although an initial set of design requirements had been specified, the NEMO BS solution does not satisfy the majority of them and leaves their support for an extended

*Table 1. Mapping between basic network mobility design requirements and demanded key features for network mobility support for the ITS scenario*

| Basic network mobility support design requirements | Demanded key features for NEMO support in the ITS scenario |
|---|---|
| Wide-Area Mobility | Global reachability |
| Transparency | Efficiency |
| Minimum signaling overload | |
| Scalability | |
| Optimal Routing | |
| Nested Mobility | Nested Mobility |
| | Mobility support inside the mobile network |
| Mobile CN | Mobile CN |
| Security | Security |
| Backward compatibility | Backward compatibility and adaptability |
| Minimum impact on existing protocols and infrastructure | Minimum impact on existing protocols and infrastructure |
| - | MR multihoming |
| - | Multiple MR support |

NEMO management version. As a result of this, manifold proposals have been published throughout the literature with the aim of enhancing the NEMO BS protocol and tackling different design requirements. It is worth pointing that up to now, no solution covering those open issues has been standardized. Furthermore, the particularities of the ITS context are not considered by NEMO BS but a generalist solution is provided.

## A TAXONOMY OF NEMO MANAGEMENT PROTOCOLS

In this section a taxonomy of the different approaches suggested so far to cover the open issues of NEMO BS is presented. The proposals are classified in two groups: approaches that aim at enhancing MIPv6 and approaches that are based on alternative host mobility management protocols and aim at optimizing diverse aspects that stem from the base protocol design.

## MIPv6 Based NEMO Management Solutions

Main research efforts of NEMO BS have the goal to improve its inherent suboptimal routing due to its crucial importance. Several works exist where a taxonomy of the existing NEMO solutions is carried out (Ng, Zhao, Watari, & Thubert, 2007), (Shahriar et al., 2010). In this chapter the gap found in analyzing network mobility solution space is filled out focusing on approaches that address other shortcomings of NEMO BS.

As safety and security are the rule in the ITS scenario, this chapter concentrates on the security provisioning and on RO schemes supporting security of NEMO management solutions. It is worth mentioning that security support has not been studied as widely as RO support. In fact, main works related to security analyze the interaction of NEMO with an AAA (Authentication, Authorization and Accounting) framework. Next, major contributions of security support in NEMO management solutions suggested so far are described.

## Security Support in NEMO Management Solutions

Bournelle et al. (2006) studied how to combine network mobility and access control mechanisms from an operator's point of view. In this work, authors present three possible NEMO deployment scenarios: a MR-pan, where the personal NEMO includes a MR and a set of MNNs; a MR-bus, where the MR belongs to the operator; and a MR-pan in the MR-bus, which is a combination of the first two scenarios, i.e. a nested mobility scenario. Authors in the analysis assume that the PANA protocol is used between the clients and the Network Access Server, and RADIUS or Diameter are considered as the AAA protocol. One of the issues highlighted in the article is that only the egress interface of the MR is authenticated, and consequently, the traffic coming from the mobile network can access the internet while it has not been explicitly authorized. On the other hand, authors point that the nested NEMO case presents challenges related to the authorization of users connected through the secondary MRs. This approach aims to solve authentication by means of integrating NEMO BS and an AAA scheme but a security solution for the NEMO BS protocol is not completely defined.

Phang, Lee, and Lim (2007) also proposed an AAA framework and a firewall traversal solution. The proposed framework solves access control between nodes in the mobile network and the service providers, and firewall traversal issues. Authors state that the proposed framework does not affect the smooth operation of NEMO, but no validation of this statement if shown in the article. In addition, authors mainly focus on the nested NEMO scenario and the security threats related to the miss of authentication methods between MRs. Consequently, authentication support is one of the solved security properties together with firewall traversal challenges, but other security features required in NEMO communications are not addressed. On the other hand, the advantage of this solution resides in the fact that it does not require any changes in the NEMO BS protocol.

Fathi et al. (2006) brought out that the use of IPsec to secure NEMO procedures does not prevent from a critical security issue: leakage of stored secrets. Leakage of stored secret or private keys causes serious flaws that are enough to breakdown the overall security of the system. The potential of stored secrets is not negligible due to computer viruses, system misconfigurations or even lost/stolen mobile devices (Shin, Kobata, & Imai, 2005). Private keys or secrets are usually stored in tamper-resistant modules, but in some situations, these types of modules are not possible to be used, mainly because of their cost. Based on this issue, Fathi et al. (2006) join NEMO with AAA and add a unique mechanism robust against leakage of secrets and other attacks, such as spoofing or eavesdropping. Authors propose in this work a new handover procedure to be performed by MRs and visiting mobile networks. This handover procedure is based on the Leakage Resilient-Authenticated Key Establishment (LR-AKE) protocol. The LR-AKE protocol is an authenticated key exchange protocol based on RSA (a public key cryptography algorithm for signing and encryption) and it is used for establishing secure communications relying on a short secret that can be remembered by humans to avoid leakage from devices. The architecture presented by authors in this work comprises an AAA server (AAAF if the MR is in a foreign domain and AAAH if the MR is its home domain), the HA, the MR and the VMN. For this last entity, the CN is also considered in the handover process because being a visiting node it can have a session established with a CN, and the VMN and the CN need to generate session keys used for encrypting the messages to be exchanged between them, via their respective HAs. In summary, the LR-AKE based AAA solution for NEMO secures the MR-HA, VMN-HA, VMN-CN and VMN-MR communications by means of LR-AKE exchanges.

The main advantage of this approach is the guarantee of the leakage of stored secrets. On the other hand, involved entities have to be upgraded with the LR-AKE protocol, and which is worse, the integration and interaction of the LR-AKE protocol and the mobility management protocol (NEMO BS) is required. Consequently, a complex framework results from this approach. It is worth pointing that, being the LR-AKE protocol a key exchange protocol, proper message encryption apart from authentication should be achieved. However, this scheme aims at providing security to the signaling plane of NEMO BS and leaves aside the data plane security. Authors have extended the LR-AKE based protocol for nested scenarios (Fathi et al., 2007).

Kin and Samsudin (2007) stated that the use of IPsec is expensive in terms of signaling and may impact in the performance of the system, which is worsen when nested scenarios are present and several tunneling levels are required. More specifically, the motivation behind their proposal is that the process of establishing and getting the SAs synchronized is costly and that IPsec may not be the best solution to be deployed. Following this idea, authors propose the use of a centralized system to simplify the security management. This centralized system consists on a Certificate Authority (CA) managed by the Internet Service Provider (ISP) and a trusted database; that is, a Public Key Infrastructure (PKI) is considered in this approach mainly for authentication purposes. Authors argue that a tradeoff between encryption (achieved by IPsec) and computing power is needed to adopt the CA-PKI process they propose. In addition, they underline that there is no need to re-establish and re-agree the SAs once the trust has been established by means of the PKI, unlike in IPsec, and therefore, their CA-PKI approach will result in less computation overloading, but no validation has been presented. The main disadvantage of this approach is that it considers a centralized security management, which is managed by a certain ISP using centralized policies opposed to maintaining the SAs by the tunnel endpoints that can have local policies, but does not consider the scenario where the mobility management involves different ISPs each with its own security policies and its own PKIs. Furthermore, the need for rekeying to reduce vulnerabilities related to the key usage time is not considered. Referring to user data security, this cannot be confidentiality and integrity protected end-to-end, but only between the CN and the MR, because the symmetric key is shared between these two parties.

Jiang et al. (2006) focused on the security provisioning of group communication in NEMO scenarios and proposed a multiple key sharing and distribution scheme. Three parties are usually involved in group communications: the CA, a trusted party in charge of issuing the certificates to group members; members, which can be mobile users or mobile routers; and the Network Center (NC), in charge of processing the messages from members, reconstruction, renewing and distributing the keys to members. The approach proposed by Jiang et al. (2006) is a key sharing and distribution scheme for exchanging the secret between the parties that comprise the group. The scheme is based on the multisecret sharing and the modular square root techniques. The solution also offers unique procedures to renew the keys of members dynamically joining the group, and automatic key refreshment when these have expired. The proposed solution supports the security requirements of group communications: conversation privacy (confidentiality), member identity anonymity and no tracking (user privacy), fraud prevention (mutual authentication), replay attack prevention and forward and backward secrecy[iii]. These group communication security requirements are extensible to one-to-one communications, but, the scheme should be modified to be applicable to one-to-one communications.

Table 2 resumes the security properties supported by the aforementioned proposals.

*Table 2. Analysis of the security support of approach that improve NEMO BS security*

| Desired security related key features | | Bournelle *et al.* (2006) | Phang *et al.* (2007) | Fathi *et al.* (2006) (LR-AKE) | Kin & Samsudin (2007) (CA-PKI) | Jiang *et al.* (2006) |
|---|---|---|---|---|---|---|
| *Authentication* | | ✓ | ✓ | ✓ | ✓ | ✓ |
| Confidentiality | *Signaling plane* | X | X | X | X | X |
| | *Data plane* | X | X | X | X | ✓ |
| Integrity protection | *Signaling plane* | X | X | ✓ | X | X |
| | *Data plane* | | | | | ✓ |
| *Infrastructure based* | | ✓ | ✓ | ✓ | ✓ | ✓ |
| Other properties | | Access Control | Firewall traversal / Access Control | Protection against leakage of stores secrets | - | Group communications are considered |

Apart from studying the security provisioning, Table 3 shows whether the proposals satisfy other features required by the ITS context.

As can be regarded in the approaches described herein demanded key features for the ITS context are hardly fulfilled. Regarding security, authentication support is one of the main challenges solved by the majority of them. However, other security properties like end-to-end confidentiality or integrity protection remain unsolved or are relied on IPsec by most of the proposals. It is worth pointing also the strong dependency on infrastructure which can be a performance constraint in real deployments.

*Table 3. Evaluation of the adaptability of demanded key features in the ITS context by the security enhancing approaches*

| Demanded key features for NEMO support in the ITS scenario | Bournelle *et al.* (2006) | Phang *et al.* (2007) | Fathi *et al.* (2006) (LR-AKE) | Kin & Samsudin (2007) (CA-PKI) | Jiang *et al.* (2006) |
|---|---|---|---|---|---|
| *Global Reachability* | ✓ | ✓ | ✓ | ✓ | ✓ |
| *Efficiency (RO)* | X | X | X | X | X |
| *Nested Mobility* | ✓ | ✓ | ✓ | X | X |
| *Mobility support inside the mobile network* | X | X | X | X | X |
| *Mobile CN* | X | X | X | X | X |
| *Backward compatibility and adaptability* | ✓ | ✓ | ✓ | ✓ | ✓ |
| *Minimum impact on existing protocols and infrastructure* | X | X | X | X | X |
| *MR multihoming* | X | X | X | X | X |
| *Multiple MR support* | X | X | X | X | X |

## Security Support in Route Optimization Schemes

Apart from solutions that aim at solving the suboptimal routing problem or improving the security properties offered by NEMO BS, several approaches have been proposed so far to provide at the same time security services and route optimization. Baldessari et al. (2009) and Eddy et al. (2009) underlined the following security properties as mandatory in a NEMO RO scheme to be applied in the ITS context:

- *The RO scheme must permit the receiver of a BU to validate the ownership of the CoA claimed by the MR.* This requirement can be interpreted as the need of authentication mechanisms between the MR and the CN. In addition, as proof of ownership is needed a Certification Authority (CA) should certify that the claimed address is owned by the MR. Consequently, besides authentication procedures, a certificate validation process is demanded.
- *The RO scheme must ensure that only explicitly authorized MRs are able to exchange BUs for a specific mobile network prefix.* This requirement is related to the needed procedures to complete the aforementioned address ownership validation. Therefore, authentication mechanisms and a valid certificate are demanded.
- *The RO scheme must not further expose the mobile network prefix on the wireless link.* In other words, BU messages are required to be confidentiality protected not to expose its context to other nodes but to the intended CNs.

Apart from these mandatory security properties, features like data integrity, non-repudiation, reliability, privacy or accountability are desired. Although little work exists addressing security and RO issues, the most relevant approaches are

described and analyzed herein because the support of security and optimal routing are fundamental in the ITS context.

The introduction of security in RO scenarios is commonly known as the most challenging scenario. The main goal of NEMO solutions providing security in RO scenarios is to ensure address ownership and proxying rights, which is achieved with Cryptographically Generated Addresses (CGA) and its variants. Calderón et al. (2005) studied the approaches to secure RO and outline how to adapt host mobility management protocols in the NEMO scenarios. The conclusion stemming from this work is that delegation of signaling rights between MNNs and the MR covers the required address ownership and proxying rights proof. Furthermore, authors suggest directions on how to delegate this rights but a complete RO proposal is not suggested. Three main delegation approaches are studied in the article: delegation using PKI certificates, delegation using self-signed certificates and implicit delegation. Delegation based on PKI certificates is the most flexible solution but requires strong infrastructure support. On the other hand, implicit delegation is the simplest solution but has limitations like incompatibilities with other security mechanisms like SEND (SEcure Network Discovery) (Arkko, Deverapalli, & Dupont, 2005). Consequently, the delegation of signaling rights based on self-signed certificates is highlighted as the trade-off between complexity and flexibility. Authors do not select any of the studied solutions but this decision is outlined as operator preference dependant.

Jo and Inamura (2008) proposed a secure route optimization mechanism based on a variant of CGA: MCGA (Multi-key CGA). The MCGA is used to proof address ownership and demonstrate signaling rights authorization. The MNN configures its CoA by the MCGA mechanism. In order to secure the MCGA computation process, a modified SEND mechanism is considered in this approach. Using the MCGA, the MNN shares the address ownership with the MR and provides authorization to the MR to update the binding information. More precisely, the MNN sends to its HA the computed MCGA, and the MR sends a Proxy Binding Update (PBU) message to the HA where the MCGA and the MR's CoA are included. Based on the MCGA, the HA associates both nodes, the MNN and the MR. Therefore, whenever the MR handoffs, it sends a PBU message to the HA including the new CoA and the MCGA. Therefore, the location of the MNNs is updated by the MR.

Even being a proposal that aims at solving security challenges of the NEMO scenario, no formal security validation neither a meta-reasoning validation are described in the article. Consequently, leap of faith is required for the security provisioning of this approach. However, authors have carried out a real implementation to analyze the performance of the solution where the effect of RO has been analyzed, but the MCGA has not been implemented.

Although the strength of the solution proposed by Jo and Inamura (2008) resides in the simplicity of its implementation, the security goals are mainly related to authentication and proof of address ownership, while other features like end-to-end confidentiality and integrity protection are not considered in the paper.

Koo et al. (2010) proposed a solution to support authentication in RO schemes based on CGAs named CARO. One of the major weaknesses of CARO is that it assumes that IPsec ESP tunnels are established between the MNN and its MR, between the MR and its HA, between the CN and its MR and between this MR and its HA, but no further security mechanisms are described prior establishing the IPsec ESP tunnels. Notice that this approach considers that the CN may be placed in another mobile network.

CARO assumes that if each of the IPsec ESP tunnels is secure, the communication between MNNs and a CN will be secure. In fact, the pre-established tunnels are used for exchanging information with the goal of generating a session

key ($K_{MG}$) shared by the MR and the CN. These nodes are authenticated on the $K_{MG}$ key. Whether the authentication has been successful, BU messages are exchanged between the MR and the CN. It is worth pointing that the MR is in charge of processing the messages on behalf of the MNN. Furthermore, assuming that the CN is located in another mobile network, the BU message will be sent to the HA of the peer MR. Consequently, signaling messages for updating the location of the mobile network are not exchanged between the endpoints, but third parties are required to forward and process them. In fact, RO is only achieved in terms that the CN can bind the MNN's CoA (obtained considering the new mobile network prefix) with the HoA, but the signaling required for being this completed traverses both HAs (MR's and CN's). Consequently, the CARO solution can face performance constraints.

Authors have demonstrated the security properties of CARO by means of a meta-reasoning verification. The performance in terms of handover delay and computation cost is also described to provide a comparison table where NEMO BS and the proposal (Jo & Inamura, 2008) are placed. Results show that the CARO protocol does not outperform the other solutions. In addition, although authors assume that steady IPsec ESP tunnels are established between the involved entities, (MNN, MRs, HAs and CN) it is envisioned that this is not a realistic approach as it is likely to have several communications established between the MNNs and different CNs, which are not necessarily long term communications. Therefore, whenever a new communication between a certain MNN and a different CN is requested, a high amount of signaling is required to establish a new IPsec ESP tunnels and to agree the required session key for securing the BUs.

Kukec et al. (2010) presented CRYPTRON (CRYptographic Prefixes for Route Optimization in NEMO) focusing on the aeronautics scenario. CRYPTRON presents two specific goals: a new security tool (Crypto Prefixes) and a RO solution

*Table 4. Analysis of the security provisioning and number of signaling messages of solutions supporting RO and security*

| Desired key features for NEMO RO schemes in the ITS context | | Jo & Inamura (2008) | Koo *et al.* (2010) (CARO) | Kukec *et al.* (2010) (CRYPTRON) |
|---|---|---|---|---|
| The RO scheme must permit the receiver of a BU to validate the ownership of the CoA claimed by the MR. *(Authentication and Certification)* | | ✓ (MCGA) | ✓ (CGA) | ✓ (Crypto Prefixes) |
| The RO scheme must ensure that only explicitly authorized MRs are able to exchange BUs for a specific mobile network prefix. *(Authorization-Certification)* | | ✓ (MCGA) | ✓ (CGA) | ✓ (Crypto Prefixes) |
| The RO scheme must not further expose the mobile network prefix on the wireless link. *(Confidentiality)* | | X | ✓ | X |
| Integrity protection | *Signaling plane* | ✓ (signature) | ✓ (signature) | ✓ (Crypto Prefix signature) |
| | *Data plane* | X | X | X |
| N° of signaling messages required for handoff | | 6 | 15 (CN is in a mobile network) / 12 (CN is not in a mobile network) | 2 |

based on these Crypto Prefixes. Crypto Prefixes are IPv6 prefixes that contain embedded cryptographic information (a hash of a public key). The basic idea of this proposal resides in the ability to proof ownership by means of the cryptographic nature of the crypto prefixes, hence, not requiring any additional infrastructure. In fact, the prefix ownership is claimed by proving the knowledge of the corresponding private key. Using the cryptographic prefix, communications between the mobile network and the CN are secured. Therefore, CRYPTRON is a solution for securing the binding establishment between the MR's CoA and the CN, and it focuses on the authentication mechanism between the involved parties in the RO procedure.

Authors have also analyzed how CRYPTRON covers the NEMO RO requirements specified by the IETF for the aeronautics and space exploration mobile networks. They also assure that CRYPTRON outperforms other solutions (Calderón, Bernardos, & Bagnulo, 2005), (Wakikawa & Watari, 2004) and (Na, Cho, Kim, Kee, Kang, & Koo, 2004) in terms of number of signaling messages, handover delay and packet overhead, but no formal validation is presented in the work.

Table 4 resumes how the solutions that gather security and RO provide required features for the ITS context related to security and number of signaling messages, which in turn provides insight of their performance.

Besides aforementioned aspects, Table 5 shows the other demanded key features for being these solutions applicable in the ITS context.

The security provisioning in NEMO RO environments is mainly focused on supporting authentication of entities, but an overall security framework that addresses not only authentication but also other crucial features like confidentiality is not covered by the proposals. Indeed, to the best of our knowledge, no efficient MIPv6 based NEMO solution addressing basic security features (authentication, confidentiality and integrity) in RO scenarios is published throughout the literature.

*Table 5. Evaluation of the adaptability of demanded key features in the ITS context by the NEMO RO schemes supporting security*

| Demanded key features for NEMO in the ITS context | Jo & Inamura (2008) | Koo *et al.* (2010) (CARO) | Kukec *et al.* (2010) (CRYPTRON) |
|---|---|---|---|
| *Global Reachability* | ✓ | ✓ | ✓ |
| *Efficiency (RO)* | ✓ | ✓ | ✓ |
| *Nested Mobility* | X | X | X |
| *Mobility support inside the mobile network* | X | X | X |
| *Mobile CN* | X | ✓ | X |
| *Backward compatibility and adaptability* | ✓ | ✓ | ✓ |
| *Minimum impact on existing protocols and infrastructure* | X | ✓ | ✓ |
| *MR multihoming* | X | X | X |
| *Multiple MR support* | X | X | X |

Apart from network mobility solutions based on MIPv6, there are several works that consider HMIPv6, FMIPv6 or PMIPv6 to address NEMO management. HMIPv6 has been designed to reduce the signaling overhead, hence, the NEMO solution based on HMIPv6 inherit this property. Most relevant HMIPv6 based NEMO protocols are those defined by Tsai (2009). The mobility management by the network has been defined in the PMIPv6 protocol. Hence, PMIPv6 based NEMO protocols proposed by Pack et al. (2009) and Woo et al. (2010) addresses this scenario.

On the other hand, as FMIPv6 has the goal of enhancing the MIPv6 handover process, FMIPv6 based NEMO solutions intend to take benefit from this issue. FMIPv6 has been considered by Mussabir et al. (2007) to solve mobility of NEMO contexts. There are additional proposals that consider enhancing the handover process to fulfill application QoS requirements. The approach named FNEMO (Mitra et al. 2011) brings in the concept of IP pre-fetching and advance registration to acquire the care-of-address to be used in the new cell. The analytical evaluation carried out shows that FNEMO supports higher vehicle velocities compared to FMIPv6. Lee et al. (2008) proposed a QoS integrated cross-layer hierarchi-cal architecture to manage network mobility. This approach, named HiMIP-NEMO, demonstrates the advantages of combining route optimization techniques with resource allocation mechanisms to improve application performance. This solution supports higher vehicle velocities because the network part is in charge of managing the handover process.

## NEMO Management Solutions Based on Alternative Mobility Management Protocols

In general, the research community has been focused on introducing required mechanisms to the NEMO BS protocol to solve the open issues and enhance its properties. Consequently, little work exists where alternative protocols to MIPv6 are considered as the basic procedures to support NEMO management. The common aspect of the solutions grouped herein is that they consider novel design protocols that address from scratch some NEMO support requirements. In this way, it is straightforward to have those requirements satisfied in the NEMO solution. That is, the alternative protocols to MIPv6 that had been considered in the different approaches present characteristics like route optimization that are directly exploitable in NEMO scenarios.

This section describes the main approaches suggested in this area and provides an evaluation of the advantages and drawbacks each solution presents, highlighting whether the demanded key features in the ITS context are satisfied.

## Session Initiation Protocol Based NEMO Solutions (SIP-NEMO)

SIP is one of the most outstanding protocols considered to provide NEMO support. The IETF standardized the SIP protocol (Rosenberg et al., 2002) which was first designed for multimedia session management. However, thanks to its simplicity and flexibility other usages like mobility and/or location management have been proposed.

SIP is an application layer control protocol that consists on sending invitation messages to establish multimedia sessions. These invitation messages contain session descriptions to agree on the media types. SIP utilizes proxy servers to route requests to user's location, authenticate and authorize users and offer features to users. Thanks to the SIP registration function, users can upload their current location. These functionalities allow the utilization of SIP as a mobility management protocol.

Huang et al. (2006) proposed a SIP based NEMO management scheme (SIP-NEMO). In this work, authors describe a network mobility management solution that considers SIP as the base mobility management protocol. The main feature of the SIP-NEMO protocol is the inter-NEMO and intra-NEMO RO. The performance of SIP-NEMO is evaluated through simulation concluding that SIP-NEMO takes less time than NEMO-BS for reachability recovery and for data transmission, even having a complex nesting level. The following design goals are described before presenting the solution:

- *Global Reachability.* All SIP clients have to be reachable regardless the location of the NEMO in the internet.
- *Movement Transparency.* SIP clients should not detect the movement of the mobile network.
- *Backward compatibility.* SIP-NEMO has to be compatible with the previous SIP architecture. Any node supporting SIP should be able to enter a SIP-NEMO scenario and vice versa.
- *Route Optimization.* Packets between a node located inside the mobile network and a node located in the outside network should traverse the shortest path and avoid the *dog-leg routing*.
- *Signaling Reduction.* Not to diminish the quality of the exchanged data, signaling messages should be reduced.

SIP-NEMO defines a new architecture with three new types of SIP back-to-back user agents: SIP-NMS (SIP Network Mobility Server), which is an analogue entity to the MR; SIP-HS (SIP Home Server), which is the entity where SIP-NMS are registered and their current location is stored; and SIP-FS (SIP Foreign Server), which is in charge of forwarding the invite messages based on the SIP identifier. In the same way the functionalities of the HA defined in NEMO BS can be divided into the procedures defined for the SIP-HS and the SIP-FS. Figure 4 shows the SI^-NEMO architecture.

It can be observed that several design requirements specified by Ernst (2001) for NEMO BS are specified also by SIP-NEMO, but other crucial characteristics like security are not considered in this solution. It is worth mentioning that authors point out security provisioning as future work.

Huang et al. (2009a) had also further suggested additional properties to the SIP-NEMO solution. This solution presents multihoming support for the SIP-NEMO solution, solving the interface selection issue using the IEEE 802.21

*Figure 4. SIP-NEMO architecture*

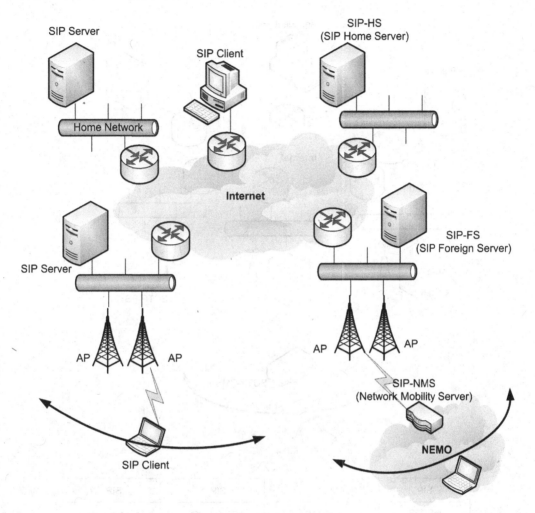

Media Independent Handover (MIH) protocol (IEEE Standards [IEEE], 2008). In this work, an analysis on how to extend SIP-NEMO to support multihoming is provided focusing on the *(1,\*,\*)* case; that is, a single MR, one or more HAs and one or more mobile network prefixes. Authors evaluate through simulation that SIP-NEMO outperforms NEMO BS, and that multihoming does not result in suboptimal routes, but packets are routed through the shortest path between the node inside the NEMO and the CN. Compared to SIP-NEMO, the multihomed SIP-NEMO solution requires modifications in the SIP-NMS node as the MIH functionalities have to be introduced. In

addition, in order to indicate the correspondent egress path, a new field is introduced in the SIP re-registration message to manage the multihoming scenario. Consequently, modifications in the SIP-HS nodes are also required.

As a step ahead to the multihoming support, Huang et al. (2009b) also presented a multiple MR solution. In this work, new extensions to SIP-NEMO are proposed. On the one hand, a new field to indicate the priority level of each egress path is included. On the other hand, new procedures are included to redirect SIP sessions between the different SIP-NMSs.

*Figure 5. SIP SCTP-NEMO architecture*

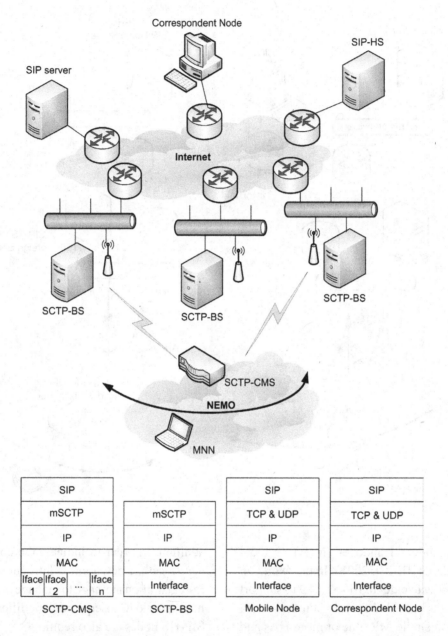

From the study of the SIP-NEMO solutions, it can be observed that a different set of functionalities is required to support each of the basic NEMO design requirements, increasing the complexity level of the solutions. As the complexity level of the solution increments with the supported requirements, a trade-off between the needs of the application scenario and the design of a full operative solution is fundamental. Consequently, it is interesting to have as much as possible properties supported in the basic host mobility management protocol specification.

Regarding the application of SIP-NEMO solutions in the ITS context, it is important to point out that, despite the rest of the design requirements are supported, security properties are not considered in the solutions but are only mentioned in the future works.

Apart from the proposals that consider the SIP protocol, another SIP based NEMO solution which includes the SCTP protocol has been proposed (Leu, 2009). The design goal of SIP-SCTP NEMO is to have minimal packet loss and minimal handover delay variation to support QoS in voice and multimedia applications. Authors in this work present an architecture where the following new entities are included: SCTP-CMS (SCTP Central Management Server), which acts as the gateway that can have several interfaces providing connectivity to the mobile network and speaks SIP and mSCTP protocol; SCTP-BS (SCTP Base Station), which is in charge of establishing and maintaining communication links with SCTP-CMSs, it is located in the access network and speaks the mSCTP protocol; and SCTP-HS (SCTP Home Server), which is the node where SIP users are registered. Figure 5 shows the SIP SCTP-NEMO architecture.

The SCTP-CMS and the SCTP-BS communicate by means of mSCTP control messages, and SIP is used for the signaling between the SCTP-CMS and the remote SIP clients. One of the most significant drawbacks of SIP-SCTP NEMO is that it assumes that data is forwarded from the old SCTP-BS to the new SCTP-BS. This scenario is challenging, because SCTP-BSs are likely to belong to different access providers and because these nodes can be using different wireless access technologies, and consequently, although authors demonstrate through simulation that the architecture is capable to ensure zero packet loss, data forwarding may be done through the internet backbone, increasing the delay of the communication. It is worth pointing that in this work, security issues are also considered for future works.

## Location Independent Networking for IPv6 (LIN6) Based NEMO Solutions (LIN6-NEMO)

LIN6 is a host mobility management protocol that aims at solving the tunnel and suboptimal routing header overhead of MIPv6. LIN6 separates the identifier and the locator of nodes assigning a unique *generalized identifier* to the node. This generalized identifier is used to identify the nodes in the transport layer and above. The generalized identifier is composed of the LIN6 prefix (a constant value) and the LIN6 identifier (a 64 bit identifier). As its name suggests, this generalized identifier is used for identifying the node. The routing to the node is done using the LIN6 address which is composed of a network prefix and the LIN6 identifier. Another significant feature of LIN6 is that the LIN6 prefix is translated into the network prefix and vice versa for outgoing and incoming packets correspondingly. In order to store the relation between the LIN6 address and the generalized identifier of a node the LIN6 protocol defines a Mapping Agent.

Oiwa et al. (2003) proposed a NEMO solution based on LIN6. The design goals of LIN6-NEMO are route optimization and minimum signaling overload. In addition, the protocol allows mobile nodes to move inside or outside the mobile networks.

LIN6-NEMO proposes an architecture where a LIN6 capable MR and an extended Mapping Agent are defined. In order to inform the Mapping Agent about the new location of the mobile network, a new procedure for location registration is described, where the option to indicate whether the registration belongs to an isolated mobile node or to a mobile network is introduced. Figure 6 shows the LIN6-NEMO architecture.

One of the main disadvantages of LIN6-NEMO is that a signaling storm is caused when the mobile network moves, because mobile nodes inside the mobile network must send the Mapping Update to the communicating nodes even not needing to

*Figure 6. LIN6-NEMO architecture*

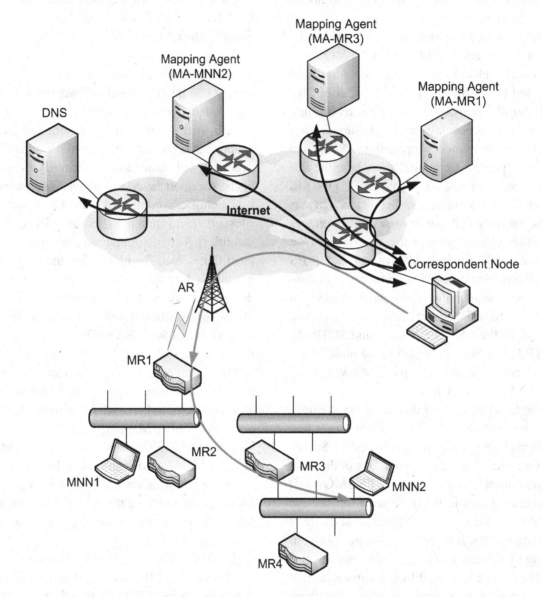

register the new location in the Mapping Agent. The signaling storm caused is worsened in a nested mobile scenario. The evaluation carried out over LIN6-NEMO demonstrated that the prefix overwriting process and the queries of the CNs to the Mapping Agents to obtain the current location of the mobile node might not cause serious delays. On the other hand, even authors point that security provisioning is indispensable; LIN6-

NEMO does not define any mechanism to support security.

Banno and Teraoka (2006) proposed an approach named *v*LIN6. This approach supports host mobility with optimal routing without header overhead, and network mobility with a single packet relay in the Mapping Agent. The advantage of this solution is that it only requires a single relay regardless the nesting level of the solution. Compared to NEMO BS, main differences reside in the header overhead and packet routing. On the one hand, as a result of rewriting the address by the Mapping Agent, LIN6-NEMO and *v*LIN6 do not perform tunneling and consequently header overhead is not introduced. On the other hand, in LIN6-NEMO packets are always forwarded to the mobile node and in *v*LIN6 a single relay is required regardless the nesting level. Consequently, it can be determined that the LIN6 based NEMO approaches outperform the NEMO BS protocol with respect to route optimization and packet overhead.

## Host Identity Protocol Based NEMO Solution (HIP-NEMO)

In the same way as LIN6, the cornerstone of HIP is the separation of node's location and identifier by means of introducing a new namespace for identifying the nodes, named Host Identity (HI). This identifier is a public-private key pair, which in practice a hashed value of this HI, i.e. Host Identity Tag (HIT) is utilized. Hence, HIP is understood as a new layer between network and transport layers. Meanwhile, HIP tackles security as a basic part of its design due to utilizing public key cryptography and other security mechanisms.

HIP defines a base exchange (BEX) to establish Security Associations (SA) relying on the Diffie-Hellman protocol (D-H) (Rescorla *et al.*, 1999) to generate keying material for the association. In order to map the identifier and the locator a Rendezvous Server (RVS) (Laganier *et al.*, 2008) where the IP addresses and the HIT of the hosts are stored, is defined. It should be noticed that no previous credential distribution is required but knowing the HIT is sufficient. Moreover, the use of a PKI (Public Key Infrastructure) is not mandatory. Notice that a flat correspondence between a RVS and a Home Agent (HA) defined in Mobile IP may be found. However, in HIP data does not pass through the RVS in Mobile IP alike, but it is used to provide global reachability of mobile nodes. Consequently, route optimization

is ensured. IPsec ESP (Kent, 2005) is adopted to transport data securely. Hence, there is no HIP specific data packet defined, but the standard IPsec ESP is used.

Main advantages that can be distinguished from the HIP protocol compared to basic MIPv6 are the following: security properties, route optimization and no need to establish tunnels. It is important to highlight that the provision of NEMO support based on HIP is not straightforward. One of the main issues when covering NEMO scenarios using end-to-end mobility management protocols is the manageability. That is, managing independently all end-to-end associations between the MNNs and CNs will result in an increased complexity and a significant overhead, as already mentioned in this chapter. However, because HIP gathers security and route optimization mechanisms, the most outstanding HIP based NEMO protocols are stressed and provide a detailed overview of them.

Melen et al. (2009) worked out a HIP based NEMO solution which was proposed to the IETF. In general terms, the solution makes use of asymmetric authentication and symmetric authorization based on delegation of signaling rights to provide end-to-end secure mobility management.

MNNs register with the MR through an extended HIP BEX. In order to authorize the MR to perform signaling exchange on behalf of them, MNNs will delegate their signaling rights to the MR after the registration phase. The security of this delegation is based on sharing a key between the MNN and MR inferred from a portion of the keying material generated during the end-to-end HIP BEX between the MNN and CN. With this key, the MR can create protected messages on behalf of the MNN; hence, the MR is able to represent the MNN by means of symmetric cryptography. The MNN sends a ticket to the MR with the new key to protect messages on behalf of the MNN. Additionally, this ticket may include ancillary data such as ticket lifetime or authorized actions that can be encrypted end-to-end. Figure 7 clarifies the signaling right delegation process:

*Figure 7. Delegation of signaling rights in the approach of Melen et al. (2009)*

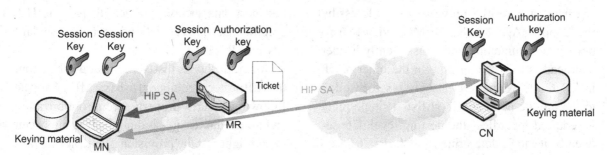

As IP addresses of the MNNs are considered to be non-publicly available addresses, the MR is in charge of carrying out address mappings when a MNN is communicating with a CN. More precisely, this approach recommends owning link-local subnet addresses or unique local sub-network addresses by the MNN. Hence, the MR establishes a binding state and tracks each active session. In this approach, a Security Parameter Index multiplexed Network Address Translation (SPINAT) (Melen *et al.*, 2008) state, mapping MNN's IP address together with the SPI value to corresponding public values is created. As SPI values and ESP headers are integrity protected, the MR cannot transparently translate SPI values, but it should notify the endpoints about this translation. Notice that from the CN's point of view, the MNN is reachable at the MR's locator. That is, the CN may know the IP address of the MR, but additional information to determine the communication endpoint should be provided. It is worth pointing that how MNN's are accessed from the outside network for the first time is not defined in this approach.

Regarding NEMO mobility management in this solution, the MR generates HIP association update messages to every CN that MNNs are connected with. This update includes the new IP address of the MNN and authentication information extracted from the ticket, while being signed using the authorization ticket obtained from the MNN.

This solution also tackles end-to-end update messages, thereby, end-to-end association rekeying and/or MNNs locator changes. The end-to-end update is not only for updating the state in the CN but also for creating a state in the MR. This state creation is required as the MR should know which is the current locator of the MNN in order to carry out the required SPINAT process and send the packets to the correct destination. It should be noticed that in order to avoid unnecessary signaling end-to-end updates should not take place because MNNs are located in the MR's IP address. The protocol entire flow chart is shown in Figure 8.

As this approach recommends avoiding renumbering (the action of every MNN configuring a new IP address from the new network prefix advertised by the MR inside the mobile network), non-globally routable addresses are suggested to be used by MNNs. The cost of this optimization is having a state per each HIP session in the MR to translate the IP addresses of MNNs into the globally routable IP address of the MR itself. Moreover, the solution does not specify entities or procedures to first contact a MNN, but limits the solution to the cases where the MNNs begin the communication.

Yitalo et al. (2008) extended the approach proposed by Melen et al. (2008) to minimize the mobility management signaling sent over the wireless hop, together with the renumbering events, not sacrificing RO advantages. In order to do so, this approach proposes utilizing on the path

*Figure 8. Flow chart of the Melen et al. (2009) approach*

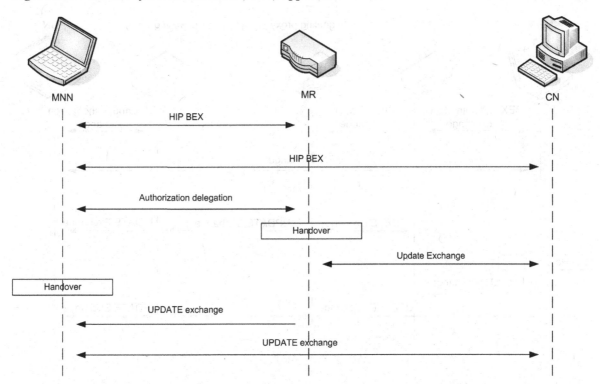

signaling proxies located in the fixed network to avoid wireless bandwidth consumption with the signaling storm and prevent ongoing sessions from QoS degradation.

In this approach, when the MNN enters the NEMO, it authorizes the MR to further authorize a signaling proxy. The authorization chain consists of two certificates: one certificate for authorizing the MR, and another certificate for the MR delegating signaling rights to the signaling proxy on the path. When the MR handoffs it sends a single location update per signaling proxy, where the authorization certificate is included. In addition, CNs are notified by MR's signaling proxies about the new address of the MR. Notice that the *on-the-path* necessity of signaling proxies on the fixed network requires a complex security management framework because of the mobile nature of the NEMO and/or the CNs.

This approach also considers the scenario where CN delegates signaling rights to a proxy

in the fixed network as peers may also use wireless access technologies. CNs include HIs of the proxies in the first end-to-end exchange in order to make MNNs learn to whom signaling right delegation has been authorized. When the MN changes its locator or decides to rekey, it sends update messages to the signaling proxies of the MNNs instead of the CNs directly. These location messages include association information to locate the CNs; therefore, the CN's signaling proxy informs the CN about the end-to-end association update requested by the MNN.

When the MR handoffs, it sends a location update message to its proxy, who forwards this message to the proxies where CNs have delegated their rights. In order to know which these signaling proxies are, the CN should include information in the HIP association establishment run between the MNN and the CN. In addition, CNs are informed about the new location of the MR. Figure 9 shows the described flow chart.

*Figure 9. Flow chart of the Yitalo et al. (2008) approach*

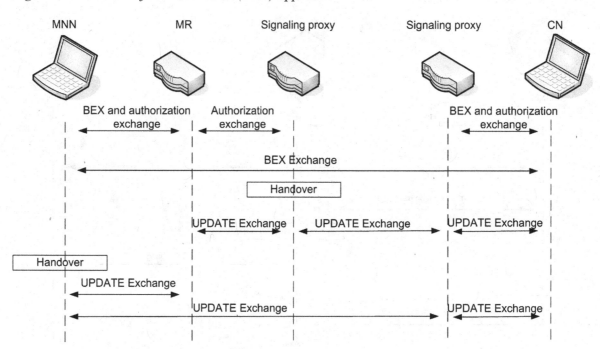

This approach can be understood as a signaling optimization regarding mainly scalability issues as long as aggregation solutions are introduced. That is, different signaling proxies should be in charge of managing the communications coming from different mobile networks. The same applies to CN's signaling proxies. The need of being the signaling proxies on the path makes the aggregation difficult. With such optimization, the solution may end missing the RO idea and approaching to the NEMO BS solution, as these new entities may acquire HA functionalities.

Nováczki et al. (2008) proposed an approach based on defining a new entity named mRVS (mobile RVS). The mRVS has the same role of the MR and provides HIP-based services to MNNs. However, as in nature it is also a mobile entity, the mRVS itself has got a RVS in order to provide global reachability of the NEMO. In this approach, the mRVS is responsible of registering MNNs in the RVS. Consequently, MNNs will be registered in the mRVS and in the RVS.

For establishing and end-to-end association between a CN and a MNN through the RVS, it is not necessary to know the address of the MNN, but only the HIT. This is because the RVS is in charge of providing the mapping between the HIT and the IP address of the destination.

When a MNN enters the NEMO, it receives service announcements from the mRVS, which trigger the registration of the MNNs in the mRVS. In the registration process, the MNN delegates its signaling rights to the mRVS through an authorization certificate. The purpose of this registration is to have the HIP-IP mapping stored in the mRVS. The mRVS links this mapping with a globally routable and topologically correct address, which is allocated and assigned to the MNN by the mRVS. In this point, the mRVS registers the MNN in the RVS to provide global accessibility to it, and stores the HIT of mRVS.

Regarding the mobility management of the NEMO, the mRVS acquires a new address which is used as the source IP address in the outgoing communications. That is, the mRVS is in charge

*Figure 10. Flow chart of the Nováczki et al. approach*

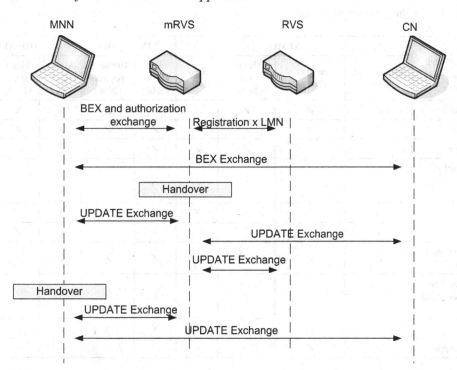

of translating the source IP address of the packet generated by the MNNs. As the mRVS has rights to perform signaling on behalf of the MNN, it is able to send an update notification containing the new prefix. Having received this packet, the RVS will update the binding between mRVS's and MNN's HIT, and MNN's address. The advantage of including the mRVS's HIT in the mapping is that all nodes indexed to the same mRVS will be updated with a single update exchange. Notice that as renumbering is avoided by means of address translation, a state per association is required in the mRVS. On the other hand, as delegation of signaling rights is present the mRVS will also update established HIP associations with CNs.

This approach does not define specific mechanisms to manage mobility and/or rekeying necessities of the MNNs, but end-to-end signaling messages are exchanged. Consequently, the advantages of delegating signaling rights are not considered in this particular scenario.

It is worth mentioning that this approach provides global accessibility to MNNs for the first time introducing the RVS, whereas does not specify mobility management solutions for MNNs. Figure 10 shows the flow chart of this approach:

These aforementioned approaches adopt the delegation of signaling rights approach (Nikander *et al.*, 2004) between the MNNs and the MR to reduce signaling overhead, optimize the signaling path and increase the manageability. An important issue regarding delegation is related to the stateless nature of the delegator, a mobile node supporting HIP. That is, if the node who is delegating its signaling rights changes its state, and this change results in requiring a signaling exchange with the peer node, the delegate must be aware of these changes. Consequently, notifications between the delegator and the delegate are required.

HIP generates confidential data (keys) during the end-to-end association establishment which has to be updated (rekeying) when mobility is

*Table 6. Analysis of the support of the demanded key features by the NEMO protocols based on alternative host mobility management protocols*

| Demanded key features for network mobility support in the ITS scenario | SIP-NEMO | | | | LIN6-NEMO | | HIP-NEMO | | |
|---|---|---|---|---|---|---|---|---|---|
| | Huang *et al.* (2006) | Huang *et al.* (2009a) | Huang *et al.* (2009b) | Leu (2009) | Oiwa *et al.* (2003) | Banno & Teraoka (2006) | Melen *et al.* (2009) | Yitalo *et al.* (2008) | Nováczki *et al.* (2008) |
| *Global reachability* | ✓ | ✓ | ✓ | ✓ | ✓ | ✓ | X | X | ✓ |
| *Efficiency - RO* | ✓ | ✓ | ✓ | ✓ | ✓ | ✓ | ✓ | ✓ | ✓ |
| *Nested Mobility* | ✓ | ✓ | ✓ | ✓ | ✓ | ✓ | ✓ | ✓ | ✓ |
| *Mobility support inside the mobile network* | ✓ | ✓ | ✓ | ✓ | X | X | ✓ | ✓ | ✓ |
| *Mobile CN* | X | X | X | X | X | X | X | ✓ | X |
| *Security* | - | - | - | - | - | - | ✓ | ✓ | ✓ |
| *Backward compatibility and adaptability* | ✓ | ✓ | ✓ | ✓ | ✓ | ✓ | ✓ | ✓ | ✓ |
| *Minimum impact on existing protocols and infrastructure* | ✓ | X | ✓ | X | X | X | X | X | X |
| *MR multihoming* | - | ✓ | ✓ | X | X | X | ✓ | ✓ | ✓ |
| *Multiple MR support* | - | ✓ | - | - | - | - | - | - | - |

present by means of signaling exchanges. Having delegated signaling rights, the MR will be in charge of maintaining the established end-to-end security association. But, this delegation of signaling rights presents challenges when it is needed to ensure the secrecy of information that should not be revealed to third parties. So, it is commonly understood that end-to-end rekeying agreements by the MR is a security vulnerability. More specifically, being the MR in charge of rekeying, end-to-end confidentiality and integrity are not ensured. Therefore, the case when the NEMO changes its point of attachment and associations are rekeyed by the MR while end-to-end integrity and confidentiality are ensured is still challenging. The straightforward solution would be to perform end-to-end signaling exchanges for updating the associations but then the advantages of delegating signaling rights would be lost. It has to be pointed out that up to now none of the HIP based NEMO protocols solve mobility including rekeying while

they take advantage of delegation of signaling rights benefits.

## Summary

A comparison between the different approaches based on alternative mobility management protocols to MIPv6 is provided in Table 6, analyzing how the requirements of NEMO solutions for the ITS context are fulfilled.

In summary, and analyzing how MIPv6 based NEMO proposals address demanded key features for the ITS context, the most interesting design approach is to consider a base host mobility management protocol that supports as much as possible requirements from scratch. This approach will mitigate compatibility problems between the different add-ons and the flaws derived from the integration of them will be decreased. In addition, the complexity of the solution will be significantly reduced.

# FUTURE RESEARCH DIRECTIONS

Internet connectivity while traveling is a flourishing demand that needs to be undertaken. However, the introduction of ICT (Information and Communication Technologies) in the ITS context is slowing down. This is mainly because there are several agents that have to cooperate in order to have a communication system deployed in the ITS context.

These agents embrace vehicle (cars, aircrafts, trains, ships,…) manufacturers; ISPs (Internet Service Providers) which have a large heterogeneity in the access network; Internet mobility operators that aim at managing the mobility of the ongoing communications; standardization bodies who are in charge of defining standardized protocols and procedures; and governmental organizations that are expected to legislate the infrastructure deployment and its utilization. Having such a variety of agents involved in the introduction of ICT in the ITS context requires strong mutual effort for decision taking and best solution design. Furthermore, this solution should be globally unique and should consider all the necessary features for optimizing the interests of every involved agent when possible. In addition, this global unique solution should fulfill international laws related to safety and security. In addition, this scenario leads to do research on new business models. Consequently, research efforts should be channeled into optimum cooperation procedures.

Currently, the validation of the different NEMO proposals is being carried out by means of mathematical modeling, simulation, formally validating the security properties through specific tools or mathematics, and small testbeds deployed in relatively controlled areas. As in any validation process, validation of NEMO protocols requires assumptions which in the ITS context usually result in not defining real deployment scenarios. Issues like the number of cars, when the ITS context refers to the automotive scenario; the impact of the buildings in the propagation of the signal and the interference of other communication systems when urban areas are considered; the velocity of the vehicle, when assuming high speed trains or aircrafts; the number of nodes located inside the mobile network, when modeling a large cruise ship, etc. have to be further studied. Consequently, work should be carried out in defining proper validation scenarios. Regarding measurements done for the NEMO protocol validation, almost no work exists in analyzing the fulfillments of user applications. Furthermore, the validation process should move to real scenarios introducing the solution to complex environments in order to ascertain the feasibility of the NEMO solutions.

Research in NEMO protocol design for the ITS context should focus on two different aspects. On the one hand, a complete NEMO protocol that addresses at the same time the most important shortcomings of NEMO BS has to be defined. On the other hand, work should be done in introducing the demanded key features for the ITS context that are less stringent but at the same time are fundamental to aid in safety purposes and to have a wide scope of services deployed for users. In addition, research should also focus on lower layers defining new propagation models in accordance with the different aforementioned ITS contexts. Consequently, all the procedures and protocols involved in the communication stack should evolve to consider specific scenario characteristics. In addition, applications to be deployed in the different ITS scenarios should be defined together with their QoS requirements. The IETF is aware of this latter need and there is ongoing work (Karagiannis et al., 2010), but further research is required in this field.

In conclusion, research on NEMO ITS should seek new design solutions that enable the development of new philosophical basis and stem valuable ideas for the definition of future network mobility systems, which will be a stepping stone for the Future Internet.

## CONCLUSION

Internet connectivity service while traveling is gaining *momentum* within the development of ITS technologies. In fact, standardization bodies like ISO or ETSI have lately considered the introduction of emerging procedures and protocols to support the demanding needs. A fundamental key characteristic required for providing Internet connectivity is the management of mobile networks, which is achieved by means of the NEMO (NEtwork MObility) protocols.

This chapter has provided an overview of the NEMO BS protocol, which is the most outstanding solution for managing the mobility of networks and the solution adopted by the standardization bodies, highlighting its main shortcomings. Being aware of these drawbacks, the international research community is actively working on improving the NEMO BS protocol. Because of this, several approaches have been proposed so far. In this chapter the most relevant proposals have been described and provide an analysis of their characteristics outlining their advantages and disadvantages. In addition, this chapter has underlined the demanded key features by NEMO protocols to be applied in the ITS context and map these key features with the features of the analyzed proposals.

As a result of the analysis, it can be concluded that the most interesting design approach is to consider a base host mobility management protocol that defines as much as possible key features from scratch. By doing this, the complexity of the resultant NEMO protocol will be controlled and problems arising from solutions that are tackled by means of add-ons will be mitigated. Therefore, apart from addressing the demanded key feature for the ITS context, the NEMO solution will have its integrity, availability, fault tolerance and deployment opportunities enhanced. Although several research efforts are being carried out by the research community, industry and standardization bodies, work should continue in order to bring forward the *anywhere-anytime-anyhow*

connectivity to Internet to the always moving globalized world.

## REFERENCES

Arkko, J., Deverapalli, V., & Dupont F. (2004). *Using IPsec to protect mobile IPv6 signaling between mobile nodes and home agents*. IETF RFC 3776.

Arkko, J., Kempf, J., Zill, B., & Nikander, P. (2005). *Secure network discovery*. IETF RFC 3971.

Baldessari, R., Ernst, T., Festag, A., & Lenardi, M. (2009). *IETF draft*. draft-ietf-mext-nemo-ro-automotive-req-02.

Baldessari, R., Festag, A., & Abeille, J. (2007a). *NEMO meets VANET: A deployability analysis of network mobility in vehicular communications*. International Conference on ITS Telecommunications (ITST). Sophia. France.

Baldessari, R., Festag, A., & Lenardi, M. (2007b). *C2C-C consortium requirements for usage of NEMO in VANETs*. IETF draft, draft-baldessari-c2ccc-nemo-req-00.

Banno, A., & Teraoka, F. (2006). *LIN6: An efficient network mobility protocol in IPv6. Information Networking: Advances in Data Communication and Wireless Networks, LNCS 3961* (pp. 3–10). Berlin, Germany: Springer-Verlag.

Bernardos, C. J., Soto, I., Calderón, M., Boavida, F., & Azcorra, A. (2007). VARON: Vehicular ad hoc route optimization for NEMO. *Computer Communications*, *30*, 1765–1784. doi:10.1016/j.comcom.2007.02.011

Bhagwat, P., Perkins, C., & Tripathi, S. (1996). Network layer mobility: An architecture and survey. *IEEE Personal Communications*, *3*(3), 54–64. doi:10.1109/98.511765

Bournelle, J., Valadon, G., Binet, D., Zrelli, S., Laurent-Maknavicius, M., & Combes, J. M. (2006). *AAA considerations within several NEMO deployment scenarios*. International Workshop on Network Mobility (WONEMO). Sendai, Japan.

Calderón, M., Bernardos, C. J., & Bagnulo, M. (2005). Securing route optimization in NEMO. *International Symposium on Modeling and Optimization in Mobile, Ad Hoc and Wireless Networks, WIOPT2005* (pp.248).

Cooperative Vehicle-to-Infrastructure System. (2006). *CVIS project*. Retrieved from http://www.cvisproject.org

Devarapalli, V., Wakikawa, R., Petrescu, A., & Thubert, P. (2005). *Network mobility (NEMO) basic support protocol*. IETF RFC 3963.

Eddy, W., Ivanvic, W., & Davis, T. (2009). *Network mobility route optimization requirements for operational use in aeronautics and space exploration mobile networks*. IETF RFC 5522.

Ernst, T. (2001). *Network mobility support in IPv6*. Doctoral Dissertation, University Joseph Fourier.

Ernst, T., & Lach. H.-Y. (2007). *Network mobility support terminology*. IETF RFC 4885.

Ernst, T., & Charbon, J. (2004). *Multi-homing with NEMO basic support*. In International Conference on Mobile Computing and Ubiquitous Computing.

ETSI EN 302 665. (2010). *Intelligent transport systems (ITS): Communications architecture.*

Fathi, H., Shin, S.-H., Kobara, K., & Chakraborty, S.-S. (2006). LR-AKE-based AAA for network mobility (NEMO) over wireless links. *IEEE Journal on Selected Areas in Communications, 24*(9), 1725–1737. doi:10.1109/JSAC.2006.875111

Fathi, H., Shin, S.-H., Kobara, K., & Imai, H. (2007). Secure AAA and mobility for nested mobile networks. In *International Conference on Intelligent Transportation Systems Telecommunications, ITST'07* (pp. 1-6). IEEE Communications Society.

Gundavelli, S., Devarapalli, V., Chowdhury, K., & Patial, B. (2008). *Proxy mobile IPv6*. IETF RFC 5213.

Huang, C. M., Lee, C. H., & Tseng, P. H. (2009a). Multihomed SIP-based network mobility using IEEE 802.21 media independent handover. In *IEEE International Conference on Communications ICC2009,* (pp. 1-5).

Huang, C. M., Lee, C. H., & Tseng, P. H. (2009b). Multiple router management for SIP-based network mobility. In *International Symposium on Computers and Communications, ISCC2009* (pp. 863-868). IEEE Communications Society.

Huang, C. M., Lee, C. H., & Zheng, J. R. (2006). A novel SIP-based route optimization for network mobility. *IEEE Journal on Selected Areas in Communications, 24*(9), 1682–1691. doi:10.1109/JSAC.2006.875113

Jemma, I. B., Tsukada, M., Menouar, H., & Ernst, T. (2010). *Validation and evaluation of NEMO in VANET using geographic routing*. In International Conference on Intelligent Transportation Systems Telecommunications, ITST. Kyoto, Japan.

Jiang, Y., Shi, M., & Shen, X. (2006). Multiple key sharing and distribution scheme with (n,t) threshold for NEMO group communications. *IEEE Journal on Selected Areas in Communications, 24*(9), 1738–1747. doi:10.1109/JSAC.2006.875114

Jo, M., & Inamura, H. (2008). Secure route optimization for mobile network node using secure address proxying. *IEEE Network Operations and Management Symposium NOMS2008, 7,* 137-143. IEEE Communications Society.

Johnson, D., Perkins, C., & Arkko, J. (2004). *Mobility support for IPv6*. IETF RFC 3775.

Jung, S., Zhao, F., Felix Wu, S., & Kim, H. (2004). *Threat analysis of network mobility (NEMO)*. Information and Communication Security, LNCS (Vol. 3264, pp. 333–337). Berlin, Germany: Springer-Verlag.

Karagiannis, G., Wakikawa, R., Kenny, J., Bernardos, C. J., & Kargl, F. (2010). *Traffic safety applications requirements*. IETF Draft. Draft-karagiannis-traffic-safety-requirements-02.

Kent, S. (2005). *IP encapsulating security payload (ESP)*. IETF RFC 4303.

Kin, T. K., & Samsudin, A. (2007). *Efficient NEMO security management via CA-PKI*. In International Conference on Telecommunications and Malaysia International Conference on Communications, ICT-MICC2007.

Koo, J.-D., Oh, S. H., & Lee, C. (2010). Authenticated route optimization scheme for network mobility (NEMO) support in heterogeneous networks. *International Journal of Communication Systems, 23*, 1252–1267. doi:10.1002/dac.1112

Koodli, R. (2009). *Mobile IPv6 fast handovers*. IETF RFC 5568.

Kukec, A., Bagnulo, M., & Oliva, A. (2010). CRYPTRON: CRYptographic prefixes for route optimization in NEMO. In *International Conference on Communications, ICC2010* (pp. 1-5).

Laganier, J., & Eggert, L. (2008). Host identity protocol (HIP) rendezvous extension. *IETF RFC 5204*.

Lee, C. W., Chen, M. C., & Sun, Y. S. (2008). *A network mobility management scheme for fast QoS handover*. In International Wireless Internet Conference. Hawaii. USA.

Leu, F. Y. (2009). A novel network mobility handoff scheme using SIP and SCTP for multimedia applications. *Journal of Network and Computer Applications, 32*, 1073–1091. doi:10.1016/j.jnca.2009.02.007

Li, F., Yang, Y., & Wu, J. (2009). *Mobility management in MANETs: Exploit the positive impacts of mobility. Guide to Wireless Ad Hoc Networks, Computer Communications and Networks Series* (pp. 211–235). London, UK: Springer-Verlag.

McCarthy, B., Edwards, C., & Dunmore, M. (2009). *Using NEMO to support the global reachability of MANET nodes*. In IEEE International Conference on Computer Communications (INFOCOM). Rio de Janeiro, Brazil.

Melen, J., Yitalo, J., & Samela, P. (2008). *Security parameter index multiplexed network address translation (SPINAT)*. IETF Draft, draft-melenspinat-01.

Melen, J., Yitalo, J., Samela, P., & Henderson, T. (2009). *Host identity protocol-based mobile router (HIPMR)*. IETF draft, draft-melen-hip-mr-02.

Mitra, A., Sardar, B., & Saha, D. (2011). Efficient management of fast handoff in wireless network mobility (NEMO). Indian Institute of Management Calcutta. *Working Paper Series WPS, N° 671*.

Mussabir, Q. B., Yao, W., Niu, Z., & Fu, X. (2007). Optimized FMIPv6 using IEEE 802.21 MIH services in vehicular networks. *IEEE Transactions on Vehicular Technology, 56*, 3397. doi:10.1109/TVT.2007.906987

Myles, A., & Skellern, D. (1993). Comparing four IP based mobile host protocols. In *Joint European Networking Conference* (pp. 191–196). Macquarie University, Sydney, Australia.

Na, J., Cho, S., Kim, C., Kee, S., Kang, H., & Koo, C. (2004). *Route optimization scheme based on path control header*. IETF Draft, draft-na-nemo-path-control-header-00.

Ng, C., Thubert, P., Watari, M., & Zhao, F. (2007). *Network mobility route optimization problem statement*. IETF RFC 4888.

Ng, C., Zhao, F., Watari, M., & Thubert, P. (2007). *Network mobility route optimization solution space analysis*. IETF RFC 4889.

Nikander, P., & Arkko, J. (2004). Delegation of signaling rights. *Security Protocols, LNCS, 2845*, 203–214. doi:10.1007/978-3-540-39871-4_17

Nováczki, S., Bokor, L., Jeney, G., & Imre, S. (2008). Design and evaluation of novel HIP-based network mobility protocol. *Journal of Networks, 3*(1), 10–24. doi:10.4304/jnw.3.1.10-24

Oiwa, T., Kunishi, M., Ishiyama, M., Kohno, M., & Teraoka, F. (2003). A network mobility protocol based on LIN6. In *IEEE Vehicular Technology Conference, vol. 3,* (pp. 1984-1988). IEEE Communications Society.

Pack, S., Kwon, T., Choi, Y., & Paik, E. K. (2009). An adaptative network mobility support protocol in hierarchical mobile IPv6 networks. In *IEEE Transcations on Vehicular Technology Conference, 58,* (p. 3627).

Petrescu, A., Olivereau, A., Janneteau, C., & Lach, H.-Y. (2004). *Threats for basic network mobility support (NEMO threats)*. draft-petrescu-nemo-threats-xx.

Phang, S. Y., Lee, H. J., & Lim, H. (2007). *A secure deployment framework of NEMO (network mobility) with firewall traversal and AAA server*. In International Conference on Coverage Information Technology. IEEE Computer Society.

Prasad, R., & Muñoz, L. (2003). *WLANs and WPANs towards 4G wireless*. Norwood, MA: Artech House.

Rescorla, E. (1999). *Diffie-Hellman key agreement method*. IETF RFC 2631.

Rosenberg, J., Camarillo, G., Johnston, A., Peterson, J., Sparks, R., Handley, M., & Schooler, E. (2002). *SIP: Session initiation protocol*. IETF RFC 3261.

Shahriar, A. Z. M., Atiquzzaman, M., & Ivanvic, W. (2010). Route optimization in network mobility: Solutions, classification, comparison and future research directions. *IEEE Communications Survey and Tutorials, 12*(1), 24–38. doi:10.1109/SURV.2010.020110.00087

Shin, S., Kobata, K., & Imai, H. (2005). A simple leakage-resilient authenticated key establishment protocol, its extensions and applications. *IECE Transactions in Fundamentals. E (Norwalk, Conn.), 88-A*(3), 248–254.

Soliman, H., Castellucia, C., El Malki, K., & Bellier, L. (2008). *Hierarchical mobile IPv6 (HMIPv6) mobility management*. IETF RFC 5380.

IEEE Std. 802.21. (2008). *IEEE standard for local and metropolitan area networks-Part 21: Media independent handover*. May 2008. ISO-21217-CALM-Architecture. (2010). *Intelligent transport systems - Communications access for land mobiles (CALM) - Architecture*.

Tsai, C. S. (2009). Designing a novel mobility management scheme for enhancing the binding update of HMIPv6 with NEMO environment. In *International Conference on Future Networks* (pp. 87-91). IEEE Communications Society.

Wakikawa, R., Thurbet, P., Boot, T., Bound, J., & McCarthy, B. (2007). *Problem statement and requirements for MANEMO*. IETF Draft, draft-wakikawa-manemo-problem-statement.

Wakikawa, R., & Watari, M. (2004). *Optimized route cache control protocol (ORC)*. IETF Draft, draft-wakikawa-nemo-orc-01.

Woo, M. S., Lee, H. B., Han, Y. H., & Min, S. G. (2010). A tunnel compress scheme for PMIPv6-based nested NEMO. In *International Conference on Wireless Communications Networking and Mobile Computing, WiCOM2010* (p. 1).

Yitalo, J., Melen, J., Samela, P., & Petander, H. (2008). *An experimental evaluation of a HIP based network mobility scheme. WWIC 2008, LNCS 5031* (pp. 139–151). Berlin, Germany: Springer-Verlag.

## ADDITIONAL READING

Aura, T. (2005). Cryptographically Generated Addresses, *IETF RFC 3972.*

Baldessari, R., Festag, A., & Lenardi, M. (2007). C2C-C Consortium Requirements for Usage of NEMO in VANETs. *IETF Draft*, draft-baldessari-c2ccc-nemo-req-00.

Bernardos, C.J., Oliva de la, A., Calderón, A., Hugo von D., & Kahle, H. (2005). NEMO: Network Mobility. Bringing ubiquity to the Internet Access. *The European Online Magazine for the Information Technologies. Profesional.*, *VI*(2), 36–42.

Car 2 Car Communication Consortium Manifesto. *Overview of the C2C-CC System*, August 2008.

Carpenter, B. (1996). Architectural Principles of the Internet. *IETF RFC 1958.*

Carracedo, J. (Ed.). (2004). *Seguridad en redes telemáticas*. McGraw-Hill/Interamericana de España, S.A.U.

Ellison, C., & Schneier, B. (2000). Ten Risks of PKI: What You're not Being Told about Public Key Infrastructure. *Computer Security Journal*, *XVI*(1), 1–7.

ETSI EN 302 665. Intelligent Transport Systems (ITS); Communications Architecutre, September 2010.

Gurtov, A. (Ed.). (2008). *Host Identity Protocol (HIP). Towards the Secure Mobile Internet*. John Wiley & Sons. doi:10.1002/9780470772898

Heikkinen, S., Kinnari, S., & Heikkinen, K. (2009). Security and User Guidelines for the Design of the Future Networked Systems, In *International Conference on Digital Society, ICDS2009* (pp. 13–19).

Henderson, T. (2003). Host Mobility for IP Networks: A Comparison. *IEEE Network*, *17*, 18–26. doi:10.1109/MNET.2003.1248657

IEEE Draft Standard for Wireless Access in Vehicular Environments (WAVE)-Networking Services. P1609.3/D8, August 2010. ISO-21210-CALM-IPv6-Networking. (2009). *Intelligent Transport Systems – Continuous air interface, long and medium range (CALM) – IPv6 Networking. ISO Draft DIS 21210.*

Kent S., & Seo, K. (2005). Security architecture for the Internet protocol. *IETF RFC 4301.*

Lach, H. Y., Janneteau, C., & Petrescu, A. (2003). Network Mobility in beyond-3G systems. *IEEE Communications Magazine*, *41*, 52. doi:10.1109/MCOM.2003.1215639

Lam, P. P., & Liew, S. C. (2007). Nested Network Mobility on the Multihop Cellular Network. *IEEE Communications Magazine*, *45*, 100. doi:10.1109/MCOM.2007.4342863

Lau, V., & Kwok, Y. (Eds.). (2007). *Wireless Internet and Mobile Computing: Interoperability and Performance* (pp. 357–375). IEEE.

Lim, H. J., Kim, M., Lee, J. H., & Chung, T. M. (2009). Route Optimization in Nested NEMO: Classification, Evaluation, and Analysis from NEMO Fringe Stub Perspective. *IEEE Transactions on Mobile Computing*, *8*, 1554. doi:10.1109/TMC.2009.76

Lin, Y. B., & Pang, A. C. (Eds.). (2005). *Wireless and Mobile All-IP Networks*. John Wiley & Sons.

Milojicic, D., Douglis, F., & Wheeler, R. (Eds.). (1999). *Mobility: Processes, Computers and Agents*. ACM Press. Addison Wesley.

Miltchev, S., Ioannidis, S., & Keromytis, A. D. (2002). A Study of Relative Cost of Network Security Protocols. In *USENIX Annual Technical Conference*.

Montavont, N., Boutet, A., Ropitault, T., Tsukada, M., Ernst, T., Korva, J., et al. (2008). Anemone: A ready-to-go testbed for IPv6 Compliant Intelligent Transport Systems. In *International Conference on Intelligent Transportation Systems-Telecommunications ITS-T*.

Montavont, N., Lorchart, J., & Noel, T. (2006). Deploying NEMO: A Practical Approach, In *International Conference on Intelligent Transportation Systems-Telecommunications ITS-T*.

Nikander, P., Gurtov, A., & Henderson, T. R. (2010). Host Identity Protocol (HIP): Connectivity, Mobility, Multi-Homing, Security, and Privacy over IPv4 and IPv6 Networks. *IEEE Communications Surveys & Tutorials*, *12*, 189–204. doi:10.1109/SURV.2010.021110.00070

Sakhaee, E., & Jamalipour, A. (2006). The Global In-Flight Internet. *IEEE Journal on Selected Areas in Communications*, *24*(9), 1748. doi:10.1109/JSAC.2006.875122

Solomon, J. D. (Ed.). (1996). *Mobile IP. The Internet Unplugged*. Prentice Hall Series in Computer Networking and Distributed Systems.

Stalling, W. (Ed.). (1995). *Network and Internetwork Security- Principles and Practice*. Prentice-Hall, Inc.

## KEY TERMS AND DEFINITIONS

**Confidentiality:** Protection by means of encryption against unauthorized access to the content of the packets.

**Integrity Protection:** Protection of packets against modifications to the content of them on the communication path.

**Nested Mobility:** When a moving network attaches another moving network to acquire connectivity to the outside network a tree topology is set up. The entire topology comprises the new *nested* mobile network where the root MR is the only node directly attached to the fixed network.

**Network Mobility:** The motion of a set of nodes that move together as a whole changing the point of attachment to the fixed network, and having continuous connectivity to the Internet.

**Route Optimization:** In NEMO scenarios, Route Optimization refers to the case where packets going from a node located inside the mobile network to a node located in the fixed network and vice versa are routed directly between the two endpoints, no third parties processes and re-routes them to the destination.

## ENDNOTES

[i] NEMO can refer to NEtwork MObility or to NEtwork that MOves depending on the context.

[ii] The terminology used in this chapter is the specified by the IETF by Ernst and Lach (2007).

[iii] Forward secrecy is that an adversary who knows the old keys is not able to generate subsequent new keys, and the backward secrecy is that an adversary who knows the current keys is not able to know the preceding keys.

# Chapter 9
# Security Framework for Mobile Agents–Based Applications

**Raja Al-Jaljouli**
*Deakin University, Australia*

**Jemal Abawajy**
*Deakin University, Australia*

## ABSTRACT

*Mobile agents have been proposed for key applications such as forensics analysis, intrusion detection, e-commerce, and resource management. Yet, they are vulnerable to various security threats by malicious hosts or intruders. Conversely, genuine platforms may run malicious agents. It is essential to establish a truly secure framework for mobile agents to gain trust of clients in the system. Failure to accomplish a trustworthy secured framework for Mobile Agent System (MAS) will limit their deployment into the key applications. This chapter presents a comprehensive taxonomy of various security threats to Mobile Agent System and the existing implemented security mechanisms. Different mechanisms are discussed, and the related security deficiencies are highlighted. The various security properties of the agent and the agent platform are described. The chapter also introduces the properties, advantages, and roles of agents in various applications. It describes the infrastructure of the system and discusses several mobile agent frameworks and the accomplished security level.*

## INTRODUCTION

Mobile agents are autonomous programs, typically written in interpreted machine-independent languages. They act on behalf of users and have some level of intelligence (Bradshaw, 1997).

DOI: 10.4018/978-1-4666-0080-5.ch009

Users can delegate mobile agents to accomplish different tasks such as access remote resources, cooperation with other mobile agents to perform complex tasks, e-trading, or filtering information autonomously from potential service providers for decision making purposes.

Mobile agents have advanced distributed computing as they exhibit special characteristics including mobility, persistence, autonomy, flexibility, cooperation, and adaptation. They control where they execute and can run on heterogeneous environments and are adaptable to changes in environments. Agents can remain stationary filtering incoming information or become mobile searching for specific information across the internet and retrieving it. The actions of an agent are not entirely pre-established and defined. The agent is able to choose what to do and in which order according to the external environment and user's requests. They can accomplish their tasks while the user might go off-line.

They traverse the internet from one platform to another through various architectures to access remote resources or even to meet, cooperate and communicate with other programs and agents to accomplish their tasks. They migrate from one platform to another based on a predefined agent itinerary or platforms dynamically allocated in response to any changes in the environment. Agents and platforms in the Mobile Agent System are not always trusted. Honest agents may run on unknown and non-trusted platforms or there might be malicious agents that concurrently execute on the platform. Moreover, agents might transfer through insecure communication channels where they might be intercepted and intruded by malicious agents during their migration. Both agent code and data are vulnerable to security threats. Conversely, a genuine platform might run non-trusted agents that might illegally get access to resources or services at the platform or breach its security.

The implementation of mobile agents in distributed computing has introduced many advantages. They overcome limitations of latency, connectivity, and bandwidth. Also, they allow a large degree of flexibility in creating computations and organize the use of distributed resources on the internet. Moreover, they allow remote software distribution and network management. Mobile agents have been used in various areas such as Forensics analysis, Intrusion detection,

E-commerce, E-health, and Resource management (Chess, Harrison, & Kershenbaum, 1997; Franklin & Graesser, 2006; Karjoth, 2000). Shopping agents in E-commerce applications are employed to search the marketplace for offers, negotiate the terms of agreements, or even purchase goods or services.

The Mobile Agent System is vulnerable to direct security attacks by malicious agents or/and non-trusted platforms. Hence, security is of concern as agents or/and platforms might handle very sensitive and critical information that should remain intact during the execution of mobile agents on agent platforms. It is essential to establish a secure framework for mobile agents to accomplish the delegated tasks successfully, and thus, gain trust of clients and service providers. Without a secure framework for Mobile agent System, the implementation of mobile agents in important application domains will be limited. Research is still ongoing for advances in securing the Mobile Agent System. The security refers to certain security properties, such as authenticity, confidentiality, integrity, etc. The security techniques can be divided into two categories: (a) preventive techniques, and (b) detective techniques. The preventive techniques hinder malicious acts to take place, whereas the detective techniques reveal the malicious acts that took place through verification processes.

In this chapter, authors address the problem of securing Mobile Agent System. First the authors present a background on mobile agents and their related properties, advantages, infrastructure, and routing. Then they discuss the role of agents in different applications and present a comparison between different frameworks of mobile agents as regards security, communication, mobility, collaboration, and other features. Next, the various security threats to Mobile Agent Systems including threats to agent code, agent data, agent platform, agent communication, agent transmission is discussed (Jansen, 2000; Nitschke, Paprzyeki, & Ren, 2006). The security properties of agent data, agent platform, and agent code are then outlined. A comprehensive taxonomy of security mechanisms is provided and mapped to the various security levels classified as preventive and detective. The drawbacks of the different security mechanisms relating to reliability, practicality and efficiency are outlined. Finally, the authors summarize the deficiencies of the existing security mechanisms and identify future research challenges.

## BACKGROUND

### Mobile Agents

Mobile agents are software entities written in interpreted machine-independent languages and can run in a distributed heterogeneous environment (Chess, Harrison, & Kershenbaum, 1997). They are small so as to improve transmission efficiency. They decide where to be executed and act on behalf of human users. They have specific internal goals and states, and can achieve the goals without human intervention.

Mobile agents possess the following properties that distinguish them from other agents (Lange, 2002; Pham & Karmouch, 1998):

- *Autonomous* – can define their goals and achieve them without human intervention.
- *Intelligent* – based on their knowledge of past actions and fixed rules they adapt to changes in the environment and to the needs of their users and hence they anticipate the most probable actions of the user.
- *Proactive* – can take initiatives as needed and exert control over their actions. They control where they execute.
- *Communicative* – can communicate with other agents, the system, and the user as needed.
- *Cooperative* – can carry out complex tasks, which they cannot handle by themselves, by cooperating with other agents.

- *Mobile* – traverse the internet from one host to another through various architectures and platforms to access remote resources or even to meet, cooperate, and communicate with other programs and agents to accomplish their tasks.
- *Flexible* – actions are not entirely predefined. They choose which actions to take and in which order according the changes in the environment.
- *Self-starting* – can decide when to start an action in response to changes in the environment.
- *Goal oriented* – receive requests from users and act to their best to satisfy the requests.
- *Individual character* – have well-defined characteristics.

Mobile agents have been implemented in several applications such as Forensic, Intrusion detection, E-commerce, Network resource management, etc. They present the following advantages (Bieszczad, Pagurek, & White, 1998; Lange, 2002; Lange & Oshima, 1999; Ylitalo, 2000):

- Use remote resources
- Cut in bandwidth usage
- Cut in communication costs
- Support off-line computations
- Facilitate distributed computing
- Distribute software across the internet
- Allow asynchronous execution of the agent on heterogeneous environments
- Manage network resources remotely
- Reduce network latency
- Use of API's rather than communication

### Mobile Agent Infrastructure

The infrastructure of the mobile agent is referred to as the Mobile Agent System. It has five components: Agent, Agent platform, Operating system, Hardware, and Communication infrastructure. The FIPA has defined the agent platform to be

*Figure 1. Mobile agent infrastructure*

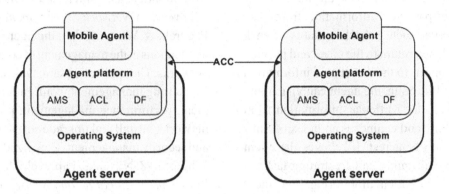

managed by a set of key agents as depicted in Figure 1 (Bellifemine, Poggi, & Rimassa, 1999; Chen, Lin, & Lien, 1999; Java World): (1) Agent Management System AMS; (2) Agent Communication Channel ACC; (3) Directory Facilitator DF; (4) Agent Communication Language ACL.

The *Agent Management System* is the agent that has control over the use and access to the agent platform. It authenticates resident agents and controls registrations. The *Agent Communication Channel* is the agent that provides the route for basic contact between agents residing on different platforms. The *Directory Facilitator* is the agent that provides the agent platform with a yellow page service. The *Agent Communication Language* specifies the standard message language including encoding, semantics, and pragmatics of the messages.

## Agent Routing

Initially, a mobile agent resides at a home host referred to as *client* or *initiator*. It allows the user to provide the agent with its request and configuration information through the Application Programmer's Interface (API). The agent is then dispatched to a remote host for execution. The accommodating host would provide suitable runtime environment for the mobile agent and is referred to as *server* or *executing host*. The mobile agent might have a restricted access to server resources

(data files, runtime library, network ports, etc.) based on security policy and/or payment for consumed resources. The server executes the mobile agent, which collects particular information, and generates runtime states and dynamic variables ready to migrate to the next server in the agent's itinerary. The process continues till the agent accomplishes its goals. The execution of an agent can be halted briefly during transport between servers and then resumed at the new server. The server should be able to execute and deactivate the agent, capture its state, transport the agent, communicate with other agent servers. The agent state includes its data and execution states. The mobile agent system is responsible of transporting the agent binary image between execution layers of executing hosts (Ng, 1999).

A *terminating agent* is an agent that returns back to the client following to its visit to a set of servers. The agent might traverse the internet on a predefined route, or a route that is set dynamically based on gathered information (Al-Jaljouli, 2006).

## Mobile Agent Applications

Mobile agents have several significant applications that include distributed information collection, searching and filtering and monitoring (Chess, Harrison, & Kershenbaum, 1997; Franklin & Graesser, 2006; Karjoth, 2000; Milojicic, 1999; Java World). In distributed information collection domain, the agent could roam the internet trying to collect particular information from each executing server such as backup status of each disk. It could then return to the client and provide a summary report. In the distributed information searching and filtering, the agent can visit many sites, searching for particular information from each visited site and composes an index of links to information of interest. It relieves the client from a time consuming and frustrating task. In distributed information monitoring, the agent can go to a particular host and waits for certain information to be available. An example, a stock agent that waits for a certain stock to hit a price limit, then sells the stock on behalf of the user. In targeted information dissemination, the agent distributes interactive news or posts ads on the internet through an information router, which would map features to interests and then routes the information to a group of users of the related interests.

Agents are also used in E-commerce (Al-Jaljouli & Abawajy, 2009), E-health (Chan, Ray, & Parameswaran, 2008), and E-negotiation (Karjoth, 2000). In E-commerce, the agent can shop for the client, make orders, and pay with the client's credit card number. The agent traverses the internet collecting offers from potential trading agents and returns to the client to make a decision on the offer that best meets the client requirements. It then makes an order and pays for it on behalf of the client. In-E-health, multi agents collect patient's data in a mobile environment and recommend on proper actions to patients and medical staff based on cooperative reasoning. In E-negotiation domain, the agent can migrate to E-marketplaces searching for an offer that satisfies the client requirement profile and maximizes utility. It collects and evaluates offers from potential trading agents and then carries out multiple negotiation rounds till it is satisfied. Finally, it congregates an agreement with the winner trader and makes a payment on behalf of the client. Another area that agent technology has been prominent is the *Net-*

*work Management*, and *Resource Allocation and Recovery*. In *Network Management* (Bieszczad, Pagurek, & White, 1998), the agent has several applications in the management of communication networks. The applications are: fault management, accounting management, configuration management, performance management, plug and play networks, install and upgrade software remotely, and security management (Bieszczad, Pagurek, & White, 1998; Nitschke, Paprzyeki, & Ren, 2006). In *Resource Allocation and Recovery* (Nitschke, Paprzyeki, & Ren, 2006), the agent can report on under-utilized network resources so as to assign it intensive processing functions. Also, it can report on over-utilized network resources so as not to restrain other jobs till it is relaxed. In intrusion detection (Wayne, 2002), multiple agents in networks are designated to perform event generation, data abstraction, or attack detection tasks. If a node is likely to be attacked, agents communicate with each other by sending messages and then a comprehensive series of tests are carried out till the attack is blocked.

## Mobile Agent Frameworks

The Mobile Agent System has to possess a set of features that are essential to support the development and execution of mobile agent applications. The features include (Silvaet al., 2001): (i) Execution Environment; (ii) Identity Management; (iii) Agent Types Management; (iv) Persistence Capability; (v) Security Framework; (vi) Migration Capability; (vii) Communication Capability; (viii) Access to external resources.

An *Execution Environment* is an important component of the system. It provides a runtime software system that is necessary for mobile agent execution. It also provides a common interface for interaction between communicating agents, or agents and respective agent platforms at which they execute. Moreover, it provides services for creation, migration and termination of agents The *Identity Management* identifies agents and execution environments distinctively, while *Agent Types Management* provides a mechanism to access and search for types or classes of agents and to grasp how agents are instantiated. The *Persistence Capability* provides storage for the agent state that could be used when agent transmits from one platform to another platform. The *Security Framework* focuses on the integrity, authentication, repudiation, access control, privacy, and control of resources consumption. The *Migration Capability* supports migration primitives (Wilhelm, 1999). Migration can be classified relating to the agent execution state as strong mobility or weak mobility. *Strong mobility* is the ability of the agent programming language to allow agents to move their code and execution states to a different server. *Weak mobility* is the ability of the agent programming language to allow agents at a host to be bound dynamically to code coming from different server. Migration can also be classified relating to the agent code as entire code migration, incremental code migration, and no code migration. The *Communication Capability* ensures agent migration do not interfere with agent communications, while *Access to external resources* provides agents with a controlled access to external resources such as file system, user interface devices, communication channels, etc.

Several mobile agent frameworks have been developed (Gupta, 2011; Suri, Bradshaw, Breedy, Groth, Hill, & Jeffers, 2000; Tripathi, Karnik, Ahmed, Singh, Prakash, Kakani, Vora, & Pathak, 2002; Brewington, Gray, Moizumi, Kotz, Cybenko, & Rus, 1999; Karnik, 1998; Silva, Romão, Deugo, & Silva, 2001) such as *Telescript, Tacoma, Aglets, D'Agents, NOMADS, Ara, Grasshopper, Odyssey, Tacoma, Concordia, Jumping Beans, AgentSpace, Mole, MOA, PageSpace, Voyagar, Ajanta, Agent Tcl,* and *ffMAIN*. A set of frameworks are discussed below. Table 1 compares different mobile agent frameworks as regards the following features (Rahimi, Angryk, Bjursell, Paprzycki, Ali, Cobb, Gibutowski, & Kolodziej, 2001; Silva et al., 2001; Gray, Cybenko, Kotz, Peterson, & Rus, 2002):

- Security: authentication, access control, and resource control
- Itinerary
- Agent collaboration
- Communication
- Persistence
- Agent language
- Agent cloning
- Remote agent creation
- Code migration
- Data migration
- Agent mobility

*Security* is a fundamental feature that should be ensured to maintain the security of the executing host. A malicious mobile agent might try attack a visited host or another agent running on the host. A host would not directly execute an incoming mobile agent. It should verify the identity of the agent owner and grants restricted permissions to local resources based on a security policy and trust in the owner of the agent.

*Itinerary* is a data structure that defines addresses of hosts to be visited. The agent might maintain it separately from the agent object. The itinerary can pre-defined before the agent is dispatched to the internet or can be dynamically composed. The itinerary control's routing of the agent and can handle unexceptional events, e.g. crash of scheduled host in agent's itinerary.

*Agent collaboration* refers to two agents or more that collaborate in purpose of accomplishing a particular task. The task is sub-divided into sub-tasks and each agent carries a sub-task. Agents communicate with each other and integrate their results upon the completion of all sub-tasks.

*Agent communication* defines how cooperating agents communicate to invoke methods or exchange data in order to accomplish a complex task.

*Persistence* refers to the ability of the network to halt the execution of a mobile agent in case of failure of the current server or the next server in agent's itinerary and to resume the agent execution upon the recovery of the server. It should

*Table 1. Features of different mobile agent frameworks*

| | Aglets | Concordia | D'Agents | Ara | Mole | NOMADS | Tacoma | Telescript | Voyager | Ajanta |
|---|---|---|---|---|---|---|---|---|---|---|
| *Authentication* | Code signed by owner; data signed by user | Signed by user | Code signed by owner | Code signed by owner; data signed by user | No | Accounts & passwords | No | Code signed by owner | Code signed by user, password authentication | Code signed by owner |
| *Itinerary* | Pre-defined | Dynamically defined | | Not supported | Not supported | Pre-defined | Not supported | Not supported | Dynamically defined | Pre-defined |
| *Access control* | ACL based on Sig. | ACL based on user | ACL based on user; enforce resource access rights | ACL based on Sig. | Sandbox security model | Java security, Pre-count policy file | Relies on OS | Permits based on user | Security manager | ACL based on sig. |
| *Resource control* | Limited control | Allowance based on agent identity | Resource manager | Allowance based on sig. | Access to resource only to immobile service agent | Aroma VM Dynamic resource control | No | Permits based on user | Security manager | Resource registry & security policy |
| *Agent collaboration* | Facilitated | Facilitated | | Not supported | Facilitated | | Not supported | Not supported | Not facilitated | Facilitated |
| *Communication* | Send object as message via proxy | Send object as event | String messages | Free-form via named points | Send object, message, RPC | Raw message | Raw message | Method invocation or events | Method call via proxy | Remote method invocation |
| *Persistence* | Yes | Yes | No | Yes | Sophisticated rollbacks | No | No | Yes | Yes | Yes |
| *Language* | Java | Java | Java, Tcl, Scheme | C, C++, Tcl | Java | Java | C+:V1.0; C:V2.0 | Telescript | Java | Java |
| *Cloning* | Supported | Supported | Not supported | Not supported | Not supported | Not supported | Not supported | Not supported | Supported | No |
| *Remote creation of mobile agent* | Supported | Not supported | Not supported | Not supported | Not supported | Not supported | Not supported | Not supported | Supported | No |
| *Code migration* | Sender push or on-demand from server | Sender push or on-demand from sender | Sender push | Sender push | On-demand from code server | Sender push | Sender push | Sender push | *MoveTo* method | Go method |
| *Data migration* | Java serialization | Java serialization | Custom | Custom | Java serialization | Custom | Manual | Custom | Java serialization | Java serialization |
| *Agent mobility* | Weak | Weak | Strong | Strong | Weak | Strong | Weak | Strong | Weak | Weak |

maintain the mobile agent on a persistent host until the system recovers.

*Agent language* refers to the programming language that a programmer (owners) uses to develop a mobile agent.

*Agent cloning* refers to the ability of a mobile agent to create a copy of itself and assign it a particular task.

*Remote agent creation* refers to the capability of creating a mobile agent at a remote host by just sending the necessary parameters.

*Code migration* can take three forms: (1) agent's code does not migrate; (2) entire agent's code migrates at once; (3) basic and specific agent's code migrates initially and remaining code migrates later on and is loaded dynamically based on demand.

*Data migration* refers to the implemented method for migrating agent's data between visited hosts.

*Agent migration* can be strong or weak. Strong agent mobility refers to the ability of the network to transport the agent's execution state together with agent's code and data as the agent migrates from one host to another, whereas, weak agent mobility refers to the ability of the network to transport agent's data and code only.

## Telescript

The *Telescript* framework (White, 1995; Wilhelm, 1999) consists of three components: *Engine, Agent Protocol*, and *Agent Language*. The *Engine* is responsible of agent execution, agent migration, and access to external resources. The *Agent Protocol* defines the exchange of agents between Telescript engines. The Telescript *Language* is an interpreted object-oriented language by General Magic for developing mobile agents. It is the first language that enabled processes to migrate to other agent platforms during execution. It has two language levels: *High Telescript* and *Low Telescript*. The *High Telescript* is the language for writing programs which would be complied

into *Low Telescript*, a postscript compatible language. *The Low Telescript* has all class information the agent needs and which can hold as it migrates. There are two types of threads that can communicate: mobile processes named agents and stationary processes named places. An agent can migrate itself by invoking the *Go* command, which interrupts agent execution, collects agent's objects and execution stack, serializes, encrypts the agent, initiates a communication channel, and resumes the agent execution from the command that follows the *Go* command on the succeeding server in the agent's itinerary. Telescript was mainly devised for E-commerce applications. It enables code mobility and asserts strong security of the system by protecting it from the attacks of malicious agents.

*Telescript* supports security, communication, persistence and mobility. Nevertheless, it has failed as being unstable, huge, and programmers have to learn a difficult programming language. Also, name and address management had no internet-based standards (Cugola, Ghezzi, Picco, & Vigna, 1997; Dömel, 1997).

## Tacoma

*Tacoma* (Cugola et al., 1997; Johansen, Renesse, & Schneider, 1995) is an extension of the Agent Tcl language that supports concurrency. A running Tcl process can request the execution of another Tcl process on another agent platform. The two processes might be executing simultaneously and can interact with each other while they are executing. It does not enforce security nor captures the execution state as the agent migrates.

## Aglets

Autonomous agents in *Aglets* (Gray, 1996; Karjoth, Lange, & Oshima, 1997; Lange & Oshima, 1998) are written in Java programming language and can migrate to other agent platforms in a network. During migration, *Aglets* transport both their code and data. They communicate through message passing. They implement event-handling methods such as a method that would alert the *Aglet* upon leaving its host and upon arriving at another platform.

*Agelts* has two layers: runtime layer and communication layer. The *runtime layer* is responsible for agent creation, managing, and dispatching the agent to a remote platform. The communication layer is responsible for transferring and receiving the serialized data of the agent between two platforms. *Aglets* mobile agent defines an entity that requests access to a system and has a unique identity the system might authenticate as *principal*. The identity includes class name and other attributes. The system defines several principals such as *Aglets principal*. Principals may enforce security policies so as to control access to their data or to audit their security actions. The *Aglets principal* can set security preferences to ensure its safety such as entities that may interact with the *Aglets* on its itinerary, maximum number of visited hosts, allowed CPU-time consumption, etc. The enforced security is limited and also it does not capture the execution state as the agent migrates.

## Concordia

Mobile agents in *Concordia* (Mitsubishi Electric, 1997; Glass, 1999; Silva et. al, 2001) are written in Java programming language. The agent is associated with particular library object that defines its itinerary and methods to be executed at each host. It supports agent mobility and class loading mechanisms. It enforces security policy to control access to resources based on the identity of the agent owner rather than the agent developer as compared to *Aglets*. It also stores the agent's initial state and recovers the agent in case of system failure. It keeps track of events and notifies concerned parities (agents and objects) of the events it receives. Moreover, it manages inbound and outbound queues to enhance reliability of agent transportation through its itinerary. The agent

state is protected using encryption protocols. However, it does not capture the execution state as the agent migrates.

## Voyager

Mobile agents in *Voyager* (Weeler, 2005) can be written in Java or Net programming languages. It supports scalability and avoids single point of failure of the mobile agent system through the utilization of a high-performance protocol. It introduces a new feature called *Space* that allows highly scalable message passing. A *Space* is composed of one or more connected subspaces and any event published to a subspace is broadcast to the entire space. It is capable of broadcasting up to 50,000 events per sec to a maximum of six clients. It also supports interoperability by providing common universal API. It supports object mobility and security.

## Ajanta

*Ajanta* (Karnik, 1998; Tripathi, Ahmed, Pathak, Carney, & Dokas, 2001) is a Java-based framework that addresses the security of mobile agent system and supports strong mobility. The agent is executed in an isolated protection domain at the agent server to protect the agent from malicious agents or non-trusted servers. The agent is assigned non-forgeable credentials including agent's name and agent's owner. The server implements a security policy that controls the agent access to its local resources based on the identity of the agent creator. It has control over its remote agents including execution termination or agent call back. Agents at different servers can communicate with each other through *Remote Message Invocation* (RMI), if the hosting server trusts the remote server. The agent server has five components: Server Interface, Domain Registry, Agent Environment, Agent Transfer, and Resource Registry. The function of each component is described below.

- *Server Interface*: supports queries about agents residing on a server.
- *Domain Registry*: keeps record of the current executing agents.
- *Agent Environment*: provides an interface between incoming agents and the server.
- *Agent Transfer*: authenticates transfer of an agent from one host to another.
- *Resource Registry*: controls access of executing agents to local resources.

## MOBILE AGENT SECURITY THREATS

Mobile agents are expected to run in partially unknown and untrustworthy environments. They transport from one host to another host through insecure channels and may execute on non-trusted hosts. The executing host would have full control over the agent. Thus, they are vulnerable to direct security attacks of intruders and non-trusted hosts (Farmer, Guttman, & Swarup, 1996; Al-Jaljouli, 2006; Jansen & Karygiannis, 2000).

The security of mobile agents relies on two components: (i) *Program code and static data*. (ii) *Dynamic data* (Maggi & Sisto, 2002). The dynamic data comprises three types of data (Fischer, 2003) as follows:

- Fixed size changeable, e.g. global variables.
- Dynamically allocated static and are commonly referred to as execution results
- Dynamically allocated changeable, e.g. register contents and execution stack.

## THREATS TO MOBILE AGENT

Intruders and non-trusted hosts can perform any of the following malicious acts (Farmer, Guttman, & Swarup, 1996; Al-Jaljouli, 2005; Karnik, 1998; Loureiro, Molva, & Roudier, 2000; Nitschke, Paprzyeki, & Ren, 2006):

- Modify the program code to have it perform malicious acts or unable to accomplish its original task.
- Delay the agent execution until the task is no more significant e.g., denial of service attack.
- Terminate the mobile agent or capture it and obstructs it from returning to the client.
- Embed a virus into the agent's code to halt its execution.
- Alter the control flow of the agent so as the agent turns out with incorrect results.
- Reinitialize the agent.

## THREATS TO AGENT DATA

Intruders and non-trusted hosts can perform any of the following malicious acts (Al-Jaljouli, 2006):

- *Truncation* of the gathered data.
- *Alteration* of the collected data. It takes place if the agent visits a host twice or two hosts collude with each other. A malicious host may send back the agent to an earlier host in the agent's itinerary. It could then truncate the data acquired at intermediary hosts and alter the data it formerly provided to the agent without being detected if it replaces the current agent's state with the state that was present when the agent firstly visited it.
- *Replacing* the gathered data with the data of similar agents. An adversary might intercept the agent. It could then replace the current agent's state with the state of a similar agent.
- *Tracing* the flow of agent execution and thus comprehend the point of execution in memory and point of storage in execution stack.

## THREATS TO THE AGENT PLATFORM

The agent platform is subject to several security threats from malicious agents. The threats include masquerading, unauthorized access, denial of service, breach privacy, alter the system, copy and replay, and harassments (Giansiracusa, 2003; Jansen & Karygiannis, 2000; Loureiro, Molva, & Roudier, 2000; Nitschke, Paprzyeki, & Ren, 2006; Ylitalo, 2000).

- *Masquerading:* A mobile agent can impersonate the identity of another agent to gain access to services and resources at an agent platform for which it is not authorized.
- *Unauthorized Access:* Remote users and agents might request resources for which they are not authorized.
- *Denial of Service:* A mobile agent can repeatedly send messages or constantly consume network connections to an agent platform so as to impede other platforms and mobile agents to have access to the platform's resources or services. Hence, the platform would not be able to provide service to agents or platforms. The attacks are tracked by the Computer Emergency Response Team (CERT) and Federal Computer Incident Response Capability (FedCIRC) (Mell, 1999).
- *Breach of Privacy:* An unauthorized agent can disclose confidential information
- *Alter the System:* An unauthorized agent can alter or demolish the confidential information it had access to it or the instruction codes of the platform.
- *Copy and Reply:* An entity can intercept an agent in transit, copy the agent or a message and then retransmit it. For instance, a malicious entity may retransmit a "purchase order" several times having an item paid for more than it is intended.

- *Harassment:* A mobile agent can annoy users with repeated attacks, such as displaying unwanted pictures or making the screen to flicker disturbing sensitive users. It can also shut-down or terminate the agent platform completely.

## THREATS TO AGENT COMMUNICATION

Complex tasks that cannot be done by a single mobile are subdivided among multi agents. Agents communicate with each other by passing messages. An intruder can intercept a communicated message and perform any of the following malicious acts: (i) Eavesdrop confidential information if they are sent in plain text and are not encrypted; (ii) Delete or manipulate the communicated message without being detected; and (iii) Falsify the communicated message or reply it at later time.

### Threats to Agent Transmission

Agents transport through unsecure communication channels, and thus intruders may intercept a migrating agent and attack it maliciously in different ways while it is in transit between successive hosts in the agent's itinerary. An intruder can *delay* the agent execution by a denial of service attack or *capture* the agent and obstruct it from returning to the client. It can also *redirect* the agent to a host not scheduled in the agent's itinerary or disturb the intended goals of the agent. The intruder might try to *eavesdrop* upon communications and read sensitive information sent as plain text within the agent. Supposedly the communicated information is encrypted then the intruder would try to *impersonate* the genuine initiator and hence breach the privacy of the gathered data. It may possibly intercept the agent that has a message signed by the initiator of an agent. It could then decrypt the signed message and sign the decrypted message with its private key impersonating the genuine initiator. Executing hosts in the agent's itinerary would encrypt the data they provide to the agent with the public key of the intruder assuming it is the genuine initiator. Hence, the intruder would be able to breach the privacy of the collected data. In addition, the intruder might *transmit* others data signed by the private key of a malicious host. Usually, executing hosts in the agent's itinerary send the data they provide to the agent signed with the corresponding private keys. An intruder might intercept the signed data. It could then decrypt the signed data and may sign the data of a particular host with its private key impersonating the genuine provider of the data.

## SECURITY REQUIREMENTS

The wide deployment of mobile agents requires reliable mechanisms that provide safety and security for mobile agents and agent servers. The mobile agent system must fulfill a set of goals (Jansen, 2000; Loureiro, Molva, & Roudier, 2000). It should protect the mobile agent code and data from tampering and disclosure, either while it is in transit or while it is executed on an agent platform, by other agents running on the agent server or a malicious agent server. It should also be able to verify integrity of agent code or data accurately. Agents should be correctly transferred to desired hosts and be provided with appropriate resources. The communicated messages between agents residing at different platforms should be protected. Cryptography is implemented to encrypt the exchanged messages for identification and authentication purposes. The agent platform should verify authenticity of agents accurately and execute authenticated agents correctly. It should prevent agents from interfering with each other or with the platform. It provides isolated domains for agents and controls their access to resources at the agent platform. The mobile agent system should be able to audit the security events taking place at the agent platform and manage agent platform

deadlocks properly. It should also protect the information, software, hardware, and resources of the agent platform from the abuse of malicious mobile agents.

## Security Properties

The security properties of an agent can be generally classified into three classes: (i) Agent's Data Security Properties; (ii) Agent's Code Security Properties; and (iii) Agent's Platform Security Properties. In this section, the various security properties (Giansiracusa, 2003; IBM Aglets (2002); Al-Jaljouli, 2005, 2006; Al-Jaljouli & Abawajy, 2009; Karnik, 1998; Ng, 1999) of mobile agents will be described in details.

### Agent's Data Security Properties

The security properties of agent data encompass integrity, non-repudiability, confidentiality and authenticity of data. The properties are described below.

1.  *Data integrity*: It requires that the results of executing the agent at $n$ visited hosts $m'_1$, $m'_2$, ..., $m'_n$ and are returned to the initiator $i_0$ match the genuine execution results $m_1$, $m_2$, ..., $m_n$. Whenever, the two sets of execution results differ it implies that the genuine execution results had been tampered with. The execution results the agent stores $m_j$ for $(1 \leq j \leq n)$ should be maintained intact during its transportation through the internet or execution at $n$ visited hosts, or at least any tampering with the execution results can be detected by the initiator upon the agent's return. The data integrity property can be classified into the following classes of protection:
    - *Insertion resilience*: Data can only be appended to the execution results $m_1$, ..., $m_j$ for $(j < n)$.

◦ *Deletion resilience:* Deletion of any data from the execution result $m_1$, ..., $m_n$ is prevented, or at least can be detected upon the agent's return to the initiator. If an execution result $m_j$ is deleted and the execution results cuts to $m_1$, ..., $m_{j-1}$, $m_{j+1}$, ..., $m_n$, then the initiator $i_0$ can deduce that an execution result is deleted from the chain of results.

◦ *Truncation resilience:* Truncation of the execution results $m_1$, ..., $m_j$, ..., $m_n$ at host $i_j$ and reducing it to $m_1$, ..., $m_j$ for $(j < n)$ is prevented, or at least can be detected upon the agent's return to the initiator $i_0$. Verifications are carried out and particular values are computed using the returned execution results e.g. *checksums* or *message authentication code*. The values are then verified with the corresponding value the agent stores. Any difference between the two values indicates that the execution results had been truncated.

◦ *Strong forward integrity:* The execution results are maintained intact. The property necessitates that the returned execution result $m'_j$ matches the genuine execution result $m_j$ for $(1 \leq j \leq n)$.

◦ *Strong data integrity:* It requires the fulfillment of the four protection classes: insertion resilience, deletion resilience, truncation resilience, and strong forward integrity.

2. *Data non-repudiability:* The initiator $i_0$ would have an evidence on the identity of host $i_j$ that added the execution result $m'_j$ $(1 \leq j \leq n)$ to the execution results $m'_1$, $m'_2$, ..., $m'_n$.

3. *Data confidentiality:* The execution results $m_1$, ..., $m_j$, ..., $m_n$ the agent stores can only be read by the initiator $i_0$. Adversaries would not be able to learn the plaintext of the ciphered execution results that are stored within the agent. Cryptography is implemented in ciphering the execution results, which seals the plaintext and prevents unauthorized retrieval of information. It uses asymmetric keys where one public key is used for encrypting plain-text and another private key is used for decrypting cipher-text. Only authorized parties can decrypt the ciphered text and learn the plaintext. An adversary might capture the ciphered execution results, shown below.

$$\{m_1\}K_{i_1}, ..., \{m_j\}K_{i_j}, ..., \{m_n\}K_{i_n}$$

However, she/he would not be able to deduce the plaintext $m_j$ for $(1 \leq j \leq n)$, as the decryption keys are private and thus the system is considered secure. It is essential to maintain the privacy of the encryption keys during the run of a protocol session.

4. *Data authenticity:* Upon the agent's return, the initiator $i_0$ can deduce the identity of host $i_j$ that appended $m'_j$ to the execution results $m'_1, m'_2, ..., m'_n$. The initiator $i_0$ can be certain that the execution result a host claims was indeed provided by that host. The property does not allow an adversary to impersonate a host.

## Agent's Platform Security Properties

The security properties of the agent platform encompass accountability, agent authentication, and origin-confidentiality.

1. *Accountability:* The agent at a platform should be held responsible for their actions. In order to ensure accountability, the agent platform implements a security policy that maintains an audit log of security relevant events taking place at the platform. It keeps information about events such as name of

user or agent, time of event, type of event, status of event as failed or succeeded. The audit logs have to be protected from unauthorized access, manipulation, or loss. *Accountability* is essential to build trust between different entities: agent-platform, agent-agent running at a platform, and user-platform. It also enables identification of malicious act/s of an entity so as to bring it to legal courts, identification of entities being affected by the malicious act, and recovery from security breach/s. Authentication techniques provide accountability for actions of entities.

2. *Agent authentication*: The platform should be able to verify that an agent is genuinely who it declares itself to be.

3. *Origin-confidentiality*: The identity of host $i_j$ for $(1 \leq j \leq n)$ that generated and added the execution result $m_j$ to the chain $m_1, \ldots, m_n$ can only be known at host $i_0$. An executing host $i_{j+1}$ should not be able to deduce the identity of a previously visited host $i_j$ for $(1 \leq j < k)$ from the agent. Though, it is possible for a malicious host $i_{j+1}$ to get the identity of host $i_j$ where the execution result $m_j$ was generated by analysing the agent's dynamic data just before and after the agent visited it. Also, the identity of host $i_{j-1}$ will possibly be revealed on the network layer. This can be prevented by using anonymous connections (Syverson, Goldschlag, & Reed, 1997) which hide the identity of the previously visited host.

## Agent's Code Security Properties

The Security properties of the agent code encompass anonymity, accountability, integrity, availability, and agent platform authentication. The properties are described below.

1. *Anonymity*: The real identity of the agent owner should be hidden except for legislative authorities and when necessary. However, credit agencies would not accept to allow entities to have access to their credit without revealing their true identities through passwords, biometrics, or smart cards.

2. *Accountability*: The agent platform should be held responsible for their interactions with agents.

3. *Integrity*: The agent should be maintained intact and not altered by any malicious entity.

4. *Availability*: The agent platform should ensure the availability of its resources and data to authorized agents. The platform should be able to manage deadlocks, recover from system failures, and allocate resources fairly, control access to resources, and provide access to shared data. Ensuring authenticity, integrity, and accountability restraint the availability of resources and data to agents due to delays caused by encryption of messages and generation of audit logs.

5. *Agent platform authentication*: The agent should be able to verify that a platform is genuinely who it declares itself to be.

## Security Mechanisms Taxonomy

The security mechanisms relate to two types of entities: (1) agent, and (2) agent platform. The security mechanisms can be classified into two levels of security as protection and detection levels. The prevention level makes it impossible or very hard for an entity to implant a malicious attack, whereas the detective level can identify malicious attacks of malicious entities through verification processes. They may also be able to identify who should be held responsible for the act. The security mechanisms assume that the client (initiator) behaves non-maliciously and prevents any security violation. Also, it assumes that the mobile agent trusts the client from where it is instantiated.

*Figure 2. Classification of prevention mechanisms for protecting agents*

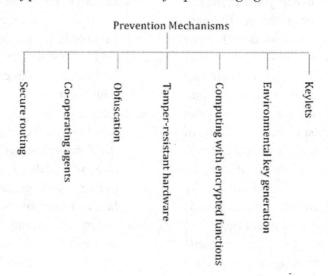

## Protecting Agent

The agent protection mechanisms that are presented in the literature (Al-Jaljouli, 2006; Karnik, 1998; Loureiro, Molva, & Roudier, 2000; Nitschke, Paprzyeki, & Ren, 2006; Ylitalo, 2000) are outlined and described below.

## Prevention Mechanisms

The prevention mechanisms include the following mechanisms: Secure routing (Agent routing security policy), Co-operating agents, Obfuscation (Time-limited black box security), Tamper-resistant hardware, Computing with encrypted functions, Environmental key generation, and Keylets. In general, the prevention mechanisms can be classified as shown in Figure 2.

### Secure Routing (Agent Routing Security Policy)

It implements a routing security policy that restricts the agent itinerary to trusted servers only. Thus, agent platform has control over the migration of the agent and the agent would not be allowed to execute on non-trusted platforms. An agent creator or owner would not have to select a trusted server.

It can send the mobile agent to gateways, which provide directory services to agents and guide them where to go to get executed

The approach prevents tampering with the agent's code and data. The drawback of the approach is the limited roaming of the agent that would restrain the potentials of mobile agents.

### Co-Operating Agents

The agent's critical data and functions are split among two or more agents that cooperate with each other. Agents execute on carefully selected disjoint platform domains to avoid the likelihood of platforms to conspire in an attempt of colluding attacks. Agents have to integrate their results to accomplish the main goal. Two agents might be cooperating to accomplish a task, a stationary agent and a migrating agent (Roth, 1999). The migrating agent traverses the internet and the stationary agent resides at a trusted platform. They can access shared data securely and store data remotely. It is assumed that a malicious agent or non-trusted platform would not be able to attack the agent individually.

The approach prevents non-trusted platforms from performing a malicious act. Platforms have to conspire to attack the agent which is not possible as the selected platform domains do not allow agents to execute on non-trusted platforms simultaneously. The approach have two drawbacks: (1) Difficulty in selecting disjoint platform domains that avoid the simultaneous execution of agents on non-trusted platforms, and (2) Confidentiality is not addressed.

## Obfuscation (Time Limited Black Box)

The mobile code is jumbled so platforms would find it very hard to analyze or even attack. Polynomials and rational functions are used to obfuscate code. The Induced code should be functionally identical to the original code. The code is called black box and has a validity interval. The interval is set so that a malicious entity would not be able to reveal sensitive data or tamper with the agent data, code or state. If the validity interval expires, then the agent code and data are considered invalid.

The approach protects the entire agent temporarily and is effective in protecting short-term sensitive data. The drawbacks of the approach are: (1) It cannot be implemented to long-validity data such as credit card expiry date, (2) The protection interval the agent owner sets might be shorter than the time the agent needs to accomplish its intended task or it might long enough to be tampered by a malicious entity, (3) The security related libraries the agent encloses have to be obfuscated, and (4) The obfuscated code can be reversed.

## Tamper-Resistant Hardware

It is a physically sealed environment that contains sensitive data and performs a task. Access to the internals is through a restricted interface the tamper-resistant hardware controls. Any violation to the rules would result in invalidation of the sealed sensitive information. Tamper resistant hardware devices are such as smart cards in payment systems and access control systems in secured communications. The drawback of the approach is that an attack might be able to violate the physical protection given enough time and resources.

## Computing with Encrypted Functions

It uses cryptography to conceal functions so the agent platform would not be able to know the task a function completes. The function is encrypted using a particular encryption key and is then dispatched to the agent platform for execution. The platform would then execute the encrypted function for an input value and then gets an encrypted result. The result is dispatched to the agent owner that only can interpret the encrypted result using a particular decryption key. The drawbacks of the approach are: (1) An attacker might be able to recover an encryption function through iterative execution of the encrypted function, and (2) Interaction with the executing host is possible as results of functions are encrypted. The approach prevents intentional tampering of malicious entities unless the enforced penalties are not too high.

## Environmental Key Generation

The approach (Jansen & Karygiannis, 2000) implements conditional decryption of the executable agent code. Upon specific environmental changes, a key is generated so the ciphered agent code can be decrypted. The condition is ciphered using one way cryptographic hash function or public key encryption so as an adversary would not be able to reveal the condition. The drawbacks of the approach are: (1) The agent platform such as Java based system would not allow the execution of the ciphered code which it cannot trust, (2) The agent platform can capture the agent or may tamper with the ciphered code and thus restrains its intended goals, and (3) The agent platform can trigger the condition and thus read the code.

## Keylets

The approach (Giansiracusa, 2003; Kim & Moreau, 2001) is based on partitioning the agent code, data, and state into self-contained compo-

*Figure 3. Classification of detection mechanisms for protecting mobile agents*

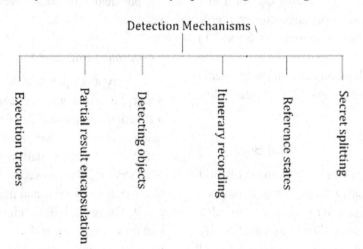

nents and then encrypting the components with different symmetric keys. The ciphered code would be neither executable nor interpretable. The symmetric keys should be propagated to the agent platform so as to decrypt the components and execute the agent properly. The symmetric keys should be ciphered before being propagated so as to preserve the integrity of the symmetric keys. It is proposed to encrypt the symmetric keys using the public key of the succeeding platform in the agent's itinerary. The agent owner might handle partitioning, encryption, and propagation tasks, nevertheless its platform might become into a bottleneck state and a focal point. It is advisable that the owner delegate a third party of key generation and propagation. The drawbacks of the approach are: (1) It introduces a considerable increase in communication loads, task duration and computation complexity, (2) The agent platform may not be willing to support the complex approach, (3) It requires a third party for key generation and propagation, and (4) Risk of incomplete key propagation.

## Detection Mechanisms

The detection mechanisms can be classified into the following detection levels based on the stage at which a threat was detected during the lifetime of the mobile agent:

- Detection after the agent execution is terminated
- Detection during the agent execution
- Instantaneous detection that require an action

The detection mechanisms include the following mechanisms: Execution traces, Partial results encapsulation, Detecting objects, Itinerary recording, Reference states and Secret splitting. In general, the mechanisms can be classified as shown in Figure 3.

### Execution Traces

The agent platform creates a log of the agent execution including the lines of code executed at the platform and the external values read by the agent while it is running on the platform. The platform computes a hash of the log using cryptographic hash function and then sends the agent with the computed hash and the agent's intermediate state both signed by its private key to the succeeding platform in the agent's itinerary. The signature acts as attestation on the platform's responsibility for the change in the agent's state as a result of the proper execution of the agent. Each platform in the agent's itinerary creates a log of the agent execution and encloses it with the previous logs. It would then send cumulative logs together with the mobile agent to the succeeding host in the agent's itinerary. The process continues till the agent returns to the client.

The agent owner can verify execution traces by simulating the agent execution and based on the agent's initial state. The agent owner can then detect if any platform has executed the agent improperly.

The approach can detect any malicious act of non-trusted platforms after the agent execution is terminated and identify malicious entities. It can also hinder malicious platforms from executing agents improperly due to the risk of being caught and penalized. Execution traces can stand as legal proofs at courts. However, they have some drawbacks. They result in heavy communication loads and necessitate large amounts of storage resources and thus are too expensive and impractical. They also need time synchronization among the executing platforms and a secure protocol for the transfer of cryptographic hashes and agent's states.

### Partial Result Encapsulation

The results of executing the mobile agent at an agent platform are enciphered in an offer to preclude other platforms from learning or deriving the actual results. The offer would also incorporate the execution results at the preceding platform and the identity of the succeeding platform so as to ensure integrity of execution results. The offer may also incorporate a cryptographic hash of a random nonce that uniquely identifies a protocol run, and timestamp. It can also incorporate the mobile code and/or a set of new platforms appended to the agent's itinerary. The offer might be signed with the private key of the platform and would then be encrypted with the public key of the agent owner. The approach implements a combination of the following security methods:

- The *public key encryption* uses the public key of the initiator to cipher the execution results at visited hosts so as to achieve secrecy of results. Also, it uses the public key of the succeeding host to encrypt particular verification terms, such as a cryptographic hash of a nonce.
- The *digital signature* is used when a host signs the results it provides with its private signing key so as to be authenticated as the initiator of the execution results.
- The *message authentication code* incorporates the identity of the succeeding host and the characteristics of the previous execution results into the execution result at a host.
- The *backward chaining* incorporates characteristics of the previous execution results into the execution result at a host.
- The *one-step forward chaining* incorporates the identity of the succeeding host into the execution result at a host.
- The *code-result binding* binds the signed code to the execution results so as to ensure that the returned results belong to the code of concern.

The approach can detect tampering with execution results after the execution of the agent is terminated. The agent owner decrypts the execution results and verifies the integrity of the execution results. The accuracy of detection is based on the parameters included in the offer. The approach preserves privacy, non-repudiation and authenticity of execution results. It can also detect any tampering with the agent code as the agent returns home and the agent owner verifies the returned code with the original one. Moreover, the approach keeps a trace of visited platforms and thus the agent owner can verify if trace actually correspond to scheduled agent's itinerary. The drawback of the approach is that it cannot maintain strong integrity of execution results. It mostly fails to detect two types of attacks: data truncation by colluding platforms and fake data insertion.

### Detecting Objects

It inserts dummy data into the execution results database that would not be included in any computations and should stay intact until the agent execution is terminated. A malicious entity that attempts to tamper with the execution results may alter the dummy data. Upon the agent's return, the agent owner can suspect tampering with the execution results if any of the dummy data items is altered or deleted.

The approach can detect tampering with execution results. The drawbacks of the approach are that the dummy data items have to be instantiated for each particular type of mobile codes and they should not be distinguished such that a malicious entity would surely attack. It is advisable to perform timely updates of the dummy data items so a malicious entity would not be able to comprehend the objective of such data items.

### Itinerary Recording

The agent platform signs the results of executing the agent with its private key and binds it to the identity of the succeeding platform in the agent's itinerary. Thus, the agent owner upon the agent's return can assemble the agent's itinerary and compares it to the itinerary it deduces from the signed execution results. If they do not match then it detects tampering with the execution results by a truncation attack.

The approach can detect tampering with execution results as the agent execution is terminated. However, the approach does not preserve the confidentiality of execution results. Also, it cannot detect truncation of execution results by colluding attack.

### References States

Reference states are agent states that resulted from executing the agent at a trusted host. The approach uses reference states to detect any tampering with the agent states. The execution results are instantaneously and absolutely verified at the next platform in the agent's itinerary.

The approach can detect tampering with the agent states spontaneously. However, it cannot detect truncation of execution results by colluding attack and does not preserve confidentiality of execution results. Moreover, it prolongs the agent execution time and doubles computation loads as compared to the normal loads of executing mobile agents.

### Secret Splitting

The agent task is split over multi-agents that cooperate to accomplish the intended task. Agents have shares of a secret key for remote digital signing. The results of agents are integrated without the need of reassembling the key.

The approach prevents a meaningful tampering with execution results. However, it does not preserve privacy of execution results and cannot protect the agent code. It also introduces additional communication and coordination loads.

## Protecting Agent Platform

The agent platform protection mechanisms that are presented in the literature (Giansiracusa, 2003; Nitschke, Paprzyeki, & Ren, 2006) are outlined and described below.

## Prevention Mechanisms

Prevention mechanisms try to make it unfeasible to perform unauthorized or illegal act to the agent platform, such as unauthorized access to confidential information or libraries or manipulation of system files. The prevention mechanisms include the following mechanisms: Access management, Security manager, Sandbox, State appraisal, Authentication and Authorization credentials (Code signing), Proof carrying code, Trust management,

*Figure 4. Classification of prevention mechanisms for protecting agent platform*

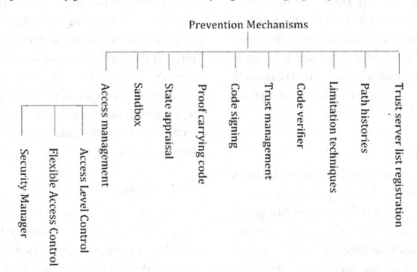

Code verification, Limitation techniques, Path histories and Trusted server list registration. In general, the prevention mechanisms can be classified as shown in Figure 4.

## Access Management

This mechanism includes three classes: (1) Access level control (Nitschke, Paprzyeki, & Ren, 2006); (2) Flexible access control (Cheng-Zhong & Song, 2003); (3) Security manager (Ylitalo, 2000). In the Access level control approach, the agent platform creates a protection domain for program execution, which restricts the agent access privileges. In Java 1.2, the granted access privileges are based on the identity of the code creator. Privileges includes: (1) read environmental variables, (2) connect, accept, or listen to host, IP, domain, or port, and (3) read, write, execute, or delete a system file. In Java 2.0, it implements Java Authentication and Authorization Services (JAAS) that grants privileges based on the identity of the client. In Aroma Virtual Machine (AVM), an attempt is made to avoid the denial of service attack by imposing additional restrictions on access to resources that includes: CPU allowed usage, transfer rate, and disk/ network quota for the agent. In Flexible access control mechanism, the agent

platform implements a security mechanism that may dynamically change access privileges given to a mobile agent as necessary. The platform may implement a Security manager that controls agent's access to local resources as follows:

- Check trusted host list
- Set access to local resources
- List permitted operations
- Monitor resource access
- Change access control list as necessary

The Security manager creates an internal Access Control List (ACL) that contains pair-wise values of objects, i.e. resource name and allowed actions for different mobile codes. The Security manager signals a user whenever a migration request is received from unknown host or the migrating mobile code is unsigned, who in turn may accept or reject the migration request.

## Sandbox

It is a domain at the agent platform that includes the non-security sensitive operations such as math and logic operations. The security sensitive operations such as the system file operations are outside the sandbox, which agents are not allowed to access or can access in a strictly controlled manner. A reference monitor consults the security policy to determine the level of access to grant a mobile agent to system resources and services agent based on the creator identity, operation requested, and access rules. It places restrictions on agent's honorable goals.

### State Appraisal

The agent creator or the agent owner produces appraisal functions such as the privileges it needs at a platform and encloses them within the agent's code. Upon the agent arrival at a platform, the platform evaluates the appraisal function based on the state of the agent and verifies the operations the agent requests. The platform will not execute the agent if it is insecure. The approach prevents security breaches by malicious or tampered agents.

### Code Signing

The agent creator or the agent owner signs the code to indicate who produced the code. The approach is also referred to as authentication and authorization credentials. A common practice to guarantee the integrity of the code is to compute a cryptographic hash of the agent's code and then encrypt it with the private key of the signer. The agent's code, message digest and public key certificate would then be sent to the agent platform, which would verify the source, integrity and authenticity of the code. The security administrator should have a list of public keys of trusted sites to decide on executing the agent. The digital signature mechanism does not ensure that the code is free of errors or illegal instructions.

### Proof Carrying Code

The agent creator builds a formal proof that the code adheres to a predefined set of safety rules, referred to as safety policy of the agent platform. The code and the proof are sent together to the agent platform, which in turn generates a safety predicate from the code to verify that the code

corresponds to the proof and satisfies the safety policy. Any tampering with the code would be detected in the verification process as a verification error. If the verification passes then the platform executes the code. The drawbacks of the technique are that proofs might be potentially large in size, and safety policies have to be formally standardized. Moreover, the generation of proofs has to be automated.

### Trust Management

The administrator only accepts mobile agents from gateways, trusted hosts, or initiated by a gateway. The administrator establishes a pre-defined list of trusted hosts based on her/his knowledge and expertise. Hosts are specified by their names, or IP addresses, or wildcards. Administrators control the level of access to non-trusted hosts.

### Code Verification

The agent platform scans the binary image of the mobile agent to check if the code is valid or not. If it includes any illegal instructions, it rejects to execute of the agent. In JAVA runtime, the byte-code verifier check for any illegal statements such as modification of system files or writing out memory space and decides if to execute the agent or not. The drawbacks of the technique are that it is slow and may not always be used. Normally, the byte-code verifier does not check any code that comes from trusted hosts and places it in a special space at the platform called *Classpath*. Though, the code is still subject to tampering by other agents running at the platform and may harm the platform.

### Limitation Techniques

The life-time of an agent can be monitored and controlled by setting time and range limits. The time limit sets a time interval for running the agent on an execution layer. If the limit has passed, then the agent is destroyed or returned to its client. The range limit pre-defines an itinerary to which the agent is restricted. The techniques may not allow the agent to fulfill its goals if limitations are not appropriately selected.

### Path Histories

It provides a visited platform with an authentic record of the previously visited platforms. Each platform the agent visits should add a signed term to the path history indicating the identity of the next host to be visited and the previous computation result. The next visited platform can determine whether it trusts the previously visited platforms and hence decides on the execution of the agent and the access level to local resources.

### Trusted Server List Registration

The agent platform is initialized with a list of trusted identity names and a list of actions an agent is allowed to a database. When a migration request is received from a host, the identity name of the host is checked against the list of trusted names. If the name is in the list, the agent platform returns *TRUE* and hence allows the agent only to perform respective actions outlined in the list. Otherwise, the agent platform returns *FALSE*, and hence it would reject the migration request as being from a non-trusted host.

## Detection Mechanisms

Audit logging is a detective mechanism. The platform records all mobile agent activities in separate audit logs. Auditors would then inspect the audit logs to detect the agent past activities. The agent would be identified and be held liable whenever any illegal activity detected. Nevertheless, the logs are susceptible to tampering or corruption by malicious agents. Also, they might have sensitive information to which auditors should not have access. Thus, audit logs can be must be protected by either implementing strong one-way cryptographic hash functions or storing logs in Tamper Resistant Hardware. Only authorized auditors should given access to audit logs through access control and search capability. However, audit logs would be invalidated given that an adversary has illegally got access to the logs.

## FUTURE ENHANCEMENT

Various security mechanisms have been proposed that assert the security of mobile agent code and data. The agent preventive mechanisms do not provide a satisfactory protection level. They are impractical, have limited application, restrain potentials of the mobile agent, or can be violated. The agent detective mechanisms do not also provide a satisfactory level as they may not be able to detect truncation attacks, introduce additional communication and computation loads, or do not accomplish all the required security properties.

The security mechanisms that assert the security of the agent platform have different protection levels. Some mechanisms might restrain the goals of honored agents, others might be breached. The most efficient mechanisms that can surely protect the agent platform are such as Security manager and Trusted server registration list. The major risks of mobile agent system are that the executing host may not execute the agent correctly or completely and it may not transmit it to the scheduled succeeding host in agent's itinerary.

The development of secure mobile agent systems still needs more research advances that balance security, practicality, reliability, and efficiency. The achievement of satisfactory advances would lead to wide deployment of mobile agent applications and gain more trust of business partners. There is a substantial need for more focus on the security of mobile agent code, especially, detection mechanisms that reveal agent termination or capture attacks.

## CONCLUSION

In this chapter authors have presented properties, advantages, and applications of mobile agent. The security threats to which the Mobile Agent System is vulnerable are explained. The authors also described the security properties a secure mobile agent system has to fulfill. The authors surveyed the security mechanisms that have been presented in the literature.

## REFERENCES

Aglets, I. B. M. (2002). *Aglets documentation home page*. Retrieved January 2, 2011, from http://www.trl.ibm.com/aglets/documentation_e.htm

Al-Jaljouli, R. (2005). Boosting m-business using a truly secured protocol for information gathering mobile agents. In *Proceedings of 4th International Conference on Mobile Business*. IEEE Computer Society Press.

Al-Jaljouli, R. (2006). *A proposed securityprotocol:security* protocol*protocol* for data gathering mobile agents. Master's Dissertation, University of New South Wales, School of Computer Science and Engineering, Australia.

Al-Jaljouli, R., & Abawajy, J. (2009). *Agents based e-commerce and securing exchanged information. Pervasive Computing: Innovations in Intelligent Multimedia and Applications, Computer Communications and Networks Series* (pp. 383–404). Berlin, Germany: Springer-Verlag.

Bellifemine, F., Poggi, A., & Rimassa, G. (1999). Jade - FIPA - complaint agent framework. In *Proceedings of 4th International Conference on Practical Applications and Intelligent Agents and Multi-Agents (PAAM)*, (pp. 97-108).

Bieszczad, A., Pagurek, B., & White, T. (1998). Mobile agents for network management. *IEEE Communication Surveys and Tutorials*, *1*(1), 2–9. doi:10.1109/COMST.1998.5340400

Bradshaw, J. (1997). An introduction to software agents. In Bradshaw, J. M. (Ed.), *Software agents* (pp. 3–46). AAAI Press.

Brewington, B., Gray, R., Moizumi, K., Kotz, D., Cybenko, G., & Rus, D. (1999). Mobile agents for distributed information retrieval. In Klusch, M. (Ed.), *Intelligent agents* (pp. 355–395). Springer-Verlag. doi:10.1007/978-3-642-60018-0_19

Chan, V., Ray, P., & Parameswaran, N. (2008). Mobile e-health monitoring: An agent based approach. *Telemedicine and e-Health Communication Systems*, *2*(2), 223–230.

Chen, W., Lin, C., & Lien, Y. (1999). A mobile agent infrastructure with mobility and management support. In *Proceedings of Workshop on Parallel Processing*, (pp. 508–513).

Cheng-Zhong, X., & Song, F. (2003). Privilege delegation and agent-oriented access control in Naplet. In *Proceedings of IEEE ICDCS Workshop on Mobile Distributed Computing (MDC)*, (pp. 493-497).

Chess, D., Harrison, C., & Kershenbaum, A. (1997). Mobile agents: Are they a good idea? *Mobile Object Systems: Towards the Programmable Internet, LNCS, 1222*, 25–47. Springer-Verlag.

Cugola, G., Ghezzi, C., Picco, G., & Vigna, G. (1997). Analyzing mobile code languages. *Mobile Object Systems: Towards the Programmable Internet, LNCS 1222*, Springer-Verlag, 93-110.

Dömel, P. (1997). Interaction of Java and Telescript agents. *Mobile Object Systems: Towards the Programmable Internet, LNCS, 1222*, 295–314. Springer-Verlag.

Electric, M. (1997). Concordia: An infrastructure for collaborating mobile agents. In *Proceedings of 1ˢᵗ International Workshop on Mobile Agents (MA 97)*.

Farmer, W., Guttman, J., & Swarup, V. (1996). Security for mobile agents: Issues and requirements. In *Proceedings of 19ᵗʰ National Information Systems Security Conference*, (pp. 591-597).

Fischer, L. (2003). *Protecting integrity and secrecy of mobile agents on trusted and non-trusted agent places*. Thesis Dissertation, University of Bremen, Germany.

Franklin, S., & Graesser, A. (2006). Is it an agent, or just a program? A taxonomy for autonomous agents. *Intelligent Agents III Agent Theories, Architectures, and Languages. LNCS, 1193*, 21–35. Springer-Verlag.

Giansiracusa, M. (2003). *Mobile agent protection mechanisms and the trusted proxy server architecture (TAPS). Technical Report*. Information Security Research Centre, Queensland University of Technology.

Glass, G. (1999). *Object space voyager core package technical overview*, (pp. 611-627). Mobility, ACM Press/Addison-Wesley Publishing Co.

Gray, R. (1996). Agent TCL: A flexible and secure mobile-agent system. In *Proceedings of 4ᵗʰ Annual TCL/TK Workshop (TCL)*.

Gray, R. S., Cybenko, G., Kotz, D., Peterson, R. A., & Rus, D. (2002). D'Agents: Applications and performance of a mobile-agent system. *Journal of Software - Practice and Experience, 32*(6), 543-573.

Gupta, R. (2011). A SURVEY of comparative study of mobile agent platforms. *Journal of Engineering Science Systems, 3*(3), 1943–1948.

Jansen, W. (2000). *Countermeasures for mobile agent security*. Computer Communications, Special Issue on Advances in Research and Applications of Network Security.

Jansen, W., & Karygiannis, T. (2000). Mobile agent security. *Special publication 800-19 of National Institute of Standards & Technology (NIST)*, 1-43.

Johansen, D., Renesse, R., & Schneider, F. (1995). Operating system support for mobile agents. In *Proceedings of 5ᵗʰ IEEE Workshop on Hot Topics in Operating Systems (HotOS-V)*, (pp. 42-45).

Karjoth, G. (2000). Secure mobile-based merchant brokering in distributed marketplaces. *Agent Systems, Mobile Agents, and Applications. LNCS, 1882*, 427–441. Springer-Verlag.

Karjoth, G., Lange, D., & Oshima, M. (1997). A security model for Aglets. *IEEE Internet Computing, 1*(4), 68–77. doi:10.1109/4236.612220

Karnik, N. (1998). *Security in mobile agent systems*. PhD dissertation, Department of Computer Science and Engineering, University of Minnesota.

Kim, H., & Moreau, T. (2001). Mobile code for key propagation. In *Proceedings of 1ˢᵗ International Workshop on Security of Mobile Multi-Agent Systems (SEMAS'2001)*.

Lange, D. (2002). Introduction to special issue on mobile agents. *Autonomous Agents and Multi-Agent Systems, 5*(1), 5–6. doi:10.1023/A:1013400230894

Lange, D., & Oshima, M. (1998). *Programming and developing Java mobile agents with Aglets*. Addison-Wesley.

Lange, D., & Oshima, M. (1999). Seven good reasons for mobile agents. *Journal of Communications of ACM, 42*(3), 88–89. doi:10.1145/295685.298136

Loureiro, S., Molva, R., & Roudier, Y. (2000). Mobile code security. In *Proceedings of ISYPAR (4ème Ecole d'Informatique des Systems Parallèles et Répartis)*. Retrieved January 2, 2011, from http://agentlab.swps.edu.pl/agent_papers/NATO_2006.pdf

Maggi, P., & Sisto, R. (2002). Using SPIN to verify security properties of cryptographic protocols. In *Proceedings of 9th International SPIN Workshop on Model Checking of Software (SPIN), LNCS 2318*, (pp. 187-204). Springer-Verlag.

Mell, P. (1999). Understanding the world of your enemy with I-CAT. In *Proceedings of 22nd National Information System Security Conference*, (pp.432-443).

Milojicic, D. (1999). Trend wars - Mobile agent applications. *Concurrency - IEEE, 7*(3), 80 – 90.

Ng, S. (1999). Protecting mobile agents against malicious hosts by intension of spreading. In *Proceedings of International Conference on Parallel and Distributed Processing Techniques and Applications (PDPTA)*, (pp.725-729).

Nitschke, L., Paprzyeki, M., & Ren, M. (2006). Mobile agent security. In Thomas, J., & Essaidi, M. (Eds.), *Information assurance and computer security* (pp. 102–123).

Pham, V., & Karmouch, A. (1998). Mobile software agents: An overview. *Journal of IEEE Communications, 36*(7), 26–37. doi:10.1109/35.689628

Rahimi, S., Angryk, R., Bjursell, J., Paprzycki, M., Ali, D., Cobb, M., et al. (2001). Comparison of mobile agent frameworks for distributed geospatial data integration. In *Proceedings of 4th AGILE Conference on Geographic Information Science*, (pp. 643-655).

Roth, V. (1999). Mutual protection of co-operating agents. In Vitek, J., & Jensen, C. D. (Eds.), *Secure Internet programming: Security issues for mobile and distributed objects* (pp. 275–285).

Silva, A. R., Romäo, A., Deugo, D., & Silva, M. M. (2001). Towards a reference model for surveying mobile agent systems. *Journal of Autonomous Agents and Multi-Agent Systems, 4*(3), 187–231. doi:10.1023/A:1011443827037

Suri, N., Bradshaw, J. M., Breedy, M. R., Groth, P. T., Hill, G., & Jeffers, R. (2000). Strong mobility and fine-grained resource control in NOMADS. In *Proceedings of 2nd International Symposium on Agent Systems and Applications and 4th International Symposium on Mobile Agents (ASA/MA)*, (pp. 2-15).

Syverson, P., Goldschlag, M., & Reed, M. (1997). Anonymous connections and onion routing. *IEEE Symposium on Security and Privacy*, (pp. 44-54).

Tripathi, A., Ahmed, T., Pathak, S., Carney, M., & Dokas, P. (2001). *Paradigms for mobile agent-based active monitoring of network systems*. Technical Report, Department of Computer Science, University of Minnesota. Retrieved January 2, 2011, from http://www.cs.umn.edu/Ajanta

Tripathi, A. R., Karnik, N. M., Ahmed, T., Singh, R. D., Prakash, A., & Kakani, V. (2002). Design of the Ajanta system for mobile agent programming. *Journal of Systems and Software, 62*(2), 123–140. doi:10.1016/S0164-1212(01)00129-7

Wayne, A. (2002). Jansen: Intrusion detection with mobile agents. *Journal of Computer Communications, 25*(15), 1392–1401. doi:10.1016/S0140-3664(02)00040-3

Weeler, T. (2005). *Voyager: High volume transaction processing in enterprise applications*. White Paper. Recursion Software, Inc. Retrieved January 2, 2011, from http://www.recursionsw.com/About_Us/inc/White_papers/Voyager_RSI/2005-02-16-Voyager_High-Volume-Transactions_RSI.pdf

White, J. (1995). *Mobile agents*. Technical Report, General Magic, Inc.

Wilhelm, U. (1999). *A technical approach to privacy based on Mobile agents protected by tamper resistant hardware.* PhD Dissertation, École Polytechnique Fédérale De Lausanne (EPFL), Switzerland.

World, J. Solutions for Java Developer. (2011). *Solve real problems with aglets, a type of mobile agent.* Retrieved January 2, 2011, from http://www.javaworld.com/javaworld/jw-05-1997/jw-05-hood.html?page=1

Ylitalo, J. (2000). *Secure platforms for mobile agents.* Technical Report, Department of Computer Science, Helsinki University of Technology, Finland. Retrieved January 2, 2011 from http://www.tml.tkk.fi/Opinnot/Tik-110.501/1999/papers/mobileagents/mobileagents.html

## ADDITIONAL READING

Al- Saadoon, G. (2009). Flexible and reliable architecture for mobile agent security (TR). *Journal of Computer Science, 5*(4), 270–274.

Bordini, R., Dastani, M., Dix, J., & Seghrouchni, A. (2009). *Multi-agent programming: Languages, Platforms and Applications.* Dordrecht, New York: Springer.

Dadhich, P., Dutta, K., & Govil, M. (2010). Security issues in mobile agents. *International Journal of Computers and Applications, 11*(4), 1–7. doi:10.5120/1574-2104

Deng, R., Qing, S., Bao, F., & Zhou, J. (2002). Lecture notes in Computer Science: *Vol. 2513. Information and Communications Security.* Heidelberg: Springer-Verlag. doi:10.1007/3-540-36159-6

Eassa, M., Salama, A., & Saleh, B. (2008). A secure integrated model for mobile agent migration. *Egyptian Computer Science Journal (ECS), 30*(2).

Garrigues, C., Migas, N., Buchanan, W., Robles, S., & Borrell, J. (2009). Protecting mobile agents from external replay attacks. *Journal of Systems and Software, 82*(2), 197–206. doi:10.1016/j.jss.2008.05.018

Garrigues, C., Robles, S., Borrell, J., & Navarro-Arribas, G. (2010). Promoting the development of secure mobile agent applications. *Journal of Systems and Software, 83*(6), 959–971. doi:10.1016/j.jss.2009.11.001

Genco, A. (2007). *Mobile Agents: Principles of Operation and Applications* (Genco, A., Ed.). WIT Press. doi:10.2495/978-1-84564-060-6

Ghosh, A., & Swaminatha, T. (2001). Software security and privacy risks in mobile E-commerce. *Communications of the ACM, 44*(2), 51–57. doi:10.1145/359205.359227

Hacini, S., Guessoum, Z., & Boufaïda, Z. (2007). TAMAP: A new Trust-based Approach for Mobile Agent Protection. *Journal in Computer Virology, 3*(4), 267–283. doi:10.1007/s11416-007-0056-y

Haiyang, H., & Hua, H. (2007). A reliable and configurable E-commerce mechanism based on mobile agents in mobile wireless environments. In *Proceedings of International Conference on Intelligent Pervasive Computing (IPC).*

Ismail, L. (2008). A secure mobile agents platform. *The Journal of Communication, 3*(2), 1–12.

Jean, E., Jiao, Y., & Hurson, A. (2007). Addressing mobile agent security through agent collaboration. *Journal of Information Processing Systems, 3*(2), 43–53. doi:10.3745/JIPS.2008.3.2.043

Jennings, N., & Wooldridge, M. (Eds.). (1998). *Agent technology: Foundations, Applications, and Markets.* Springer-Verlag Berlin Heidelberg New York.

Khosrow-Pour, M. (Ed.). (2004). *E-commerce security: Advice from experts.* USA: CyberTech Publishing.

Kou, W. (2003). Payment Technologies for E-Commerce., Kou, W. (Ed.), Springer-Verlag, 344 p.

Lin, C., & Varadharajan, V. (2010). MobileTrust: A trust enhanced security architecture for mobile agent systems. *International Journal of Information Security*, *9*(3), 53–178. doi:10.1007/s10207-009-0098-x

Muñoz, A., Maña, A., & Serrano, D. (2009). Protecting agents from malicious hosts using TPM. *International Journal of Computer Science & Applications*, *6*(5), 30–58.

Patel, R., & Garg, K. (2003). A new paradigm for mobile agent computing. In *Proceedings of 2nd WSEAS International Conference on Electronics, Control and Signal Processing*.

Schoeman, M., & Cloete, E. (2004). Cloete: Design concepts for mobile agents. *South African Computer Journal*, *32*, 13–24.

Stalling, W. (2003). *Cryptoraphy and Network Security* (3rd ed.). Upper Saddle River, NJ: Prentice Hall.

Tsai, J., & Ma, l. (2006). *Security modeling and analysis of mobile agent systems*. Electrical and Computer Engineering Empirical College, London.

Wang, T., Guan, S., & Chan, T. (2002). Integrity protection for code-on-demand mobile agents in E-Commerce. *Journal of Systems and Software*, *60*(3), 211–221. doi:10.1016/S0164-1212(01)00093-0

Wong, J., & Mikler, A. (1999). Intelligent mobile agents in large distributed autonomous cooperative systems. *Journal of Systems and Software*, *47*(2-3), 75–87. doi:10.1016/S0164-1212(99)00027-8

Xu, H., Zhang, Z., & Shatz, S. (2005). A Security based model for mobile agent software systems. *International Journal of Software Engineering and Knowledge Engineering*, *15*(4), 719–746. doi:10.1142/S0218194005002518

## KEY TERMS AND DEFINITIONS

**Agent Platform:** Provides the execution environment that supports the language in which the mobile agent was conceived, constrains its access to the allocated resources, and provides the access to the runtime library.

**Agent State:** Runtime (execution) state of the agent that includes program counter and execution stacks.

**Application Programming Interface:** A set of rules and specifications that enables software programs to interact with other.

**Asymmetric Keys:** A pair of keys, one public key used for encrypting plain-text and another private key for decrypting cipher-text.

**Credentials:** A store for secure credentials of an entity such as keys and digital signature, which a Certifying Authority (CA) needs to authenticate the entity in Public Key Infrastructure (PKI).

**Cryptographic Hash Function:** A function used for securing data transmitted over the internet, which transforms a message ($m$) into a fixed-sized string called message digest or cryptographic hash $h(m)$.

**Cryptographic Hash:** A fixed-sized string that represents a block of input data.

**Encryption:** Process of converting data (plain-text) into an encrypted data (cipher-text) using algorithms that make it unreadable by unauthorized people.

**Execution Stack:** A data structure that retrieves data in a reverse order of the order in which there were pushed onto the stack as a function is called.

**Fault Management:** Network management techniques that assure the survival of mobile agent despite the possible failure of remote agent hosts and associated platforms.

**Foundation for Intelligent Physical Agents:** An IEEE Computer Society standards organization that supports agent-based systems and develops interoperability of its standards with other systems.

**Interoperability:** The ability of diverse systems to communicate information and meaningfully interrupt the exchanged information.

**Mobile Agent System:** A system that is based on the implementation of mobile agents.

**Nonce:** A randomly generated number that uniquely identifies a protocol run.

**Protocol:** Defines the sequence of exchanged messages during a communication session.

**Runtime Library:** Contains functions that are needed for agent code translation during agent runtime (execution).

**Scalability:** Capability of a system to perform efficiently under heavy or growing workloads.

**Serialization:** Process of converting an object to a series of bits to be stored in a file or memory buffer.

**Strong Mobility:** Ability of the agent programming language to allow agents to move their code and execution state to a different server.

**Symmetric Key:** A shared key only known to communicating parties that is used for encrypting/decrypting exchanged messages.

**Timestamp:** A time generated for a code that identifies the time at which it was generated.

**Weak Mobility:** Ability of the agent programming language to allow agents at a remote host to be bound dynamically to code coming from different server.

# Chapter 10
# Mobile Agents Security Protocols

**Raja Al-Jaljouli**
*Deakin University, Australia*

**Jemal Abawajy**
*Deakin University, Australia*

## ABSTRACT

*Mobile agents are expected to run in partially unknown and untrustworthy environments. They transport from one host to another host through insecure channels and may execute on non-trusted hosts. Thus, they are vulnerable to direct security attacks of intruders and non-trusted hosts. The security of information the agents collect is a fundamental requirement for a trusted implementation of electronic business applications and trade negotiations. This chapter discusses the security protocols presented in the literature that aim to secure the data mobile agents gather while searching the Internet, and identifies the security flaws revealed in the protocols. The protocols are analyzed with respect to the security properties, and the security flaws are identified. Two recent promising protocols that fulfill the various security properties are described. The chapter also introduces common notations used in describing security protocols and describes the security properties of the data that mobile agents gather.*

## INTRODUCTION

Mobile agents are autonomous programs that can run in heterogeneous environments. They act on behalf of users and have some level of intelligence. They have one or more goals and can control where they execute. They traverse the Internet and get

DOI: 10.4018/978-1-4666-0080-5.ch010

executed on various architectures and platforms to access remote resources or even to meet, co-operate and communicate with other programs and agents to accomplish their goals. Hence, their functionality is not affected by the limitations of latency, connectivity, and bandwidth. Also, they support off-line computations and allow the use of distributed resources on the Internet. Agents can be stationary filtering incoming information or migrating searching for particular information across the Internet and analyzing it. The actions of an agent are not entirely pre-established and defined. They are able to take initiative in an autonomous way and exert control over their actions. The agent is able to choose what to do and in which order according to the external environment and user's requests. Mobile agents traverse the Internet and transport from one host to another host. They are expected to transport through insecure channels and execute on partially unknown and untrustworthy environments. Thus, they are vulnerable to various security threats of intruders and non-trusted hosts.

Several cryptographic protocols were presented in the literature asserting the security of data which agents gather while traversing and searching the Internet. Formal verification of the protocols reveals unforeseen security flaws, such as truncation or manipulation of gathered data, breaching the privacy of gathered data, sending others data under the private key of a malicious host, or replacing gathered data with data of similar agents. The detection of a security flaw implies that the protocol is not satisfactorily secure.

The chapter outlines the security requirements of mobile agent applications and analyzes the security flaws in the existing security protocols. It also discusses two recent security protocols that implement new security techniques on top of the existing techniques. The chapter is organized into six sections. The first section presents a background to mobile agents as regards real-world applications, types of data contained within mobile agents, and protocol's common notations. The second section discusses mobile agent security related issues including threats, properties, and techniques. The third section describes various mobile agent security protocols that are presented in the literature and the respective aimed security properties. It then analyzes the detected security problems and identifies the security properties a respective protocol has failed to accomplish. It also proposes an additional set of security techniques

that would rectify the detected security problems and establish protocols that are truly free of security flaws. The fourth section presents two recent mobile agent security protocols that implement new security techniques and are proved to be free of the security flaws by formal methods of verification. They are capable of preventing or detecting the security attacks the existing protocols have even failed to detect. The fifth section outlines the fundamental security specifications a security protocol should fulfill. The last section presents a conclusion that summarizes the security problems and the new security techniques that would rectify the security problems. It also emphasizes on the necessity of modeling and verifying security protocols to identify any security problem that might exist using formal methods.

## BACKGROUND

### Application Domains of Mobile Agents

Mobile agents are deployed in a wide range of real-world applications (Mobach, 2007; Outtagarts, 2009; Chen, Cheng, & Palen, 2009; Kok, Warmer, & Kamphuis, 2005; Paurobally & Jennings, 2005; James, Cohen, Dodier, Platt, & Palmer, 2006; Al-Jaljouli & Abawajy 2010). Several applications as listed below and particular applications are then discussed.

- Forensic management
- Network management
- Traffic management
- Distributed resource management
- Control in electricity infrastructure
- Web service negotiation and agreement
- Stock management
- Mobile healthcare (E-healthcare)
- Transportation planning
- E-commerce

- Distributed energy management (efficiency & metering)
- Grid computing and grid services
- Distributed data mining
- E-learning
- Climate environment and weather
- Wireless multimedia sensors

In *Forensic management applications* (Gray, 2000), a team of soldiers might be in the field observing terrorists in a suspected building and communicate with the military database at the headquarters. The communication links might be unreliable or of low bandwidth. So, they send mobile agents with description of people entering the building and log of intercepted calls to make queries on suspects in the military database that resides at the main network while connections goes off. Mobile agents return with pictures of people to be arrested that match suspects in the military database. Soldiers might also use network-sensing and planning agents to detect impending network disconnections and plan the best route in the network. Mobile agents would complete the task more efficiently as compared to remote message passing that would be susceptible to routing loops as passing through several soldiers' nodes and message queuing in case of network failure. They also reduce data transmission through communication links. Each soldier is supposed to have a laptop with GPS unit and two wireless Ethernet cards with different transmission frequencies.

In *Traffic management applications* (Chen et al., 2009), conditions of the roads are changeable and incidents are unforeseen. Multi-agents are used in the large scale distributed system for several tasks: (1) manage traffic, (2) detect congestion, (3) regulate routing on freeway and arterial, (4) control urban traffic signal. They are distributed geographically and interact with each other. Stationary agents are inflexible and inadaptable to changes in environment. Therefore, mobile agents are deployed that can be dynamically initiated at runtime in response to unforeseen events and

conditions. They can quickly diagnose incidents, dynamically implement new algorithms and reconfigure the system, and accordingly take ad hoc actions. They also reduce data transmission over the network. Mobile agents can also be deployed in railway and airway management applications. Choy, Srinivasan and Cheu (2003) presented a traffic management system that consists of three types of agents:

- Intersection Controller Agents (ICA)
- Zone Controller Agents (ZCA) that controls the pre-assigned (ICA) agents
- Regional Controller Agent (RCA) that controls all (ZCA) agents

Chen et al. (2009) presented a traffic detection and management system that consists of five types of stationary agents and particular type of mobile agents as follows:

- *Laser Detector* agents *(LRD)* and *Loop Detector* agents *(LPD)*: they track real-time traffic, estimate density on roads, and detect incidents and send alerts to *(TMC)* agents.
- *Video Camera Detector* agents *(VCD)*: they control various cameras to photograph incidents and provide detailed incident information including location.
- *Traffic Detection Execution System* agents *(TDES)*: they control lower level agents and maintain services for agents in a sub-network. They also process data from lower level agents and pass it to TMC agent.
- *Transportation Management Center* agents *(TMC)*: they provide an interface for interaction with traffic personnel and analyze requests into tasks. They assign tasks to mobile agents which they initiate and dispatch to incident locations. They also analyze data from mobile agents and pass reports to traffic personnel.

- *Mobile* agents: they move to incident location, analyze the problem locally and take appropriate actions independently even if the connection with the source network goes off.

In *Network management applications* (Bieszczad, White, & Pagurek, 1998), mobile agents can perform any of the following functions:

- Fault management
- Accounting management
- Configuration management
- Performance management
- Security management

Mobile agents can instantly diagnose points of network failure, analyze the problem, and fix it. They would eliminate control messages for routing and redundant alert messages and reduce network delay. There are two types of discovery agents: *Deglet* and *Netlet*. *Deglet*s detect over-utilized nodes and construct a model of nodes with utilization that exceed a threshold limit. They can collect more detailed information about nodes of concern and recover the problem by assigning some services to under-utilized nodes. Mobile agents can also perform remote network maintenance. They discover faults in remote devices and try to repair devices by incorporating machine learning techniques. *Deglet*s are dispatched to network nodes to perform a particular task and can be terminated internally based on predefined constraints, e.g. number of visited nodes. Conversely, *Netlet*s reside at a sub-network and detect changes in network configuration. Mobile agents would inform network manager if automatic repair is not feasible and human intervention is needed.

Mobile agents can handle the setup of provisioning services. For example, they can establish Provisioning Virtual Circuits (PVC) in ATM network between two ATM switches. *Deglet*s exchange information between ATM switches through Virtual Machine Components (VMC)

they incorporate. They use a particular ontology that generalizes the setup between the two ATM switches and select the best deal for the setup.

New components may be added or removed from a network. Usually, the network manager handles the configuration of new components; however, the process is tedious as the network topology is changeable and software may not be available. Mobile agents can automatically install added network devices by implementing Plug-and-Play capabilities. For example, agents can detect a new device that needs a printer driver to be installed. They download the latest version of the printer driver from vendor's web page and run the installation. They can also register with the vendor for any new version of the printer driver. Whenever, a new version is sent, the agent will automatically install the driver.

Mobile agents can detect deteriorated performance of a network and handle network delays more efficiently than stationary monitoring agents. They migrate to components of concern, diagnose the problem locally, and execute instant recovery routines. Mobile agents can also detect security threats and take recovery appropriate actions. They may implement security policies that control access to system resources.

In *e-Commerce applications* (Al-Jaljouli & Abawajy, 2010), mobile agents act on behalf of clients and traverse the Internet in search of goods/ services that meet the client requirement's profile. They may carry out multiple negotiation rounds with potential vendors of competitive offers trying to achieve best deal in terms of multiple constraints including price, warranty period, installment plan, etc. They evaluate the market and make decision based on eagerness of the client for goods/ services, vendors' reputation record, achieved utility, and negotiation deadlines. Mobile agents autonomously handle search and evaluation of marketplaces, negotiation and agreement settlement, and transaction processing. They enhance client's utility, broaden search/negotiation space, and shorten search and negotiation time and costs.

*Figure 1. Common notations in protocols' description*

| | |
|---|---|
| $\Pi$ | Program code and the static data |
| $i_0$ | Initiating host |
| $i_0, i_1, i_2, \ldots, i_n, i_0$ | Itinerary of a terminating agent |
| $i_j \rightarrow i_k: m$ | Transfer of data $(m)$ from host $i_j$ to host $i_k$ |
| $\{m\} S_{i_n}^{-1}$ | Signing data $(m)$ with the private key of host $i_n$ |
| $\{m\} K_{i_n}$ | Encrypting data $(m)$ with the public key of host $i_n$ |
| $h(m)$ | Hashing data $(m)$ |
| $m_1 \| m_2$ | Concatenation of data $(m_2)$ to data $(m_1)$ |

They work independently of connection with the source network that might goes off.

The data mobile agents gather and transport are very sensitive and should not be illegitimately revealed to non-correlated parties. They should remain protected from any malicious attacks; otherwise they would not complete their mission successfully.

## Data Types

The agent encompasses the two types of data: (i) *Static data*. (ii) *Dynamic d*ata (Maggi & Sisto 2002b). The dynamic data comprises the following data types, as in Fischer (2003):

- *Fixed size changeable* are the data that are set at the initiation of the agent but to which changes are authorized, such as global variables.
- *Dynamically allocated static* are the data that are acquired during the life cycle of the agent but to which no later changes are authorized. Commonly, they are referred to *execution results*.
- *Dynamically allocated changeable* are the data that are acquired during the life-cycle of the agent and to which changes might be authorized, such as register contents and stack.

## Protocol's Common Notations

In describing the security protocols of mobile agents, common notations are used (Fischer, 2003; Maggi & Sisto, 2003; Roth, 2001). The chapter focuses on the security of the *execution results* of terminating mobile agents. Figure 1 shows the common notations used in describing protocols. A terminating agent is an agent that is initiated by host $(i_0)$ and returns back to the initiator $(i_0)$ following to its visit to the $(n)$ executing hosts. The agent gets executed at each of the visited hosts in the agent's itinerary. The agent stores the result of execution at host $(i_j)$ for $(1 \leq j \leq n)$ as $(m_j)$. The execution results are modeled as a chain of: $\{m_1, m_2, \ldots, m_n\}$ respective to the order of the executing host in the agent's itinerary, which is stored within the agent. The agent, following its visit to $(n)$ executing hosts, returns the chain of results to the initiator $(i_0)$. The returned chain is expressed as: $\{m'_1, m'_2, \ldots, m'_n\}$. The returned execution result might differ from the genuine execution result $m_j$ for $(1 \leq j \leq n)$ due to tampering acts of adversaries. Hence, it is denoted as $m'_j$ for $(1 \leq j \leq n)$. Through the agent's migration, the transfer of data $(m)$ from host $(i_j)$ to host $(i_k)$ is denoted as $i_j \rightarrow i_k: m$, where $(i_j)$ and $(i_k)$ are executing hosts in the agent's itinerary.

The program code and the static data are both denoted as $(\Pi)$. The encryption of plaintext $(m)$ into a ciphertext is written as $(m) K_{i_n}$, where $K_{i_n}$ is the public key of host $(i_n)$ which encrypted $(m)$ data. A digital signature is written as an encryption with a private signing key $S_{i_n}^{-1}$. The bare signature is the digital signature on the computed hash of the data and is written as $S_{i_n}^{-1}(m)$. It is assumed that it is possible to deduce the identity of the signer from a signature. The concatenation of two data $(m_1)$ and $(m_2)$ is denoted as $m_1 \| m_2$, where concatenation refers to appending data $(m_2)$ to another data $(m_1)$. The hash function is denoted as $h$, and $h(m)$ stands for the hashing of data $(m)$.

# MOBILE AGENT SECURITY

## Security Threats

During migration and execution, mobile agents are vulnerable to direct security attacks of intruders and non-trusted hosts, where intruders and non-trusted hosts can perform any of the following malicious acts:

1.  *Truncation of the gathered data.* It is the common flaw in the existing security protocols and is the most difficult to deter. An attack can take place when two malicious hosts co-operate with each other to truncate the data acquired at intermediary hosts or substitute new data for the data they had previously provided to the agent. If a host conspires with a preceding host in the agent's itinerary and sends the agent back to it, then the preceding host would be able to truncate the data acquired at the intermediary hosts without being detected by replacing the agent's dynamic data that is current with the data that the agent had when it firstly visited its host, if the host had already stored the dynamic data including the register contents and stack. Truncation might result in money fritter, particularly, in the case of shopping agents that search for the lowest priced goods where the offer of the most competing host has been truncated.

2.  *Alteration of the collected data.* It takes place if the agent visits a host twice or two hosts collude with each other. The alteration of collected data can take different forms as described below.

    a.  A malicious host may send back the agent to an earlier host in the agent's itinerary. It could then truncate the data acquired at intermediary hosts and alter the data it formerly provided to the agent without being detected if it replaces the current agent's state with the state that was present when the agent firstly visited it. The host can then replace its initial data with a new data.

    b.  An adversary may truncate collected data and appends a fake stem at its discretion, if the input to all previous chaining relations is known and the validity of the chaining relation is maintained.

    c.  The adversary may truncate data and then sends the agent to hosts of its choice and repeats the process until it is satisfied with the collected data.

    The initiator would never detect data alteration. The alteration of the collected data might result in money fritter, particularly, in the case of shopping agents that search for the lowest priced goods where the offer of the most competing host has been altered.

3.  *Impersonating the genuine initiator and hence breaching the privacy of the gathered data.* An adversary may possibly intercept a message signed by the initiator of an agent. It could then decrypt the signed message and sign the decrypted message with its private key impersonating the genuine initiator. Executing hosts in the agent's itinerary would encrypt the data they provide to the agent with the public key of the adversary

assuming it is the genuine initiator. Hence, the adversary would be able to breach the privacy of the collected data.

4. *Transmitting others data signed by the private key of a malicious host.* Usually, executing hosts in the agent's itinerary send the data they provide to the agent signed with the corresponding private keys. An adversary might intercept the signed data. It could then decrypt the signed data and may sign the data of a particular host with its private key impersonating the genuine provider of the data.

5. *Replacing the gathered data with the data of similar agents.* An adversary might intercept the agent. It could then replace the current agent's state with the state of a similar agent.

## Security Properties

The security properties of the *execution results* of mobile agents are defined below (Fischer, 2003; Maggi & Sisto, 2003).

1. *Data integrity*: During the migration of the agent or during its execution at visited hosts, tampering with the already stored execution results $(m_j)$ for $(1 \leq j \leq n)$ is prevented, or at least any tampering will always be detected by the initiator upon the agent's return. The data integrity requires that the chain of execution results: $\{m'_1, m'_2, ..., m'_n\}$ which is returned to the initiator $(i_0)$ matches the genuine chain of execution results: $\{m_1, m_2, ..., m_n\}$. Otherwise, the initiator $(i_0)$ has a proof that the genuine execution results had been tampered with. The data integrity property refers to the following classes of protection:

   ○ *Insertion resilience*: Data can only be appended to the chain: $\{m_1, ..., m_j\}$ for $(j < n)$.

   ○ *Deletion resilience:* Deletion of an execution result $(m_j)$ for $(1 \leq j \leq n)$ from the chain of results: $\{m_1, ..., m_n\}$ is prevented, or at least is detected upon the agent's return to the initiator. If an execution result $(m_j)$ is deleted and the chain of execution results reduces to: $\{m_1, ..., m_{j-1}, m_{j+1}, ..., m_n\}$, then the initiator $(i_0)$ has a proof that an execution result is deleted from the chain of results.

   ○ *Truncation resilience:* Truncation of the chain: $\{m_1, ..., m_j, ..., m_n\}$ at host $(i_j)$ and reducing it to the chain: $\{m_1, ..., m_j\}$ for $(j < n)$ is prevented, or at least is detected upon the agent's return to the initiator $(i_0)$. The computed verification terms, e.g. message authentication code would indicate inconsistency with the returned chain of execution results.

   ○ *Strong forward integrity*: None of the execution results in a chain can be modified. The property necessitates that the returned execution result $(m'_j)$ matches $(m_j)$ for $(1 \leq j \leq n)$.

   ○ *Strong data integrity* requires the four classes of protection: insertion resilience, deletion resilience, truncation resilience, and strong forward integrity.

2. *Data non-repudiability:* The initiator $(i_0)$ can build a proof about the identity of host $(i_j)$ that added the execution result $(m'_j)$ for $(1 \leq j \leq n)$ to the chain of results: $\{m'_1, m'_2, ..., m'_n\}$.

3. *Data confidentiality*: The chain of execution results: $\{m_1, ..., m_j, ..., m_n\}$ stored within the agent can only be read by the initiator $(i_0)$. No one else is permitted to learn the plaintext of the ciphered execution results that are stored within the agent. Therefore, the unauthorized retrieval of information is prevented and the chain of the ciphered execution results should not reveal information about its contents to unauthorized entities.

Note that an adversary may of course see the chain of ciphered execution results, shown below.

$$\{m\}K_{i_1}, ..., \{m\}K_{i_j}, ..., \{m\}K_{i_n}$$

Nevertheless, as long as he is not able to get hold of the decryption keys, he is still unable to deduce the plaintext $(m_j)$ for $(1 \leq j \leq n)$ and the system is still deemed secure. Thus, it is fundamental to keep the encryption keys confidential during the run of a protocol session.

4. *Data authenticity*: Upon the agent's return, the initiator $(i_0)$ can determine for sure the identity of host $(i_j)$ that appended $(m'_j)$ to the chain of execution results: $\{m'_1, m'_2, ..., m'_n\}$. The initiator $(i_0)$ can be sure that the results that purport to be from a certain host were indeed provided by that host. Thus, an adversary should not be able to impersonate a host.

5. *Origin-confidentiality*: The identity of host $(i_j)$ for $(1 \leq j \leq n)$ that generated and added the execution result $(m_j)$ to the chain: $\{m_1, ..., m_n\}$ can only be known at host $(i_0)$. An executing host $(i_k)$ should not be able to deduce the identity of a previously visited host $(i_j)$ for $(1 \leq j < k)$ from the agent. Though, it is possible for a malicious host $(i_k)$ to get the identity of host $(i_{k-1})$ where the execution result $(m_j)$ was generated by analyzing the agent's dynamic data just before and after the agent visited it. Also, the identity of host $(i_{k-1})$ will possibly be revealed on the network layer. This can be prevented by using anonymous connections (Symbolic Model Prover, 2000) which hide the identity of the previously visited host.

## Security Techniques

In the literature the security of the execution results, particularly the integrity of the results, is built based on the implementation of one or more of the following security techniques (Corradi, Montanari, & Stefanelli, 1999b; Hannotin, Maggi, & Sisto, 2001; Karnik & Tripathi, 1999; Karjoth, Asokan, & Gülcü, 1998; Maggi & Sisto, 2002a; Yao, Foo, Peng, & Dawson, 2003): public key encryption, digital signature, message authentication code, backward chaining, one-step forward chaining, and code-result binding. The methods are described in details in (Aziz, Gray, Hamilton, Oehl, Power, & Sinclair, 2001).

- The *public key encryption* uses the public key of the initiator to cipher the execution results at visited hosts so as to achieve secrecy of results. Also, it uses the public key of the succeeding host to encrypt particular verification terms, such as a hashed nonce.
- The *digital signature* is used when a host signs the results it provides with its private signing key so as to be authenticated as the initiator of the execution results.
- The *message authentication code* incorporates the identity of the succeeding host and the characteristics of the previous execution results into the execution result at a host.
- The *backward chaining* incorporates characteristics of the previous execution results into the execution result at a host.
- The *one-step forward chaining* incorporates the identity of the succeeding host into the execution result at a host.
- The *code-result binding* binds the signed code to the execution results so as to ensure that the returned results belong to the code of concern.

## MOBILE AGENT SECURITY PROTOCOLS

Several cryptographic protocols have been presented in the literature (Karnik & Tripathi, 1999; Corradi et al., 1999b; Al-Jaljouli, 2005, 2006 ;

Karjoth et al.,1998; Maggi & Sisto, 2002a, 2003; Hannotin et al., 2001; Yao et al., 2003; Rhazi, Pierre, & Boucheneb, 2007) that aim to secure the execution results of mobile agents.

## Existing Mobile Agent Security Protocols

In this section, the security protocols listed below are discussed, which aim to preserve the confidentiality, authenticity, non-repudiation and/or integrity of data gathered by mobile agents. Each protocol is analyzed for the aimed security properties, respective security problems are highlighted, and security techniques that would rectify the problems are proposed. The discussion is intended to show that the security techniques discussed in the previous section are not sufficient to protect data contained within mobile agents that traverse through insecure channels and execute on non-trusted hosts.

- Targeted state protocol (Karnik & Tripathi, 1999)
- Append only Container Protocol (Karnik & Tripathi, 1999)
- Publicly Verifiable Chained Digital Signature Protocol (Karjoth et al., 1998)
- Chained Digital Signature Protocol with Forward Privacy (Karjoth et al., 1998)
- Chained MAC Protocol (Karjoth et al., 1998)
- Publicly Verifiable Chained Signature Protocol (Karjoth et al., 1998)
- Configurable Mobile Agent Data Protection Protocol (Maggi & Sisto, 2003)
- Mobile Agent Integrity Protocol (Hannotin et al., 2001; Maggi & Sisto, 2002a)-
- Improved Forward Integrity Protocol (Yao et al., 2003)
- Secure protocol based on sedentary agent for mobile agent environments (Rhazi et al., 2007)

- Data gathering security protocol (Al-Jaljouli, 2006)
- Enhanced multi-hops protocol (Al-Jaljouli, 2005)

## Targeted State Protocol

The *Targeted State Protocol* (Karnik & Tripathi, 1999) is proposed to ensure the confidentiality of data carried by a mobile agent. It is based on encrypting the data that should only be available to a trusted host using the public key of the host. The initiator may intend to transmit confidential data to a number of trusted hosts, so that a trusted host would only be able to learn the confidential data that is intended for it. The initiator encrypts each confidential data with the public key of the host for which the data should only be revealed, and then signs it with its private key. The targeted state would be as follows:

$$i_n \rightarrow i_{n+1} : \left\{ \{m_1\}K_{i_1}, ..., \{m_n\}K_{i_n} \right\} S_{i_o}^{-1}$$

The security flaw is that an adversary can strip off the initiator's signature from the targeted state, and then copy the targeted state into an agent of its own. Next, the adversary signs the targeted state with its private key impersonating itself as the genuine initiator. Next, the adversary sends its own agent to executing hosts: $\{i_1, ..., i_n\}$. Each host inspects the targeted state, decrypts the ciphertext it can decrypt using its private decryption key and makes the chain of plaintext: $\{m_1, ..., m_n\}$ available to the agent. The agent migrates back to the adversary carrying the plain-text. Subsequently, the adversary possesses the text that is supposed to be confidential and the initiator would never detect such breach of privacy. The attack is illustrated by considering the agent's targeted state to contain a single plaintext $(m_1)$ encrypted with the public key of host $(i_1)$. The initiator $(i_0)$ sends out the agent code $(\prod_0)$ and its targeted state to $(i_1)$, as follows:

$$i_0 \rightarrow i_1 : \left\{ \Pi_0, \{m_1\}K_{i_1} \right\} S_{i_o}^{-1}$$

An adversary ($i_a$) intercepts the communication, and then strips off the initiator's signature. Next, it copies the targeted state $\left\{ m_1 \right\} K_{i_1}$ into an agent ($\Pi_a$) of its own, and then signs the targeted state with its own signature. Next, it sends out the agent to host $i_1$ as follows:

$$i_a \rightarrow i_1 : \left\{ \Pi_a, \{m_1\}K_{i_1} \right\} S_{i_a}^{-1}$$

Host $i_1$ innocently decrypts the ciphertext using its private key hence it makes the plaintext ($m_1$) available to the adversary. The agent migrates back to the adversary as follows:

$$i_1 \rightarrow i_a : \Pi_a, \{m_1\}K_{i_a}$$

## Append Only Container Protocol

The *Append Only Container Protocol* (Karnik & Tripathi, 1999) is proposed so that new objects can be appended to a container of objects in an agent but any later modification or deletion of an already contained object can be detected by the initiator ($i_0$) upon the agent's return. Also, the insertion of a new object within the contained objects can be detected. The security of protocol is based on an encrypted checksum ($C_n$). The initial value of the checksum ($C_0$) is a nonce ($r$) that is chosen randomly by the initiator and is encrypted with its public key so having $\{r\}K_{i_o}$. The initiator has to maintain the nonce ($r$) confidential during the protocol run and would use in the verification of the protocol upon the agent's return. The agent migrates to ($n$) hosts. Each host executes the agent, signs the execution results ($m_n$) with its digital signing key ($S_{i_n}^{-1}$), computes a new checksum ($C_n$) using the checksum it received from the preceding host in the agent's

itinerary ($C_{n-1}$) and the signed results ($m_n$), appends the execution results to the chain of objects, and then sends out the agent with the new chain to the succeeding host in the agent's itinerary. The *Append Only Container Protocol* is defined as follows:

$$i_n \rightarrow i_{n+1} : \left\{ \{m_1\}S_{i_1}^{-1}, ..., \{m_n\}S_{i_n}^{-1}, C_n \right\}$$

The checksum ($C_n$) is to be updated as follows:

$$C_n = \{C_{n-1} \;\|\; S_{i_n}^{-1}(m_n)\}K_{i_o}$$

Upon the agent's return, the initiator successively decrypts the checksums, and then extracts the signatures of executing hosts. Next, it verifies the extracted signatures with the corresponding objects in the container. The last verified checksum must be equal to the initial nonce ($r$).

The security flaw is that an adversary may collude with a host ($i_j$), which the agent had previously visited, and then learn the checksum ($C_j$) at ($i_j$). Next, the adversary can truncate the container up to the $j^{th}$ object without being detected if it replaces the most recent checksum ($C_n$) with the checksum ($C_j$) that was present when the agent firstly visited host ($i_j$) for ($1 \leq j < n$). Moreover, the adversary can replace the initially provided execution result ($m_j$) with a new execution result ($m'_j$) if it computes a new valid checksum based on the learnt ($C_{j-1}$). The same type of attack can take place if the agent visits a malicious host more than once. The attack violates the data integrity property. Suppose the agent visits a malicious host ($i_j$) twice during its life cycle, then host ($i_j$) is able to perform any of the following attacks without being detected:

- Truncates the chain at ($m_j$) so it is reduced to: $\{m_1, ..., m_j\}$ and then updates the checksum accordingly. Next, it can send out the agent with the reduced chain to a new set of executing hosts $(i'_{j+1}, ..., i'_n)$ and replaces the chain of results: $\{m_1, ..., m_j, ..., m_n\}$ with an altered chain of results: $\{m_1, ..., m_j, m'_{j+1}, ..., m'_n\}$.

- Replaces the initially provided execution result $(m_j)$ with a new execution result $(m'_j)$, and then updates the checksum accordingly. Next, it dispatches the agent again to hosts $(i_k)$ for $(j < k \leq n)$.

## Multi-Hops Protocol

The *Multi-hops Protocol* (Corradi et al., 1999b) has the same purpose as the Append only Container Protocol. The protocol uses: (1) a hash chain $(\gamma_n)$, (2) a message authentication code $(\mu_n)$, (3) a static part $(\prod)$ that includes: program code, and static (initialization) data, (4) a chain of execution results in plaintext $(M_n)$, and (5) a chain of encapsulated execution results $(P_n)$. The chain of execution results $(M_j)$ at host $(i_j)$ is the concatenation of the execution results $(m_j)$ for $(1 \leq j \leq n)$ at visited hosts, and the corresponding hosts' identities $(i_j)$. The protocol binds the static part $(\prod)$ to the terms: $(M_n, P_n, \gamma_n,$ and $\mu_n)$. At initialization, $(\gamma_n)$ is set to $\gamma_0 = h(r)$ where $(r)$ is chosen randomly by the initiator $(i_0)$, and the terms: $\{\mu_n, M_n, P_n\}$ are left empty. At each executing host, the agent updates the terms $M_n, P_n, \gamma_n,$ and $\mu_n$ to incorporate the execution result at the host.

The message authentication code $(\mu_n)$ that is computed at host $(i_n)$ acts as a chaining relation that incorporates: (1) the hash chain computed at the preceding host $(\gamma_{n-1})$, (2) the message authentication code computed at the preceding host $(\mu_{n-1})$, which summarizes all execution results previously obtained by the agent, (3) the execution result at the current host $(m_n)$, and (4) the identity of next host $(i_{n+1})$ in the agent's itinerary. The protocol data $(P_n)$ is a chain of the signed message authentication codes $(\mu_n)$ for $(1 \leq j \leq n)$.

The protocol is described as follows:

$$\gamma_n = h(\gamma_{n-1})$$
$$\mu_n = h(m_n, \gamma_{n-1}, \mu_{n-1}, i_{n+1})$$
$$P_n = P_{n-1} \| S_{i_n}^{-1}(\mu_n)$$
$$M_n = M_{n-1} \| m_n \| i_n$$
$$i_n \rightarrow i_{n+1} : (\Pi, M_n, P_n), \{\gamma_n\}K_{i_{n+1}}, \mu_n$$

The security flaw is that an adversary may collude with host $(i_j)$ for $(j < n)$, which the agent had previously visited. The adversary would send back the agent to host $(i_j)$. It could then truncate the data gathered following the first visit of the agent to its host, and replace its offer with a new valid offer choosing a new successor and using the values of $(\gamma_{j-1})$ and $(\mu_{j-1})$, which it might still store. Another security flaw can take place in case the agent visits host $(i_j)$ twice and the host stores $(\gamma_{j-1})$ and $(\mu_{j-1})$. The host can replace the data it provided to the agent during the first visit $(m_j)$ with any of the data gathered following the first visit of the agent to its host $(m_x)$ for $(j < x)$, and then compute new valid values of $(\gamma_j)$ and $(\mu_j)$. Next, it signs $(\mu_j)$ with its private key masking itself as the genuine provider of the data, and then truncates the data gathered following the first visit of the agent to its host. The initiator would never detect the data truncation or replacement.

## Publicly Verifiable Chained Digital Signature Protocol

The *Publicly Verifiable Chained Digital Signature Protocol* (Karjoth et al., 1998) aims to preserve the confidentiality and integrity of data acquired by mobile agents. The protocol uses a hash chain $(C_n)$ and a chain of encapsulated execution results: $\{M_1, ..., M_n\}$. The hash chain $(C_n)$ binds the encapsulated execution result at the preceding host $(M_{n-1})$ to the identity of the next host in the agent's itinerary $(i_{n+1})$. The encapsulated execution result $(M_n)$ incorporates the execution result at the current host $(m_n)$, the randomly selected nonce $(r_n)$, and the hash chain $(C_n)$. The nonce $(r_n)$ prevents an adversary from attacking the encryption. The $(m_0)$ is a dummy data provided by the initiator $(i_0)$. The protocol is defined as follows:

$$M_n = \left\{ \{m_n, r_n\} K_{i_o}, C_n \right\} S_{i_n}^{-1}$$
$$C_n = h(M_{n-1}, i_{n+1})$$
$$M_0 = \left\{ \{m_0, r_0\} K_{i_o}, C_0 \right\} S_{i_o}^{-1}$$
$$C_0 = h(r_0, i_1)$$
$$i_n \rightarrow i_{n+1} : \{M_0, ..., M_n\}$$

The security of the protocol is based on the assumption that an attacker does not change the last element $(M_n)$ in the chain. The protocol does not aim to preserve forward privacy, though it is noted that there is a possibility of an adversary to intercept the agent and then deduce the identities of signing hosts from the encapsulated execution results $\{M_0, ..., M_n\}$.

The security flaw is that an adversary can truncate chain elements and can grow a fake stem, since the input to all previous chaining relations is known. Elements can be appended to the chain at the discretion of the adversary, though the validity of the chaining relation is maintained. The adversary sends the agent with a chain of execution results, e.g. $\{M_0, ..., M_{j-1}\}$ to host $(i_j)$ of its own choice and repeats the process until it is satisfied with the collected elements. Then, the adversary chooses an element and pastes it into agent and sends the agent to $(i_{j+1})$. Another security flaw is that an adversary can append arbitrary objects, generated to the terms of the adversary rather than the initiator, to the agent without being detected.

## Chained Digital Signature Protocol with Forward Privacy

The *Chained Digital Signature Protocol with Forward Privacy* (Karjoth et al., 1998) has the same purpose as the *Publicly Verifiable Chained Digital Signature Protocol* as well as forward privacy/origin confidentiality. It proposes a change

in the order of encrypting and signing the execution results so as to accomplish forward privacy. The execution results at a visited host are firstly signed by the host and then are encrypted with the public key of the initiator. Hence, no one other than the initiator can decrypt the ciphered execution results. The protocol uses a hash chain $(C_n)$ and a chain of encapsulated execution results $(M_1, ..., M_n)$. The hash chain $(C_n)$ binds the encapsulated execution result at the preceding host $(M_{n-1})$ to the identity of the next host in the agent's itinerary $(i_{n+1})$ and a random nonce the host selects $(r_n)$. The encapsulated execution result $(M_n)$ incorporates the execution result at the current host $(m_n)$, the randomly selected nonce $(r_n)$, and the hash chain $(C_n)$. The nonce $(r_n)$ prevents an adversary from attacking the encryption. The $(m_0)$ is a dummy data provided by the initiator $(i_0)$. The protocol is defined as follows:

$$M_n = \left\{ \{m_n\} S_{i_n}^{-1}, r_n \right\} K_{i_o}, C_n$$
$$where \ C_n = h(M_{n-1}, r_n, i_{n+1})$$
$$M_o = \left\{ \{m_0\} S_{i_o}^{-1}, r_0 \right\} K_{i_o}, C_0$$
$$where \ C_0 = h(r_0, i_1)$$
$$i_n \rightarrow i_{n+1} : \{M_0, ..., M_n\}$$

The problem with the protocol is that executing hosts would not be able to know the identity of the initiator of agent, since the signature of the initiator is encrypted within $(M_0)$. The security flaws of the protocol are the same as the flaws of the *Publicly Verifiable Chained Digital Signature Protocol*.

## Chained MAC Protocol

The *Chained MAC Protocol* (Karjoth et al., 1998) aims to preserve confidentiality, integrity, and forward privacy of data acquired by mobile agents. It does not provide authenticity. The protocol uses a hash chain $(C_n)$ and a chain of encapsulated execution results $(M_0, ..., M_n)$. The executing host ($i_n$) computes the hash chain of the succeeding host ($C_{n+1}$). The hash chain ($C_{n+1}$) binds the identity of the succeeding host ($i_{n+1}$) to the hash chain ($C_n$), execution results ($m_n$), and a random nonce ($r_n$) generated at the current host ($i_n$). The encapsulated execution result at host ($M_n$) binds the random nonce ($r_n$) and the execution results ($m_n$) generated at the host to the identity of the succeeding host ($i_{n+1}$). The protocol is defined as follows:

$$M_n = \left\{ r_n, m_n, i_{n+1} \right\} K_{i_o} \qquad for \ n \geq 0$$
$$C_{n+1} = h(C_n, r_n, m_n, i_{n+1}) \qquad for \ n \geq 1$$
$$C_0 = h(r_0, m_0, i_1) K_{i_o}$$
$$i_n \rightarrow i_{n+1} : \{M_0, ..., M_n\}, C_{n+1} \qquad for \ n \geq 0$$

The problem with the protocol is that executing hosts would not be able to know the identity of the initiator of agent. The security flaw of the protocol is that an adversary can collude with host ($i_j$), which the agent had previously visited and had stored the hash chain ($C_j$), and send the agent back to it. The host can then truncate the gathered data just after the data of host ($i_j$) and replace its initial encapsulated execution result ($M_j$) with a new encapsulated execution result ($M'_j$). Next, it sends the agent with the updated hash chain ($C'_{j+1}$) to host/s of its selection. Hence, the initiator would never detect the data truncation or replacement. The other security flaws of the protocol are the same as the flaws of the *publicly Verifiable Chained digital Signature protocol*.

## Publicly Verifiable Chained Signature Protocol

The *Publicly Verifiable Chained Signature Protocol* (Karjoth et al., 1998) aims to preserve confidentiality and integrity of data acquired by mobile agents. The protocol does not provide authenticity. The protocol uses temporary key pairs (private signing key, and the corresponding verification key) and a chain of encapsulated execution results. Each host generates a pair of keys (private and public). The host encloses the public key ($y_{n+1}$) it generates within the encapsulated execution result computed at its host ($M_n$) and then signs the encapsulated execution result with the private key ($\overline{y}_1$) it received from its predecessor. It provides its successor with the private key it generates ($\overline{y}_{n+1}$).

At initiation, the initiator provides the agent with a dummy data $m_0$ and an initial key pair. It encloses the public key ($\overline{y}_1$) it generates within the encapsulated execution result ($M_0$) and then signs the encapsulated execution result with its private key. It provides host ($i_1$) with the private key ($\overline{y}_1$). The hash chain ($C_n$) at a host binds the encapsulated execution results at the preceding host ($M_{n-1}$) to the identity of the succeeding host ($i_{n+1}$). The encapsulated execution result ($M_n$) incorporates the following data terms that are generated at the host: data ($m_n$), a random nonce ($r_n$), a hash chain ($C_n$), and a public verification key corresponding to the succeeding host ($y_{n+1}$).

The protocol is defined as follows:

$$M_n = \left\{ \{m_n, r_n\} K_{i_o}, C_n, y_{n+1} \right\} S_{\overline{y}_n} \qquad for \ n \geq 1$$
$$M_n = \left\{ \{m_0, r_0\} K_{i_o}, C_0, y_1 \right\} S_{i_o}^{-1}$$
$$C_n = h(M_{n-1}, i_{n+1}) \qquad for \ n \geq 1$$
$$C_0 = h(r_0, i_1)$$
$$i_n \rightarrow i_{n+1} : \{M_0, ..., M_n\}, \overline{y}_{n+1} \qquad for \ n \geq 0$$

The security flaws of the protocol are as follows:

- An adversary can intercept the agent and then decrypt the encapsulated execution result ($M_0$) with the signature verification key of the signer. Next, it signs the decrypted term of ($M_0$) with its private key impersonating the genuine initiator. Executing hosts would encrypt the data they provide to the agent with the public key of the adversary believing that it is the genuine initiator. Hence, the adversary can breach the privacy of the gathered data.

An adversary can collude with host $(i_j)$, which the agent had previously visited and had stored the private key it had received from the preceding host $(\overline{y}_n)$, and send the agent back to it. The host can then truncate the gathered data just after the data of host $(i_j)$ and replace its initial encapsulated execution result $(M_j)$ with a new encapsulated execution result $(M'_j)$. Next, it sends the agent to host/s of its selection. Hence, the initiator would never detect the data truncation or replacement.

An adversary can truncate chain elements and can grow a fake stem, since the input to all previous chaining relations is known. Elements can be appended to the chain at the discretion of the adversary, though the validity of the chaining relation is maintained. The adversary sends the agent with a chain of execution results: $\{M_0, \ldots, M_{j-1}\}$ to host $(i_j)$ of its own choice and repeats the process until it is satisfied with the collected elements. Then, the adversary chooses an element and pastes it into agent and sends the agent to $(i_{j+1})$.

An adversary can append arbitrary objects, generated to the terms of the adversary rather than the initiator, to the agent without being detected.

## Configurable Mobile Agent Data Protection Protocol

The *Configurable Mobile Agent data Protection Protocol* (Maggi & Sisto, 2003) is intended to accomplish a combination of the security properties: authenticity, confidentiality, integrity, origin confidentiality/ forward privacy, and non-repudiation. The protocol can be configured for the properties of concern. The security is based on: (1) securely storing the addresses of next hosts to

be visited, and (2) binding the static part $(\Pi)$, which consists of program code and static data, to a chain of encapsulated execution results at hosts in the agent's itinerary $\{M_0, \ldots, M_n\}$ (Roth, 2001; Roth, 2002). The static part $(\Pi)$ is paired with a timestamp $(t)$ and signed by the initiator so having $\Pi_0 = \{\Pi, t\} S_{i_0}^{-1}$. The agent's initial itinerary is denoted as $(P_0)$. The set of new hosts added to agent's initial itinerary is denoted as $(P_n)$. The chain of encapsulated execution results is denoted as $(M_n)$ and is composed of two parts: $(D_n)$ and $(C_n)$. The $(D_n)$ binds the execution results at a host $(d_n)$ to $(P_n)$. The $(C_n)$ binds the static part $(\Pi_0)$ to: (1) the parameter $(C_{n-1})$ that is computed at the preceding host, (2) the identity of the succeeding host $(i_{n+1})$, (3) the execution result $(d_n)$ at host $(i_n)$, and (4) the addresses of new hosts $(P_n)$ added to agent's initial itinerary $(P_0)$.

The protocol configuration for data-authenticity, data-confidentiality, and data-integrity properties is as follows:

$$i_n \rightarrow i_{n+1} : \Pi_0, \{M_0, \ldots, M_n\}$$
$$Where, \Pi_0 = \{\Pi, t\} S_{i_0}^{-1}$$
$$M_n = D_n \parallel C_n$$
$$D_n = \begin{cases} P_0 & if\ i_n = i_0 \\ \{d_n\} K_{i_o}^+, P_n & otherwise \end{cases}$$
$$C_n = \begin{cases} S_{i_o}^{-1}(P_0, \Pi_0, i_1) & if\ i_n = i_0 \\ \{S_{i_n}^{-1}(d_n, P_n, \Pi_0, C_{n-1}, i_{n+1})\} K_{i_o}^+ & otherwise \end{cases}$$

Two security flaws are revealed, which are: (i) an adversary $(i_a)$ may intercept the communication between $(i_1)$ and $(i_0)$, and then strip off the initiator's signature from $(\Pi_0)$ and learn the pair $(\Pi, t)$ in plaintext. Next, it extracts the tuple $(P_0, \Pi_0, i_1)$, and then appends its signature to the tuple so having $C'_0 = S_{i_a}^{-1}(P_0, \Pi_0, i_1)$. Next, it signs the pair $(\Pi, t)$ with its private key so having $\Pi'_0 = \{\Pi, t\} S_{i_a}^{-1}$, and computes $C'_0 = S_{i_a}^{-1}(P_0, \Pi_0, i_1)$. Next, it sends the agent with the fake terms: ($\Pi'_0$), and ($C'_0$) rather than the original terms $\Pi_0$, and $C_0$ to host ($i_1$) impersonating the genuine initiator. Host ($i_1$) would receive the agent and incorrectly authenticate ($i_a$) as the genuine initiator of the agent. Thus, it would encrypt its own data with the public key of the adversary ($i_a$) rather than that of the genuine initiator ($i_0$). The agent continues migrating till all hosts defined in the set: $\{P_0, ..., P_n\}$ are visited. Next, the adversary intercepts the agent and spies out the gathered data. Next, the adversary ($i_a$) sends the agent with the original terms: $\Pi_0$, $D_0$, $C_0$ as a fresh protocol instance. The attack would result in erroneous authenticity to ($i_a$) and breach of privacy of the gathered data. Actually, host ($i_a$) sends two instances of the protocol. The first instance with ($D_0$) and the fake terms: ($\Pi'_0$) and ($C'_0$), and the second instance with original terms: $\Pi_0$, $D_0$, and $C_0$. As a result, the adversary would possess the data that should only be revealed to the genuine initiator. The initiator would never detect such breach of privacy. (ii) An adversary can truncate the data acquired at hosts visited between the first and the second visits of the agent to its host hence can alter the data it has provided to the agent in the first visit maintaining the consistency of checksums of the gathered data.

## Mobile Agent Integrity Protocol

The *Mobile Agent Integrity Protocol* (Hannotin et al., 2001; Maggi & Sisto 2002a) aims to preserve the integrity of data gathered by mobile agents. The protocol is based on the chain of execution results ($AD_n$), a message integrity code ($MIC_n$), and hash chain of a nonce ($C_n$). The ($AD_n$) is a chain of execution results at hosts in the agent's itinerary as $\{D_1, ..., D_n\}$. The ($MIC_n$) binds the execution result ($D_n$) and hash chain of the nonce ($C_n$) computed at the current host to the ($MIC_{n-1}$) computed at the preceding host. Upon the agent's return, the message integrity code is verified to detect any tampering with the already gathered

data. At initialization, ($C_n$) is set to $C_0 = r$ where ($r$) is chosen randomly by the initiator ($i_0$). The ($MIC_0$), and ($AD_0$) are left empty. The protocol is defined as follows:

$$AD_n = \{D_1, ..., D_n\} = AD_{n-1} \cup \{D_i\}$$
$$MIC_n = h(D_n, C_n, MIC_{n-1})$$
$$C_n = h(C_{n-1})$$
$$i_n \rightarrow i_{n+1} : MIC_n, \{C_n\}K^+_{i_{n+1}}, AD_n$$

Upon the agent's return, the initiator computes ($MIC'_n$) from the values of: $\{C_0, ..., C_n\}$ and $\{D_1, ..., D_n\}$ and then verifies that the computed message integrity code ($MIC'_n$) matches the message integrity code that has just been received from the agent ($MIC_n$).

The security flaw is that an adversary can read the gathered data and append arbitrary data to the already gathered data, since the terms: ($MIC_{n-1}$) and ($C_{n-1}$) that are needed to compute the message integrity code ($MIC_n$) at host ($i_n$) are known. Another security flaw is that a non-trusted host can append arbitrary objects, generated to the terms of the adversary rather than the initiator, to the agent without being detected.

## Improved Forward Integrity Protocol

The *Improved Forward Integrity Protocol* (Yao et al., 2003) is an extension to the "Publicly Verifiable Chained Signature Protocol" (Karjoth et al., 1998). It aims to preserve strong forward integrity. It uses a mechanism that restricts tampering with the data acquired to a certain extent or at least be detectable. Tampering includes modification, deletion, and insertion, even through collusion with other servers. The mechanism defends against the "colluding servers" attack to which the "Publicly Verifiable Chained Signature Protocol" is vulnerable. The protocol involves the preceding and the succeeding hosts of a particular host in signing its offer.

At initiation, the initiator constructs a dummy offer ($M_0$) that incorporates a dummy data ($m_0$), a hash chain ($h_0$), and a one-time public key ($t\_Pub_1$) it generates for its successor. Then, it signs the offer with its long-term private key ($k\_Priv_0$) and sends it along with the one-time private key it generates for its successor ($t\_Priv_1$). The hash chain ($h_0$) binds a random nonce the host generates ($r_0$) to the identity of the succeeding host in agent's itinerary ($i_1$). At execution, a host constructs a joint private key used for signing its offer ($j\_Priv_n$). It is composed of two parts: a one-time private key ($t\_Priv_n$) generated by the preceding host ($i_{n-1}$) and its long-term private key ($k\_Priv_n$). The corresponding public key ($j\_Pub_n$) can be deduced from the public keys ($t\_Pub_n$) and ($k\_Pub_n$). A host constructs an encapsulated offer ($M_n$) that incorporates a data ($m_n$), a hash chain ($h_n$), a random nonce the host generates ($r_n$), and a one-time public key ($t\_Pub_{n+1}$) it generates for its successor. Then, it signs the offer with its joint private key ($j\_Priv_n$) and sends all encapsulated offers with the one-time private key it generates for its successor ($t\_Priv_{n+1}$). The hash chain ($h_n$) binds the encapsulated offer of the preceding host ($M_{n-1}$) to the identity of the succeeding host in agent's itinerary ($i_{n+1}$). At migration, a host computes a joint verification function. If the verification is true, then the signature is authentic.

The protocol is described below:

$$M_0 = Sig k\_Priv(\{m_0, t_0\}k\_Pub_0, h_0, t\_Pub_1)$$
$$h_0 = H(r_0, i_1)$$
$$M_n = Sig_j\_Priv_n(\{m_n, r_n\}K\_Pub_0, h_n, t\_Pub_{n+1}), 0 \le n \le m$$
$$h_n = H(M_{n-1}, i_{n+1})$$
$$Ver_j\_Pub_n(M_n) = True$$
$$i_n \rightarrow i_{n+1} : \Pi, \{M_0, M_1, ..., M_n\}, \{t\_Priv_{n+1}\}K_n, 0 \le n \le m$$

The security flaw of the protocol is that an adversary can collude with host ($i_j$), which the agent had previously visited and had stored the private key it had received from the preceding host ($t\_Priv_n$), and send the agent back to it. The host can then truncate the gathered data just after the data of host ($i_j$) and replace its initial encapsulated execution result ($M_j$) with a new encapsulated execution result ($M'_j$). Next, it sends the agent to host/s of its selection. Hence, the initiator would never detect the data truncation or replacement.

## Secure Protocol Based on Sedentary Agent

The *Secure Protocol Based on Sedentary Agent* (Rhazi et al., 2007) is aimed at protecting the agent code and data from various security attacks. It focuses on the detection of various security attacks, particularly, re-execution of mobile code with different data, incorrect execution of code, tampering with agent's code or state, and denial of service. It relies on the *Cryptographic Security of Mobile Code Approach* (Algesheimer, Cachin, Camenisesh, & Karjoth, 2000), which subdivides the mobile code into fragments, and the *Cooperating Agents Approach* (Roth, 1998).

The protocol makes a replica of the critical agent code that should be protected and stores it in a Sedentary Agent (*SA*). It has the code that cares for the verification of the execution results of the Mobile Agent (*MA*) that traverses the Internet. The agent (*SA*) is dispatched to a Reliable Third Party (*RTP*) and stay there till the (*MA*) completes its itinerary. The (*MA*) agent sends intermediate execution results to (*SA*), which can detect any security attack. The (*MA*) can only have access to the (*SA*) through the communication with the (*RTP*), which preserves the confidentiality of mobile computing.

The protocol can be summarized in the following steps:

- The initiator platform sends (*SA*) to (RTP).
- The initiator platform dispatches (*MA*) to traverse the internet.
- The (*MA*) sends an Arrival() message to (*SA*) informing it of its arrival on host ($i_n$).

- Upon the receipt of the Arrival() message, (*SA*) validates agent's itinerary. If corruption has not taken place, the (*SA*) initializes a time counter (*Timeout*) that restricts the execution time of (*MA*) at host ($i_n$)

- Prior to the execution of mobile code at ($i_n$), the (*MA*) sends (*SA*) an Input() message of all input data (X) provided by ($i_n$) for the run of the agent.

- Upon the receipt of the Input() message, the (*SA*) executes the cloned agent code using (X) and gets results (R').

- Upon the complete execution of the agent code at ($i_n$), the (*MA*) sends the (*SA*) an Output() message of output results (R).

- Upon the receipt of the Output() message, the (*SA*) executes the cloned mobile code that verifies the output results. If results do not match, the (*SA*) concludes that an attack took place. If they match, it concludes that the mobile code was properly executed. The (*SA*) appends the identity of the next host ($i_{n+1}$) to agent's itinerary.

- The (*MA*) migrates to next host on its itinerary.

- Upon the completion of agent's itinerary, the (*MA*) returns to the initiator's host and the initiator requests (*SA*) to send it all data pertaining to the mobile code execution and excludes any data from attacked hosts. It then verifies the output results (R') and (R) of the (*MA*) and the (*SA*), respectively.

The messages the (*MA*) communicates to the (*SA*) can be described as follows.

$$i_n \rightarrow i_{RTP} : \left( Arrival() : i_n, S_{i_n}^{-1}(ID), i_{n-1} \right)$$
$$\rightarrow i_{RTP} : \left( Input() : i_n, S_{i_n}^{-1}(X), (X) \right) K_{RTP}$$
$$i_n \rightarrow i_{RTP} : \left( Output() : i_n, S_{i_n}^{-1}(R), (R), i_{n+1} \right) K_{RTP}$$

The input data (X) and output data (R) are signed with the private signing key ($S_{i_n}^{-1}$) of the visited host. The input and output message are encrypted with the public key ($K_{RTP}$) of the (RTP). The node code identity is denoted as (*ID*).

The initiator runs an Estimator Agent (EA) on a platform to estimate the (*Timeout*) that represents the maximum time needed to execute the mobile code and to transmit Input() and Output() messages. If the (*Timeout*) expires before (*SA*) receives an Output() message from (*MA*), then (*SA*) it concludes that the mobile code has been re-executed. The (*SA*) also initializes a supplementary waiting Interval (*SWA*) as an additional waiting time. If the (*Timeout*) expires and the (*SWA*) expires before (*SA*) receives an Output() message from (*MA*), then (*SA*) concludes that a denial of service took place. The initiator sends the estimate of (*Timeout*) the (*SA*).

The protocol has three security flaws as described below.

- It does not preserve privacy of results of code execution at visited hosts, which are contained within the (*MA*) during migration

- It is subject to data truncation attack. An intruder may intercept any of Arrival(), Input(), or Output() messages which the (*MA*) sends to the (*SA*) upon its arrival/execution at host ($i_n$) and destroy or delay it. The (*SA*) would assume that the host ($i_n$) has re-executed the code or under a denial of service attack and, hence, consider it a malicious host. Conversely, the host might provide some advantageous results as compared to that provided by other visited hosts and, thus, valuable results are excluded.

- It does not preserve origin-confidentiality as the identity of the preceding host can be deduced. The results of code execution at a host are signed with the private key of the host and are contained within the (*MA*) during its migration.

## Security Problems

The discussed security protocols have been proposed to protect mobile agents from security attacks of intruders and non-trusted hosts. Though, the security protocols should be verified for soundness using formal methods of verification. Formal methods provide rigorous analysis for the system design, and for establishing its correctness and reliability (Ma & Tsai, 2000). Thus, they help in developing error-free security protocols. Conversely, the testing of protocols is not enough to ensure the liability and correctness of their implementation, because of the unpredictable behavior and unbounded capabilities of adversaries, and the dynamic behavior of mobile agents. The analysis would not be finite since there might be infinite set of traces. Moreover, it would not be practical and feasible to test complex and large-scale systems with too huge number of traces.

Formal methods can be classified into five categories (Ma & Tsai, 2000) as follows:

- Methods based on modal logic
- Methods based on finite-state exploration
- Methods based on theorem proving
- Methods based on process algebra
- Methods based on infinite-state exploration

Modal logic method is such as BAN logic (Burrows, Abadi, & Needham, 1990). Finite-state exploration methods include: Interrogator (Millen, Clark, & Freedman, 1987), NRL protocol Analyzer (Meadows, 1996), CSP model checker FDR (Lowe, 1996; 1997), SPIN model checker (Maggi & Sisto, 2002b), and Murφ (Mitchell, Mitchell, & Stern, 1997). Theorem proving method is such as induction method (Paulson, 1997). Process algebra methods include: π-Calculus (Abadi, Blanchet, & Fournet, 2007), applied-π Calculus (Abadi & Fournet, 2001), Spi-Calculus (Abadi & Gordon

1999), Distributed-π calculus (Sewell, 1998), Seal calculus (Vitek & Gastagna, 1999), and Crypto-loc Calculus (Blanchet & Aziz, 2003). Infinite-state exploration methods include: STA (Boreale, 2001; Boreale & Buscemi, 2001; 2002a ; 2002b; STA Documentation, 2001; Boreale & Gorla, 2002; PVS & SAL (Rushby, 2006); Symbolic models (Fiore & Abadi, 2001).

Formal methods have revealed the security flaws in existing protocols (Aziz et al., 2001; Ma & Tsai 2000; Meadows, 1994). Different formal methods have been used to detect security flaws such as CSP-based tools: Casper (Lowe, 1997) and FDR (Formal Systems, 2000) in (Hannotin et al., 2001), Model checker that uses Spi-calculus (Abadi & Gordon, 1999) in (Maggi & Sisto, 2003), and STA (STA Documentation, 2001) in (Boreale & Buscemi, 2002b; Al-Jaljouli, 2005, 2006). The protocols failed to detect particular security attacks and hence failed to accomplish the aimed for security properties (Al-Jaljouli, 2005, 2006). The most common and critical attack is data truncation that violates the integrity of execution results. In Table 1, the types of security attacks the existing security protocols failed to prevent or at least detect are identified. The security attacks can be classified into the following types:

- Breach of privacy
- Data truncation
- Erroneousness authentication
- Append fake data amendment.

The security properties each protocol fulfills or fails to accomplish are identified in Table 2. It summarizes the correctness of the existing security protocols.

## Breach of Privacy

The agent's state in the *Targeted State Protocol* and agent's static part in the *Configurable Mobile Agent Data Protection Protocol* are signed with the initiator's private key. An adversary might

*Table 1. Flaws revealed in the mobile agents security protocols*

| Security protocol | Possible types of flaws |
|---|---|
| *Targeted State Protocol* | Breach of privacy |
| *Append only Container Protocol* | • Erroneous authentication<br>• Data truncation<br>• Fake data amendment |
| *Multi-hops Protocol* | • Data truncation<br>• Fake data amendment |
| *Publicly Verifiable Chained Digital Signature Protocol* | • Data truncation<br>• Fake data amendment |
| *Chained Digital Signature Protocol with Forward Privacy Protocol* | • Data truncation<br>• Fake data amendment |
| *Chained MAC Protocol* | • Data truncation<br>• Fake data appendage |
| *Publicly Verifiable Chained Signature Protocol* | • Data truncation<br>• Fake data amendment |
| *Configurable Mobile Agent Data Protection Protocol* | • Data truncation<br>• Breach of privacy |
| *Mobile Agent Integrity Protocol* | • Data truncation<br>• Fake data amendment |
| *Improved Forward Integrity Protocol* | • Data truncation |
| *Secure Protocol Based on Sedentary Agent* | • Breach of privacy<br>• Data truncation<br>• Origin confidentiality |

intercept the agent and decrypt the signed state or part. It would then send the decrypted state or part signed with its private key. Recipients would assume that the signer is the genuine initiator of the agent, and would then encrypt the data they provide to the agent with the public key of the adversary. Hence, the adversary would be able to learn the encrypted data. Whereas, the *Secure Protocol Based on Sedentary Agent* transmits the execution results contained within the (*MA*) in plain text and not encrypted with the public key of the initiator.

## Erroneous Authentication

In the *Append only Container Protocol*, the execution result at a host is just encrypted with the private key of the host. An adversary may intercept the agent and decrypt the execution results. It would then sign the execution results with private keys

of co-operating hosts, and update the checksum accordingly. Consequently, the initiator would assume that the returned data were provided by genuine executing hosts. The same flaw exists in the *Chained MAC Protocol* and the *Publicly Verifiable Chained Signature* Protocol. The two protocols do not aim to preserve authenticity, but it is just a remark.

## Appending a Fake Stem to the Agent

In some security protocols the inputs that are needed to compute an encapsulated execution result at a host are available. Hence, an adversary can append fake execution results to the agent without being detected, as follows:

An intruder may intercept the agent and append an offer to the results of the agent's execution. The intruder is a non-scheduled host in the agent's itinerary. The initiator would not detect the ma-

*Table 2. Correctness of security protocols as regards the fulfillment of the aimed security properties*

| | Targeted state protocol | Append only container protocol | Multi-hops protocol | Publicly verifiable chained digital signature protocol | Chained digital signature protocol with forward privacy protocol | Chained MAC protocol | Publicly verifiable chained signature protocol | Configurable mobile agent data protection protocol | Mobile agent integrity protocol | Improved forward integrity protocol | Secure Protocol Based on Sedentary Agent |
|---|---|---|---|---|---|---|---|---|---|---|---|
| Confidentiality | ✗ | | | ✓ | ✓ | ✓ | ✓ | ✗ | | | ✗ |
| Authenticity | | ✗ | ✓ | | | | | ✓ | | | |
| Non-repudiation | | | | | | | | ✓ | | | |
| Forward privacy | | | | | ✓ | ✓ | | ✓ | | | ✗ |
| Integrity | | ✗ | ✗ | ✗ | ✗ | ✗ | ✗ | ✗ | ✗ | ✗ | ✗ |

licious act upon the agent's return. The attack is possible in the *Append Only Container Protocol*.

In the following protocols, a non-trusted host that participates in the protocol may send the agent to a succeeding host of its selection. The succeeding host would append its offer to the execution results of the agent, though the initiator would detect the malicious act upon the agent's return since each partial execution result incorporates the identity of the scheduled succeeding host.

- Multi-hops Protocol
- Publicly Verifiable Chained Digital Signature Protocol
- Mobile Agent Integrity Protocol
- Chained MAC Protocol
- Publicly Verifiable Chained Signature Protocol
- Configurable Mobile Agent Data Protection Protocol
- Chained Digital Signature Protocol with Forward Privacy Protocol
- Improved Forward Integrity Protocol

The execution results that are returned to the agent might be generated for a different protocol run, a different initiator, or a different agent. The execution results of the existing protocols differ as follows:

- The execution results of the Append Only Container protocol do not incorporate an identifier of the protocol run of concern, the identity of the genuine initiator, or identifier of the agent of concern within the execution results.
- The execution results of the following protocols only incorporate a random nonce generated by the initiator, which identifies the protocol run, within the execution results:
  ○ Multi-Hops Protocol
  ○ Mobile Agent Integrity Protocol
  ○ Improved Forward Integrity Protocol

- The execution results of the Configurable Mobile Agent Data Protection Protocol incorporate the following within the execution results.
  - Timestamp generated by the initiator that uniquely identifies the protocol run of concern
  - Identity of the first host in the agent's itinerary
  - Identity of the initiator within the offer a host provides
- The execution results of the following protocols:
  - Publicly verifiable chained digital signature protocol
  - Chained Digital Signature Protocol with Forward Privacy Protocol
  - Chained MAC Protocol
  - Publicly Verifiable Chained Signature Protocol
  - Improved Forward Integrity Protocol
- Incorporate the following within the execution results:
  - Random nonce generated by the initiator that uniquely identifies the protocol run of concern
  - Dummy data generated by the initiator
  - Identity of the first host in the agent's itinerary
  - Identity of the initiator within the offer a host provides.

## Truncation and/or Substitution of Execution Results

An adversary might truncate the data acquired at hosts visited between the first and the second visits of the agent to its host. It could then alter the data it has provided to the agent in the first visit maintaining the consistency of the chaining relation of the gathered data. The attack requires that the non-trusted host has stored the chaining relation that was present when the agent first visited it.

## Proposed Security Techniques

A set of reliable and efficient new security techniques for the detected security flaws in the existing security protocols are proposed.

## Breach of Privacy

It would be recommended that the initiator follows the signing of a term by an encryption with the public key of the recipient. Hence, an adversary would neither be able to learn the encrypted term it might intercept nor impersonate the genuine initiator by signing the agent with its private key. It has to have the decryption key of the scheduled recipient which is private to the recipient. It is only the scheduled recipient that can decrypt and learn the term. Hence, the recipient can be sure that the received term is truly signed by the genuine initiator. Whereas, it would be recommended that the *Secure Protocol Based on Sedentary Agent encrypts the execution results contained within the migrating (MA) with the public key of the initiator's host.*

## Erroneous Authentication

It would be recommended that an executing host firstly signs the data it provides to the agent with its private key and then encrypts it with the public key of the initiator.

## Appending a Fake Stem to the Agent

It would be recommended that the terms that are necessary to compute a new checksum at a host, which is based on the checksum computed at the preceding host, be transmitted from the preceding host to the host encrypted with the public key of the host. The terms are depicted in Table 3.

*Table 3. Terms computed at the preceding host and are necessary for an adversary to perform a non-detectable data truncation/alteration*

| Security protocol | Necessary terms |
|---|---|
| *Targeted state protocol* | None |
| *Append only container protocol* | Checksum $C_{n-1}$ |
| *Multi-hops protocol* | Message authentication code $\mu_{n-1}$, hash chain $\gamma_{n-1}$ |
| *Publicly verifiable chained digital signature protocol* | Encapsulated execution results $M_{n-1}$ |
| *Chained Digital Signature Protocol with Forward Privacy Protocol* | Encapsulated execution results $M_{n-1}$ |
| *Chained MAC Protocol* | Hash chain $C_n$ computed at its predecessor |
| *Publicly Verifiable Chained Signature Protocol* | Encapsulated execution results $M_{n-1}$, and the private signature it received from its predecessor $y_{n-1}$ |
| *Configurable mobile agent data protection protocol* | Static part and a timestamp signed by initiator $\Pi_0$, chaining relation $C_{n-1}$ |
| *Mobile agent integrity protocol* | Hash chain of the nonce $C_{n-1}$, and a Message Integrity Code $MIC_{n-1}$ |
| *Improved Forward Integrity Protocol* | $M_j$ |
| *Secure Protocol Based on Sedentary Agent* | *None* |

It would also be recommended to incorporate the following terms within each encapsulated execution result.

- Random nonce or a timestamp generated by the initiator that uniquely identifies the protocol run of concern
- Dummy data generated by the initiator
- Identity of the first host in the agent's itinerary
- Identity of the initiator within the offer a host provides.

Moreover, it would be recommended to store the terms securely with an agent that is stationary at the initiator. It might be assumed that it is enough to store the terms securely in the memory of the initiator, though an adversary might tamper with the memory of initiator. Hence, the verifications of the four terms would not be accurate. The use of a secondary agent to store the verification terms would enable the initiator to trace any manipulation with the terms by the use of execution traces. Vigna (1998) recommends the agent executor to create a trace of the agent's execution. The trace contains the lines of the agent's code that were executed as well as any new values assigned to initial verification terms that were stored within the stationary agent. The trace of the agent's execution is to be stored at the executing host for a limited time. Upon the initiator's request, each executing host signs the execution trace and forwards it to the succeeding host in the agent's itinerary. The cumulative execution traces are forwarded to the initiator. It is proposed to implement the technique to store the execution trace of the secondary agent at the initiator, thus the initiator would be able to verify the terms upon the agent's return through the stored execution trace. Usually the execution traces require large amounts of resources for the storage of validating information. Conversely the execution trace of the secondary agent would be short as compared to the execution traces of the migrating agent.

## Truncation and/or Substitution of Execution Results

It would be recommended that an executing host should clear its memory from any terms acquired as a result of executing the agent before it dispatches the agent to the next host in the agent's itinerary. It is proposed to design the migrating agent in such a way that it requests an executing host to clear its memory from any terms acquired as a result of executing the agent before it dispatches the agent to the next host in the agent's itinerary. However, an executing host may not respond to the request. The denial of clearing request can be traced by implementing the execution traces technique recommended by Vigna (1998). The technique requests an executing host to create and sign the execution trace, and to store it so as to be forwarded to the initiator upon request. It would be recommended that the execution trace be limited to the line of code that requests the clearing of the memory of the executing hosts; otherwise the trace of all executable lines of the agent would be extremely long and require large amounts of resources of storage at the executing hosts. Moreover, it would lead to overburden the communication channels as traces are transmitted to the initiator upon request.

## RECENT PROMISING SECURITY PROTOCOLS

Two recent security protocols are discussed that are reliably secured. They implement new security techniques in addition to the existing techniques and are capable of preventing or at least detecting the various security attacks, particularly the flaws discussed in the chapter and the existing protocols failed at least to detect. The two protocols are: (1) Data gathering security protocol, (2) Enhanced multi-hops protocol. The protocols are verified formally and are found free of security attacks (Al-Jaljouli, 2005, 2006). The high-level system architecture is depicted in Figure 2, where a stationary agent ($A_s$) resides at the initiator's host maintaining initial verification data and a mobile agent ($A$) traverses the Internet in search of data necessary for accomplishing a particular task. It

*Figure 2. High-level system architecture of the proposed security protocols*

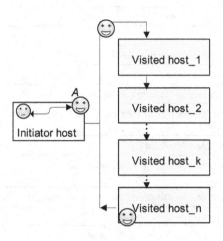

then returns to the initiator's host as it completes data gathering task.

## Data Gathering Security Protocol

The protocol (Al-Jaljouli, 2005) is derived from the *Multi-hops protocol* (Corradi et al., 1999b). New security techniques are implemented to rectify the flaws in the *Multi-hops Protocol*. The techniques are outlined below.

1.  Utilization of co-operating agents where the initial verification terms, which uniquely identify the agent and are necessary for accurate verifications of the protocol upon the agent's return, are securely stored within a secondary agent that resides at the initiator and co-operates with a major agent that traverses the Internet. The intention to store the initial verification terms within the secondary agent and not at the initiator's memory is to enable the initiator to trace any tampering with the initial verification terms through the execution traces of the secondary agent it creates and stores, which Vigna (1998) recommends.

2. Carrying out *verifications on the identity of the genuine initiator*, at the early execution of the agent at visited hosts, using a ciphertext signed by the initiator and contained within the migrating agent. The cipher securely stores the identity of the genuine initiator.

3. Scrambling the gathered offers so having the offers are arranged in a reverse order and a dummy offer, which the initiator generates, as the last offer within the chain of offers. Hence the malicious act of a non-trusted host, which has tried to delete the offer of its predecessor by deleting the last offer in the chain of offers, would be detected upon the agent's return. The initiator would check the availability of the dummy offer within the chain.

4. Clearing the memory of an executing host from any data acquired as a result of executing the agent before the host sends out the agent to the succeeding host in the agent's itinerary. An executing host may not respond to the request. The denial of the clearing request can be traced by implementing the execution traces technique recommended by Vigna (1998) recommends.

5. Verifications upon the agent's return to the initiator.

6. Transmission of cryptographic proofs of the data that agents have already gathered, along with the agent, which includes: (a) a counter that indicates the number of the actually visited hosts, (b) a chain of the encrypted offers, where each offer incorporates the following data: the data acquired at a host, an identifier of the protocol run, a counter that indicates the order of the host among the visited hosts stored as a hash chain of a random nonce, the identity of the genuine initiator, and the identity of the succeeding host, (c) a data integrity code that encapsulates the execution results at the visited hosts, and (d) a cipher that securely stores the identity of the genuine initiator.

The protocol initially creates two co-operating agents: $(A)$ and $(A_s)$. The agent $(A)$ is a *major agent* that traverses the Internet and gathers particular data. The secondary agent $(A_s)$ resides at the initiating host. It securely stores the initial verification terms and carries out a set of verifications on the execution results. At the first instance of the protocol run, the initiator $(i_0)$ generates a fresh nonce $(r)$ that identifies the protocol run. Next, the initiator dispatches the agent $(A)$ to host $(i_1)$, and then the agent is free to autonomously choose the next host to visit at each migration step during its life cycle. Each visited host provides the agent $(A)$ with the requested data. When the agent $(A)$ completes its execution at last host in the agent's itinerary, it returns to the initiator with the results of execution at the visited hosts. Upon the agent's return, the agent $(A)$ co-operates with the secondary agent $(A_s)$. The collective data from the two agents would be utilized in the detection of any malicious act performed on the execution results, and would be sufficient to identify any tampering with the execution results. The agent $(A)$ can be represented as a sequence of messages communicated between the executing hosts, with the initiating host starting the sequence of messages by sending a preliminary message and finally receiving the summary message that summarizes the execution results at different executing servers. The migration of the agent $(A)$ from one host to another is simply modeled by sending a data message. The protocol can be described as follows:

$$\gamma_j = h(\gamma_{j-1}), \quad \text{where } \gamma_0 = r$$
$$\delta_j = h(m_j, \delta_{j-1}), \quad \text{where } \delta_0 = h(i_0)$$
$$\lambda_j = \left\{ \{m_j, \delta_o, i_{j+1}, \gamma_j\} S_{i_j}^{-1} \right\} K_{i_o}^+, \quad \text{where } \gamma_0 = \{m_0\} K_{i_o}^+$$
$$i_j \rightarrow i_{j+1} : \{\lambda_j, ..., \lambda_1, ..., \lambda_0, \delta_j, \{\delta_o\} S_{i_o}^{-1}, \gamma_j\} K_{i_{j+1}}^+$$

At initiation, the protocol chooses the first host in the agent itinerary ($i_1$) and picks up a random nonce ($r$) and assigns it to ($\gamma_0$). It then generates a dummy data ($m_0$) and encrypts it with its public key ($K_{i_0}^+$). It securely stores the tuple ($m_0$, $i_1$, $r$) within the secondary agent ($A_s$), which is needed to ensure accurate verifications upon the agent's return. Then, it computes ($\delta_0$) as a hash of the identity of the initiator ($i_0$), and then signs it with its own private key ($S_{i_0}^{-1}$). The term ($\delta_0$) securely stores the identity of the genuine initiator, which would be used for the verifications at the early execution of the agent at visited hosts. Next, it encrypts the tuple $\left\{\lambda_0, \delta_0, \{\delta_0\} S_{i_0}^{-1}, \gamma_0\right\}$ with the public key of the succeeding host in the agent's itinerary ($K_{i_1}^+$) and then dispatches the agent to the next host in agent's itinerary.

At an executing host, the host ($i_j$) for ($1 \leq j \leq n$) receives a message and deduces the identity of the signer of the term $\{\delta_0\} S_{i_0}^{-1}$. It then decrypts the signed term and computes a hash of the deduced identity. It then verifies that the computed hash matches the decrypted term. If the verification passes, then the deduced identity is truly the identity of the genuine initiator and it would be used to encrypt the data host ($i_j$) provides to the agent. Next, the host provides its offer ($\lambda_j$) signed by its private key ($S_{i_j}^{-1}$) and then encrypted by the public key of the initiator ($K_{i_0}^+$). The offer incorporates the tuple $\left\{m_j, \delta_0, i_{j+1}, \gamma_j\right\}$, where ($m_j$) is the data provided by host ($i_j$). The data integrity code ($\delta_0$) is enclosed within the offer ($\lambda_j$) to indicate the identity of the initiator for which the offer is generated. If the data integrity code ($\delta_0$) that is enclosed within the received offer ($\lambda_j$) for ($1 \leq j \leq n$) does not match $h(i_0)$, then it implies that the returned data were not generated for the genuine initiator. The identity of the successor host ($i_{j+1}$) is included to record the agent's partial itinerary. The partial itineraries would be used to assemble the actual agent's itinerary. If the partial itineraries are not consistent, it implies that the acquired data have been truncated. The hash chain ($\gamma_j$) is included to indicate the actual number of visited hosts. The parameter ($\gamma_j$) is initially set to a random nonce ($r$), which uniquely identifies a protocol run. The random nonce ($r$) is expected to be hashed as many times as the number of visited hosts. If a malicious host deleted trailing offers and sent the agent to hosts of its selection, then the attack would be detected since the actual number of visited hosts deduced from the hash chain is greater than the number of returned offers. Also, if an adversary replaced the collected data with data of a similar protocol, then the attack would be detected since the nonce ($r'$) enclosed within returned offers does not match the nonce ($r$) stored within the secondary agent. It implies that the returned data were not generated for the protocol run of concern. The offer ($\lambda_j$) is appended to the chain of offers ($\lambda$) that are provided by the visited hosts ($i_k$) for ($1 \leq k \leq j-1$) and then the chain of offers are jumbled as $\left\{\lambda_j, \lambda_{j-1}, \&, \lambda_1, \lambda_0\right\}$ to deceive any adversary trying to truncate the data acquired at the preceding host. They can be arranged in a reverse order or according to a permutation chosen at random or balanced, and maintaining the dummy offer as the last offer in the chain. The agent computes the data integrity code ($\delta_j$) as a hash of the data ($m_j$) host ($i_j$) provided to the agent, and the received data integrity code ($\delta_{j-1}$) that was computed at the preceding host ($i_{j-1}$). It is used to verify if any of the data acquired at visited hosts was deleted. The agent then encloses the cumulative execution results: $\left\{\lambda_j, \&, \lambda_1, \&, \lambda_0, \delta_j, \{\delta_0\} S_{i_0}^{-1}, \gamma_j\right\}$ in a tuple and then encrypts it with the public key of the succeeding host in the agent's itinerary ($K_{i_{j+1}}^+$).

Then the agent requests the current host ($i_j$) to clear its memory from any data acquired as a result of executing the agent. Next, it dispatches the agent to the succeeding host ($i_{j+1}$) in the agent's itinerary. An intruder would not be able to append any valid offer to the chain of offers. Only the scheduled succeeding host $(i_{j+1})$ would be able to compute a valid offer.

Upon the agent's return, the agent carries the following verifications:

- Deduce identities of signers of offers and authenticate data.
- Check availability of the dummy offer.
- Count offers and verify integrity of the hash chain.
- Check integrity of the data integrity code.
- Check that the returned data have been generated for the genuine initiator.
- Check that the first visited host is the first scheduled host in the agent's itinerary.
- Check that the returned data belong to the protocol run of concern.
- Deduce accurately the actual agent's itinerary.

The protocol carries various verifications that can prevent or at least detect any malicious act on the gathered offers. The protocol truly accomplishes the intended security properties as follows.

## Ensures Strong Integrity of Gathered Data

The protocol can prevent or at least detect security attacks such as truncation, deletion, or replacement of data. It can also detect other attacks such as appending fake data, or generating data for a colluding host. It performs the following security acts:

- Securely stores the initial verification terms with a secondary agent to be verified with gathered data upon the return of the mobile agent.
- Carries out verifications on the identity of the genuine initiator at the early execution of the agent at a visited host.
- Binds the data a host provides to:

  - Data gathered at preceding hosts
  - Identify of next host
  - A unique nonce
  - Identity of initiator
  - Dummy data generated by the initiator
- Forbids an intruder from appending arbitrary data to the protocol.
- Forbids an adversary from inferring the offer of a preceding host.
- Checks that the returned data have been generated for the genuine initiator.
- Checks that the returned data belong to the protocol run of concern.
- Checks that the returned data are associated with the dummy data the initiator has generated.
- Accurately deduces the order of a host among visited hosts upon the agent's return.
- Accurately deduces the actual agent's itinerary.
- Accurately deduces the actual number of visited hosts upon the agent's return and if the number of returned offers matches the actual number of visited hosts.

## Preserves Privacy of Gathered Data

The data a host provides is encrypted with the public encryption key of the initiator.

## Preserves Origin-Confidentiality

The signed data a host provides is encrypted with the public encryption key of the initiator.

## Accurately Authenticates Gathered Data

The signed data a host provides is encrypted with the public encryption key of the initiator.

## Ensures Non-Repudiation of Gathered Data

The data a host provides is digitally signed with the private key of the host.

## The Enhanced Configurable Protection Protocol

The protocol (Al-Jaljouli, 2006) is an enhancement of the *Configurable Mobile Agent Data Protection Protocol*(Maggi & Sisto 2002a). The protocol requires the initiator to create two co-operating agents $(A)$ and $(A_s)$. The agent $(A)$ is a *major agent* that traverses the Internet and gathers particular data. The secondary agent $(A_s)$ resides at the initiating host. It securely stores the initial verification terms and carries out a set of verifications on the execution results.

The protocol can be described as follows:

$$i_n \rightarrow i_{n+1} : \Pi_{n+1}, \{M_0, ..., M_n\}$$

$$where \ \Pi_{n+1} = \left\{ \{\Pi, t\} S_{i_o}^{-1} \right\} K_{i_{n+1}}^+$$

$$M_n = D_n \| C_n \qquad for \ (0 \le n \le K)$$

$$D_n = \begin{cases} P_0 & if \ i_n = i_0 \\ \{d_n\} K_{i_0}^+, P_n & otherwise \end{cases}$$

$$C_n = \begin{cases} (P_0, \Pi_0, i_1) S_{i_o}^{-1} & if \ i_n = i_0 \\ \{(d_n, P_n, \Pi_0, C_{n-1}, i_{n+1}) S_{i_n}^{-1}\} K_{i_0}^+ & otherwise \end{cases}$$

$$where \ \Pi_0 = h(\Pi, t, h(i_0))$$

At initiation, the agent composes a pair of: (a) Agent's static part $(\Pi)$; (b) Timestamp $(t)$ that uniquely identifies a protocol run. It signs the pair:$\{ \Pi, t\}$ with its private key $( S_{i_0}^{-1} )$ and then encrypts it with the public key of the succeeding host in the agent's itinerary $( K_{i_{n+1}}^+ )$. It assigns the initial verification data: $\left\{ \{\Pi, t\} S_{i_0}^{-1} \right\} K_{i_{n+1}}^+$ to the term $(\Pi_{n+1})$ and keeps a store of the term within the secondary agent $(A_s)$ to ensure accurate verifications upon the agent's return. It also computes

a hash of the tuple: $(\Pi, t, h(i_0))$ and assigns it to $(\Pi_0)$. The agent also chooses an initial itinerary for the agent $(P_0)$ and computes $(C_0)$.

At the early execution of the agent at a visited host, the host carries out verifications on the identity of the genuine initiator. The host decrypts the cipher $(\Pi_{n+1})$ with its public key to enable the agent's execution and the inference of the initiator's identity. The agent computes a hash of the tuple: $(\Pi, t, h(i_0))$ and verifies if it matches the corresponding cipher $(\Pi_0)$ stored within the major agent. If the verification passes, then the deduced identity is truly the identity of the genuine initiator and it would be used to encrypt the data host $(i_n)$ provides to the agent. The host would provide its offer $(d_n)$ and appends new hosts $(P_n)$ to the initial agent's itinerary $(P_0)$. It then computes the tuple $\left( d_n, P_n, \Pi_0, C_{n-1}, i_{n+1} \right)$ and signs it with its private key $( S_{i_n}^{-1} )$ and then encrypts it with the public key of the initiator $( K_{i_0}^+ )$. It assigns the cipher to $(C_n)$ and then computes $(D_n)$ as $( \{d_n\} K_{i_0}^+, P_n )$. The cipher $(\Pi_{n+1})$ the host decrypts is then encrypted with the public key of the succeeding host. Next, the host requests the current host to clear its memory from any data acquired as a result of executing the agent before it dispatches the agent to the succeeding host. Hence, malicious hosts would not be able to tamper with the agent's dynamic data.

Upon the agent's return to the initiator $(i_0)$, the initiator compares the term $((\Pi_{n+1})$ returned with the migrating agent with the term received from the trusted co-operating agent. If the verification passes, it decrypts the term $(\Pi_{n+1})$ it received from the co-operating agent first with its private decryption key and then with its signature verification key. Next, it computes $(\Pi'_0)$ as a hash of the decrypted term and $h(i_0)$ and compares it with corresponding term $(\Pi_0)$, which is enclosed within $(C_n)$ for $(0 \le n \le k)$. If all the verifications pass, then it implies that the data were generated for the genuine initiator, and for the agent's code and timestamp of concern. Also, the initiator validates if the checksums that they compute are consistent with the returned checksums the agent stores.

The protocol carries various verifications that can prevent or at least any malicious act on gathered data. The protocol truly accomplishes the intended security properties as follows.

## Preserves Privacy of Gathered Data

A host signs its offer digitally, and then encrypts it with the public encryption key of the initiator. It prevents breach of privacy of collected data.

## Accurately Authenticates Gathered Data

At the early execution of the agent at a visited host, the host decrypts the term $\Pi_{n+1}$ using its decryption key and deduces the identity of the signer of the decrypted term. Then, it decrypts the term using the decryption verification key of the signer. Next, it computes a hash of $(\Pi, t, h(i_0))$ and then it verifies that the computed hash is the same as $\Pi_0$ enclosed within $C_0$, otherwise the agent execution terminate. It detects the security threat of impersonating the genuine initiator.

The offer $C_n$ is digitally signed with the private key of host $i_n$, and then encrypted with the public key of the initiator. Hence, an adversary would not be able to read the offer or send the offer under its private key. It prevents the security attack of impersonating the genuine provider of an offer.

## Ensures Non-Repudiation of Gathered Data

A host should sign the offer it provides with its private key. Hence, an adversary would not be able to append an arbitrary offer for which it is held responsible and cannot repudiate it.

## Ensures Strong Integrity of Gathered Data

Upon the agent's return, the initiator performs the following actions to detect any security threat:

- Verifies that the term $M_n$ for $(1 \leq n \leq k)$ is provided for the protocol run identified with a timestamp $(t)$. If the verification fails, then it detects the security attack of replacing the collected data with data of a similar protocol run.

- Verifies the terms $(\Pi, P_0, i_0$ and $i_1)$ stored within the agent upon its return match the corresponding terms stored within the secondary agent. If the verification fails, then it detects the security attack of replacing the collected data with data of a similar agent.

- Calls the initial offer $M_0$ which is securely stored within the secondary agent and checks the availability of the same data within the data returned with the mobile agent. If the verification fails, then it detects the security attack of data truncation and, in particular, the deletion of the initial offer $M_0$.

- Verifies that the identity of the first host in the assembled agent's itinerary matches the identity of the first host the secondary agent securely stores. If the verification fails, it then detects the security attack of data truncation and, in particular, the deletion of the first collected offer $M_1$.

- Decrypts the term $C_n$ for $(0 \leq n \leq k)$ and deduces the partial agent's itineraries, where each signed term $C_n$ includes the identity of the succeeding host $i_{n+1}$. It then verifies that the partial agent's itineraries are consistent. If assembled agent's itinerary indicates a missing connection, it then detects the security attack of data truncation and, in particular, the deletion of a collected offer.

- Verifies that the term $C_{n-1}$ enclosed within the term $C_n$ for $(0 \leq n \leq k)$ is among the stand alone terms. If a term is missing, it then detects that the offer of that host has been deleted or an adversary has replaced its previous offer $C_n$ with a new offer $C'_n$ so as to replace its original offer $d_n$ with a competitive offer $d'_n$.
- Requires that the memory of a visited host to be cleared from any terms the host may need to generate a new offer that replaces its previous offer before the mobile agent is dispatched to the next host. If a malicious host did not clear its memory from the terms, the attack is then detected.
- Encrypts the agent's code with the public key of succeeding host in the scheduled agent's itinerary. Hence, the intruder would not be able to run the agent or illegitimately insert or append data.
- Requires a visited host to sign the offer it provides with its private key. Hence, an adversary would not be able to append an arbitrary offer for which it is held responsible and cannot repudiate it. It prevents the security attack of appending fake data.

## Formal Verification of Proposed Security Protocols

The two protocols are verified using STA (Symbolic Trace Analyzer) formal method (Boreale & Buscemi, 2001, & 2002b; Boreale, 2001; STA documentation, 2001). It is a recent approach that is based on infinite-state exploration methods. It is a simple and an efficient tool for the analysis of security protocols. It is an ML (Milner, Tofte, Harper & MacQueen, 1997) based tool, where ML is basically a functional polymorphic programming language that allows parallel processing and is employed to develop verification tools (Mobility Workbench, 2004; Process Algebra Compiler, 2000; Symbolic Model Prover, 2000). The theory underlying the STA is explained in full details in

(Boreale & Buscemi, 2002a). STA tool requires the Moscow ML (Moscow, ML 2004), a compiler for the standard M. The method is used as it has the following characteristics:

- Avoids the state explosion problem, which process algebra methods suffer (Durante, Sisto, & Valenzano, 2000)
- Does not require expert guidance, which theorem proving methods suffer (Aziz et al., 2001; Durante et al., 2000)
- Does not need to model the intruder explicitly, which finite-state exploration methods suffer (Boreale & Buscemi, 2002b; Boreale & Gorla, 2002; Mitchell et al., 1997)
- Automates verification and does not need hand-written proofs, which modal logic methods suffer (Durante et al., 2000)

The formal verification showed that the two protocols are free of flaws (Al-Jaljouli, 2005, 2006) and are capable of preventing or at least detecting the security threats, which other security protocols do not.

## PROTOCOL DESIGN SECURITY FRAMEWORKS

The soundness of a protocol can be verified by the implementation of formal methods. Though, there are fundamental requirements that should be considered while designing mobile agent security protocols. The protocol has to comply with the following security framework so the outlined security properties can be accomplished.

- *Ensures strong integrity of gathered data*: The protocol satisfies the followings:
  - Accurately deduces the order of a host among visited hosts upon the agent's return.

◦ Accurately deduces the actual number of visited hosts upon the agent's return.

◦ Checks that the returned data are associated with the dummy data the initiator has generated.

◦ Checks that the returned data have been generated for the genuine initiator.

◦ Checks that the first visited host is the first scheduled host in the agent's itinerary.

◦ Checks that the returned data belong to the protocol run of concern.

◦ Carries out verifications on the identity of the genuine initiator at the early execution of the agent at a visited host.

◦ Ensures accurate verifications upon the agent's return by storing securely the initial verification terms with a secondary agent.

◦ Deduces accurately the actual agent's itinerary.

◦ Forbids an adversary from inferring the offer of a preceding host.

◦ Binds the data a host provides to identify of next host, a unique nonce/timestamp, and identity of initiator.

◦ Binds the data a host provides to data gathered at preceding hosts, and dummy data generated by the initiator.

◦ Forbids an intruder from appending data to the protocol.

◦ Provides instant check upon the agent's return. Firstly, it verifies if number of returned offers matches the actual number of visited hosts. Secondly, it verifies if returned data have been generated for the genuine initiator. Finally, it verifies if the returned data belong to protocol run of concern.

• *Preserves privacy of gathered data*: The data a host provides is encrypted with the public encryption key of the initiator.

• *Preserves origin-confidentiality*: The signed data a host provides is encrypted with the public encryption key of the initiator.

• *Ensures non-repudiation of gathered data*: The data a host provides is digitally signed with the private key of the host.

• *Accurately authenticates gathered data*: The signed data a host provides is encrypted with the public encryption key of the initiator.

## CONCLUSION

The several protocols (Corradi, et al., 1999b; Hannotin et al., 2001; Karnik & Tripathi, 1999; Karjoth et al., 1998; Maggi & Sisto, 2002a, 2003; Yao et al., 2003; Rhazi et al., 2007) presented in the literature aim to assert the security properties of mobile agent's execution results such as integrity, confidentiality, and authenticity in the presence of malicious hosts and intruders. However, they are not able to completely achieve the aimed for security properties. They do not achieve particular security properties, such as strong data integrity. It is attributed to incomplete designs, where a proper design of a security protocol should consist of: (a) precise security requirements, (b) clear assumptions, (c) various capabilities of adversaries, especially the conspiracy of non-trusted hosts, (d) formal modeling of the system and security requirements, and (e) formal verification of the security properties a protocol aims to accomplish.

The analysis showed the types of security attacks the existing protocols even failed to detect. The flaws can be classified into four types: erroneous authentication, breach of privacy, data truncation or substitution, and appending a fake stem to the execution result. The most critical flaw is data truncation, which is attributed to colluding attacks of malicious hosts.

A set of remedies that would rectify the security flaws and result in truly secured protocols are presented. The remedies include: use of execution traces an executing host may store as the initiator would request for verification purposes, encryption of the offer the host signs using the public key of the succeeding host in agent's itinerary, clearing the execution results at a visited host before the agent migrates to another host, verifying the identity of the genuine initiator at the early execution of the agent and if to proceed on the execution of the agent, carrying out a set of verifications upon the agent's return to the initiator, and incorporating initial verification data within the offer a visited host provides. The terms includes: nonce or timestamp that uniquely identifies the protocol run, dummy data, identity of the genuine initiator, identity of the first hosts in agent's itinerary.

Two recent security protocols are described that implement the security remedies and are found free of security flaws as showed by formal methods. A security framework a designer of security protocols should fulfill is discussed. Also a protocol designer should specify and verify a new protocol formally before presenting it for implementation and claiming the assertion of particular security properties. Many of the existing protocols are not yet formally described and verified. It is very beneficial to apply formal methods to existing protocols, which would help in fixing such protocols or getting a formal proof of their correctness and safety, as applicable.

# REFERENCES

Abadi, M., Blanchet, B., & Fournet, C. (2007). Just fast keying in the Pi calculus. *ACM Transactions on Information and System Security*, *10*(3), 1–59. Retrieved December 24, 2010 doi:10.1145/1266977.1266978

Abadi, M., & Fournet, C. (2001). Mobile values, new values, and secure communications. In *Proceedings of 28th ACM SIGPLAN-SIGACT Symposium on Principles of Programming Languages (POPL'01)* (pp. 104-115).

Abadi, M., & Gordon, A. D. (1999). A calculus for cryptographic protocols: The Spi-Calculus. *Journal of Information and Computation*, *148*(1), 1–70. doi:10.1006/inco.1998.2740

Al- Jaljouli. R. (2006). *A proposed security protocol for data gathering mobile agents*. Masters Thesis Dissertation. University of New South Wales, School of Computer Science and Engineering, Australia.

Al-Jaljouli, R. (2005). *Formal methods in the enhancement of the data security protocols of mobile agents* (Technical Report TR 520). University of New South Wales, School of Computer Science and Engineering, Australia. Retrieved on December 24, 2010, from http://cgi.cse.unsw.edu.au/~reports

Al-Jaljouli, R., & Abawajy, J. (2010). Negotiation strategy for mobile agent-based e-negotiation. In *Proceedings of the 13th International Conference on Principle and Practice of Information Mobile Agents (PRIMA 2010)*.

Algesheimer, J., Cachin, C., Camenisesh, J., & Karjoth, G. (2000). *Cryptographic security for mobile code*. IBM Research Report, Zurich, Switzerland.

Aziz, B., Gray, D., Hamilton, G., Oehl, F., Power, J., & Sinclair, D. (2001). Implementing protocol verification for e-commerce. In *Proceedings of International Conference on Advances in Infrastructure for Electronic Business, Science, and Education on the Internet (SSGRR 2001)*.

Bieszczad, A., White, T., & Pagurek, B. (1998). Mobile agents for network management. *Journal of IEEE Communications Surveys*, *1*(1).

Blanchet, B., & Aziz, B. (2003). A calculus for secure mobility. In *Proceedings of 18th Asian Computing Science Conference (ASIAN'03), Lecture Notes in Computer Science, 2896* (pp.188-204). Springer-Verlag.

Boreale, M. (2001). Symbolic trace analysis of cryptographic protocols. In *Proceedings of 28th International Colloquium on Automata, Languages and Programming (ICALP), Lecture Notes in Computer Science, 2076* (pp.667-681). Springer-Verlag.

Boreale, M., & Buscemi, M. (2001). *STA: A tool for trace analysis of cryptographic protocols - ML object code and examples.* Retrieved December 24, 2010, from http://www.dsi.unifi.it/~boreale/tool.html

Boreale, M., & Buscemi, M. (2002a). A framework for the analysis of security protocols. In *Proceedings of the 13th International Conference on Concurrency Theory (CONCUR)* (pp.483-498). *Lecture Notes in Computer Science, 2076* (pp. 667-681). Heidelberg, Germany: Springer-Verlag.

Boreale, M., & Buscemi, M. (2002b). Experimenting with STA, a tool for automatic analysis of security protocols. *In Proceedings of ACM Symposium on Applied Computing (SAC),* (pp. 281-285). ACM Press.

Boreale, M., & Gorla, D. (2002). Process calculi and the verification of security protocols. *Journal of Telecommunications and Information Technology – Special Issue on Cryptographic Protocol Verification (JTIT), 4,* 28-40.

Burrows, M., Abadi, M., & Needham, R. (1990). Logic of authentication. *ACM Transactions on Computer Systems, 8*(1), 18–36. doi:10.1145/77648.77649

Carnegie Melon University. (2000). *Symbolic model prover.* Retrieved December 24, 2010, from http://www-2.cs.cmu.edu/~modelcheck/symp.html

Chen, B., Cheng, H., & Palen, J. (2009). Integrating mobile agent technology with multi-agent systems for distributed traffic detection and management systems. *Journal of Transportation Research Part C: Emerging Technologies, 17*(1), 1–10. doi:10.1016/j.trc.2008.04.003

Choy, M. C., Srinivasan, D., & Cheu, R. L. (2003). Cooperative, hybrid agent architecture for real-time traffic signal control. *Journal of IEEE Transactions on Systems. Man and Cybernetics Part A: System and Humans, 33*(5), 597–607. doi:10.1109/TSMCA.2003.817394

Corradi, A., Montanari, R., & Stefanelli, C. (1999a). Mobile agents integrity in e-commerce applications. In *Proceedings of the 19th IEEE International Conference on Distributed Computing Systems Workshop (ICDCS'99),* (pp. 59-64). IEEE Computer Society Press.

Corradi, A., Montanari, R., & Stefanelli, C. (1999b). Mobile agents protection in the Internet environment. In *Proceedings of the 23rd Annual International Computer Software and Applications Conference (COMPSAC '99)* (pp. 80 - 85).

Documentation, S. T. A. (2001). *Symbolic trace analyzer.* Retrieved December 24, 2010, from http://www.dsi.unifi.it/~boreale/documentation.html

Durante, L., Sisto, R., & Valenzano, A. (2000). A state-exploration technique for Spi-calculus testing equivalence verification. *In Proceedings of the IFIP International Joint Conference on Formal Description Techniques for Distributed Systems and Communication Protocols (FORTE XIII) and Protocol Specification, Testing and Verification (PSTV XX),* (pp. 155-170). Dordrecht, The Netherlands: Kluwer Academic Publishers.

Fiore, M., & Abadi, M. (2001). Computing symbolic models for verifying cryptographic protocols. In *Proceedings of the 14th IEEE Computer Security Foundations Workshop (CSFW 2001),* (pp. 160-173). IEEE Computer Society Press.

Fischer, L. (2003). *Protecting integrity and secrecy of mobile agents on trusted and non-trusted agent places*. Diploma Dissertation. University of Bremen, Germany. Retrieved from http://www.sec. informatik.tu-armstadt.de/lang_neutral/diplomarbeiten/ docs/fischer_diplom.pdf

Formal Systems (Europe) Ltd. (2000). *Failures divergence refinement*. FDR2 user manual. Retrieved from http://www.formal.demon.co.uk/ fdr2manual/index.html

Gray, R. S. (2000). Soldiers, agents and wireless networks: A report on a military application. In *Proceedings of the 5th International Conference and Exhibition on the Practical Application of Intelligent Agents and Multi-Agents (PAAM 2000)*.

Hannotin, X., Maggi, P., & Sisto, R. (2001). Lecture Notes in Computer Science: *Vol. 2240. Formal specification and verification of mobile agent data integrity properties: A case study* (pp. 42–53). Springer-Verlag.

James, G., Cohen, D., Dodier, R., Platt, G., & Palmer, D. (2006). A deployed multi-agent framework for distributed energy applications. In *the Proceedings of the 5th International Joint Conference on Autonomous Agents and Multi-agents Systems (AAMAS 2006)*.

Karjoth, G., Asokan, N., & Gülcü, C. (1998). Protecting the computation results of free-roaming agents. *Journal of Personal and Ubiquitous Computing, 2*(2), 92–99.

Karnik, N., & Tripathi, A. (1999). *Security in the Ajanta mobile agent system* (Technical Report TR-5-99). University of Minnesota, Minneapolis.

Kok, J. K., Warmer, C. J., & Kamphuis, I. G. (2005). Multiagent control in the electricity infrastructure. In *the Proceedings of 4th International Joint Conference on Autonomous Agents and Multiagent Systems* (pp. 75-82).

Lowe, G. (1996). Breaking and fixing the Needham-Schroeder public key protocol using FDR. In *Proceedings of Tools and Algorithms for the Construction and Analysis of Systems (TACAs)* (*Vol. 1055*, pp. 147–166). Lecture Notes in Computer Science Springer-Verlag.

Lowe, G. (1997). Casper: A compiler for the analysis of security protocols. In *Proceedings of the 10th Computer Security Foundation Workshop (PCSFW)*. IEEE Computer Society Press.

Ma, L., & Tsai, J. J. P. (2000). Formal verification techniques for computer communication security protocols. In S. K. Chang (Ed.), *Handbook of software engineering and knowledge engineering*. Retrieved December 24, 2010, from ftp://cs.pitt. edu/chang/handbook/12.pdf

Maggi, P., & Sisto, R. (2002a). Experiments on formal verification of mobile agent data integrity properties. In *Proceedings of Workshop from Data to Agents (WOA)* (pp. 131-136).

Maggi, P., & Sisto, R. (2002b). Using SPIN to verify security properties of cryptographic protocols. In *Proceedings of 9th International Spin Workshop on Model Checking of Software (SPIN 2002), Lecture Notes in Computer Science, 2318* (pp.187-204). Springer-Verlag.

Maggi, P., & Sisto, R. (2003). A configurable mobile agent data protection protocol. In *Proceedings of Autonomous Agents and Multiagent Systems* (pp. 851–858). New York, NY: ACM Press. doi:10.1145/860710.860712

Meadows, C. (1994). Formal verification of cryptographic protocols: A survey. In *Proceedings of 4th International Conference on the Theory and Applications of Cryptology- Advances in Cryptography (ASIACRYPT)* (pp. 135-150).

Meadows, C. (1996). Language generation and verification in the NRL protocol analyzer. In *Proceedings of 9th IEEE Computer Society Foundations Workshop (CSFW)* (pp. 48-61).-

Millen, J. K., Clark, S. C., & Freedman, S. B. (1987). The interrogator: protocol security analysis. *Journal of IEEE Transactions on Software Engineering, 13*(2), 274–288. doi:10.1109/TSE.1987.233151

Milner, R., Tofte, M., Harper, R., & MacQueen, D. (1997). *The definition of standard ML* (rev. ed.). MIT Press.

Mitchell, J. C., Mitchell, M., & Stern, U. (1997). Automated analysis of cryptographic protocols using Mur. In *Proceedings of Symposiums on Security and Privacy* (pp. 141–153). IEEE Computer Society Press. doi:10.1109/SECPRI.1997.601329

Mobach, D. (2007). *Agent-based mediated service negotiation.* PhD Thesis, Computer Science Department, Vrije University Amsterdam.

Mobility Workbench. (2004). *A tool for manipulating and analyzing mobile concurrent systems described in the Pi.* Retrieved December 24, 2010, from http://www.it.uu.se/research/group/mobility/mwb

*Moscow, ML.* (2004). Retrieved December 24, 2010, from http://www.dina.dk/~sestoft/mosml.html

Outtagarts, A. (2009). Mobile agent-based applications. *Journal of Computer Science and Network Security, 9*(11).

Paulson, L. C. (1997). Proving properties of security protocols by induction. In *Proceedings of the 10ᵗʰ Computer Society Foundations Workshop (CSFW)* (pp. 70-83).

Paurobally, S., & Jennings, N. R. (2005). Developing agent web service agreements. In *Proceedings of the IEEE/WIC/ACM International Conference on Intelligent Agent Technology* (pp. 464-470).

Process Algebra Compiler. (2000). Verification tool. Retrieved December 24, 2010, from http://www.reactive-systems.com/pac

Rhazi, A., Pierre, S., & Boucheneb, H. (2007). Secure protocol based on sedentary agent for mobile agent environments. *Journal of Computer Science, 3*(1), 35–427. doi:10.3844/jcssp.2007.35.42

Roth, V. (1998). Mutual protection of co-operating agents. In Vitek, J., & Jansen, C. D. (Eds.), *Secure Internet programming* (pp. 26–37). Berlin, Germany: Springer-Verlag.

Roth, V. (2001). Programming Satan's agents. *Journal of Electronic Notes in Theoretical Computer Science (ENTCS), 63.*

Roth, V. (2002). Empowering mobile software agents. In *Proceedings of 6ᵗʰ IEEE Mobile Agents Conference, Lecture Notes in Computer Science, 2535* (pp. 47-63). Springer-Verlag.

Rushby, J. (2006). Tutorial: Automated formal methods with PVS, SAL, and Yices. In *Proceedings of the 4ᵗʰ IEEE International Conference on Software Engineering and Formal Methods (SEFM'06)* (p. 262).

Sewell, P. (1998). Global/ local subtyping and capability inference for a distributed calculus. Automata, languages and programming. In *Proceedings of 25ᵗʰ International Colloquium (ICALP), Lecture Notes in Computer Science, 1443* (pp.695-706). Springer-Verlag.

Vigna, G. (1998). Lecture Notes in Computer Science: *Vol. 1419. Cryptographic traces for mobile agents. Journal of Mobile Agent and Security* (pp. 137–153). Heidelberg, Germany: Springer-Verlag.

Vitek, J., & Gastagna, G. (1999). Seal: A framework for secure mobile computations. In *Proceedings of Internet Programming Language Workshop (ICCL), Lecture Notes in Computer Science, 1686* (pp. 47-77). Springer-Verlag.

Yao, M., Foo, E., Peng, K., & Dawson, E. (2003). An improved forward integrity protocol for mobile agents. In *Proceedings of the 4ᵗʰ International Workshop on Information Security Applications (WISA), Lecture Notes in Computer Science, 2908* (pp. 272-285). Springer-Verlag.

## ADDITIONAL READING

Abramowicz, W., & Mayr, H. (2007). *Technologies for Business Information Systems*. Springer, 432 p. Al- Saadoon, G. (2009). Flexible and reliable architecture for mobile agent security (TR). *Journal of Computer Science, 5*(4), 270–274.

Dadhich, P., Dutta, K., & Govil, M. (2010). Security issues in mobile agents. *International Journal of Computers and Applications, 11*(4), 1–7. doi:10.5120/1574-2104

Deng, R., Qing, S., Bao, F., & Zhou, J. (2002). Lecture Notes in Computer Science: *Vol. 2513. Information and Communications Security*. Heidelberg: Springer-Verlag. doi:10.1007/3-540-36159-6

Eassa, M., Salama, A., & Saleh, B. (2008). A secure integrated model for mobile agent migration. *Egyptian Computer Science Journal (ECS), 30*(2).

Garrigues, C., Migas, N., Buchanan, W., Robles, S., & Borrell, J. (2009). Protecting mobile agents from external replay attacks. *Journal of Systems and Software, 82*(2), 197–206. doi:10.1016/j.jss.2008.05.018

Garrigues, C., Robles, S., Borrell, J., & Navarro-Arribas, G. (2010). Promoting the development of secure mobile agent applications. *Journal of Systems and Software, 83*(6), 959–971. doi:10.1016/j.jss.2009.11.001

Genco, A. (2007). Mobile Agents: Principles of Operation and Applications. *Advances in Management Information, 6*.

Hacini, S., Guessoum, Z., & Boufaïda, Z. (2007). TAMAP: A new trust-based approach for mobile agent protection. *Journal in Computer Virology, 3*(4), 267–283. doi:10.1007/s11416-007-0056-y

Ismail, L. (2008). A secure mobile agents platform. *The Journal of Communication, 3*(2), 1–12.

Jean, E., Jiao, Y., & Hurson, A. (2007). Addressing mobile agent security through agent collaboration. *Journal of Information Processing Systems, 3*(2), 43–53. doi:10.3745/JIPS.2008.3.2.043

Jennings, N., & Wooldridge, M. (Eds.). (1998). *Agent technology: Foundations, applications, and markets*. Springer-Verlag Berlin Heidelberg New York.

Khosrow-Pour, M. (Ed.). (2004). *E-Commerce security: Advice from experts*. USA: CyberTech Publishing.

Kou, W. (2003). Payment Technologies for E-Commerce. *Advances in Information Management, 6*.

Lange, D., & Oshima, M. (1998). *Programming and Deploying Java Mobile Agents with Aglets*.

Lei, S., Zhang, R., Liu, J., & Xiao, J. (2008). A novel security protocol to protect mobile agent against colluded truncation attack by cooperation. In *Proceedings of International Conference on Cyberworlds (CW)* (pp.186-191).

Lin, C., & Varadharajan, V. (2010). MobileTrust: A trust enhanced security architecture for mobile agent systems. *International Journal of Information Security, 9*(3), 53–178. doi:10.1007/s10207-009-0098-x

Muñoz, A., Maña, A., & Serrano, D. (2009). Protecting agents from malicious hosts using TPM. *International Journal of Computer Science & Applications, 6*(5), 30–58.

Ouardani, A., Pierre, S., & Boucheneb, H. (2007). A security protocol for mobile agents based upon the cooperation of sedentary agents. *Journal of Network and Computer Applications, 30*(3), 1228–1243. doi:10.1016/j.jnca.2006.04.008

Peter Braun, P., & Rossak, W. (2005). *Mobile Agents: Basic Concepts*. Mobility Models, and the Tracy Toolkit.

Schoeman, M., & Cloete, E. (2004). Cloete: Design concepts for mobile agents. *South African Computer Journal, 32*, 13–24.

Stalling, W. (2003). *Cryptoraphy and Network Security* (3rd ed.). Upper Saddle River, NJ: Prentice Hall.

Tsai, J., & Ma, l. (2006). *Security Modeling and Analysis of Mobile Agent Systems*. Electrical and Computer Engineering, Empirical College, London.

Wang, T., Guan, S., & Chan, T. (2002). Integrity protection for Code-on-Demand mobile agents in e-commerce. *Journal of Systems and Software, 60*(3), 211–221. doi:10.1016/S0164-1212(01)00093-0

Wang, X., Xu, D., & Luo, J. (2008). A free-roaming mobile agent security protocol based on anonymous onion routing and k anonymous hops backwards. In *Proceedings of Autonomic and Trusted Computing, 5th International Conference (ATC)* (pp.588-602).

Wong, J., & Mikler, A. (1999). Intelligent mobile agents in large distributed autonomous cooperative systems. *Journal of Systems and Software, 47*(2-3), 75–87. doi:10.1016/S0164-1212(99)00027-8

Xu, D., Harn, L., Narasimhan, M., & Luo, J. (2006). An improved free-roaming mobile agent security protocol against colluded truncation attacks. In *Proceedings of 30th Annual International Computer Software and Applications Conference (COMPSAC)* (pp.309-314).

Xu, H., Zhang, Z., & Shatz, S. (2005). A Security based model for mobile agent software systems. *International Journal of Software Engineering and Knowledge Engineering, 15*(4), 719–746. doi:10.1142/S0218194005002518

## KEY TERMS AND DEFINITIONS

**Decryption:** Process of converting encrypted data (ciphertext) back into its original form (plaintext) to be readable using a particular decryption key. The key is an algorithm that reverses the work of encryption algorithm.

**Encryption:** Process of converting data (plaintext) into an encrypted data (ciphertext) using algorithms that make it unreadable by unauthorized people.

**Mobile Agent:** Software entity that have some intelligence, run on heterogeneous environments, act on behalf of clients, autonomously make decisions, and adapt to changes in environments.

**Security Attacks:** Malicious acts of intruders that are able to implant on a secured entity that break the security techniques set by the protocol.

**Security Flaws:** Failure of a security protocol to preserve one or more security properties such as integrity, privacy, authenticity, or non-repudiation.

**Security Properties:** States that assure protection or freedom from attacks. Security properties include privacy, integrity, authenticity, accountability, and origin-confidentiality.

**Security Protocol:** Security-related rules applying cryptographic techniques and specifying the sequence of messages to be sent/received during a communication session. The rules specify pre-conditions and post-conditions on global states for a particular step in the protocol.

# Chapter 11

# GSM:
## Future Viable Option for Interacting with Real-World Entities

**Srinivasa K G**
*M S Ramaiah Institute of Technology, India*

**Vijayendra R**
*M S Ramaiah Institute of Technology, India*

**Harish Raddi C S**
*M S Ramaiah Institute of Technology, India*

**Anil Kumar M**
*M S Ramaiah Institute of Technology, India*

## ABSTRACT

*The objective of this chapter is to provide an innovative system for tracking and monitoring objects using RF transmitters and receivers, and querying about these objects using mobile phones. The protocol for the system and a simulation to check the feasibility of this project before applying it to a real world scenario is presented below. In this simulation, capability of the transmitter and receiver is tested to communicate effectively and also display the data sent by the transmitter. As a result, a highly flexible wireless controlled system would enable bridging the distance between the user and the electronic device, and hence, provide easier access to real world entities, by just manoeuvring the mobile phones, without the need to be physically present near the object. Also, it is being enhanced to ensure an authenticated access to these resources, thereby taking care of the security aspect. The working of the system can be divided into two phases: phase one comprises a pair of RF transmitter and receiver coupled with a central database, and phase two is comprised of GSM modem with a unique SIM linked to the central data base via GSM network. In the first phase of the system, RF transmitters are tagged to the objects of everyday use and have the capability of transmitting signals, and the paired receiver detects the transmission of the tagged object and stores its corresponding location in the central database, which is created specifically for information maintenance of the tagged objects. In the second phase of the system, a mobile phone is used to query the location of any tagged objects based on its availability in the vicinity, by sending*

DOI: 10.4018/978-1-4666-0080-5.ch011

*a message to the SIM connected to a GSM modem. If an object is present, the GSM modem fetches the location and other relevant information from the central database and encapsulates this information into a message, which is sent back to the mobile device or phone that has requested the information.*

## INTRODUCTION

The growth in technology facilitates competitive and less expensive devices; keeping this in mind a flexible system is put forth to play a major role in future computing systems. The aim of the system is to ensure a degree of freedom for an individual to locate objects in his vicinity without the need of any physical movement. This is done by monitoring everyday physical processes or entities and keeping track of it by a central database, provides novel features to the system. What currently hinders most of the conceived parallel similar type of systems from becoming commercial applications, however, is a lack of adequate infrastructure of following types: First, a well marked sensing infrastructure must be installed in area or block wise to perform an efficient and ordered sensing task. Second, a communication infrastructure is required to transmit the data within the network of the connected devices; this involves distribution and aggregation of readings from multiple transmitters and so on. Thirdly, a commercial infrastructure is needed to manufacture and deploy the transmitting, receiving and communicating devices like GPS system, and to generate revenue from the system.

Compared to these hindrances, authors propose a system with a unique approach to establish a relationship between RF transmitters and receivers using a microchip, and link it to the GSM network, to provide an uninterrupted access to the data by the user. The incorporation of mobile phones in the system provides a unique opportunity to overcome most of the difficulties, indicating the ease of operating the system by an individual without the need of a specialized knowledge. Sensing technologies can be implemented using transmitters and a receiver and, data from the

central database can be accessed from the mobile handset via wireless mobile networks. Wide area communication is a core property of the cellular network. It enables the integration of data from many transmitters and the support of applications with backend services such as data storage in a central system. The effectiveness of the system is shown by locating important day to day entities or objects, whether moveable or fixed, and querying their information by various users irrespective of their location.

Making use of these unique properties of mobile phones, RF components and the cellular network, proposed system is mainly concerned with monitoring and locating objects by means of mobile phones, and sensing functionality carried out by the objects that are tagged with the hardware i.e. the transmitters in the vicinity, and are detected by the corresponding receivers. And the location is queried from the central database using a mobile phone by means of a GSM modem. The dual technique employed in this project is: First, object sensing using the RF transmitters and receivers. Second, user interface provided by the mobile phone aid for sensing and querying the personal items simultaneously. This system allows to identify a particular challenge that is common to many applications that makes use of large people-centric infrastructure provided by mobile phones and the cellular network. The challenge is to answer the question of whether a system based on RF transmitters and receiver can provide sufficient coverage in relatively short period of time. In an extensive evaluation, which includes a real-world experiment with the object localization prototype, analysis of properties of the coverage obtained given a wide range of different operational parameters such as the distance between transmitter and receivers, objects in the path of transmission and range of the transmitters used were performed. The study deals with the feasibility of object localization based on mobile phones and can provide valuable guidelines for the design of future people-centric sensing systems in general (Christian, Philipp, Christof, & Wolfgang , 2008).

Security aspect is an essential factor that surrounds any network based application. Strengthening the security aspect of the system strengthens its application, thereby indicating a linear relation among the two. It is important to realize and ensure a reliable secured network based application, without which the entire system and its application tend to be meaningless. Therefore security presents sacrosanct part of the system, hence needs to be addressed with due diligence. Keeping this in view, few security related operations of this system that will increase the overall credibility and reliability of the task being carried out such as Iris scan, Finger print secured access and Encrypted password access are discussed.

The chapter is organized as follows. The Background section recalls various methods which deal with interaction of real world entities. The System Architecture section deals with design of all the components such as Sensing by RF transmitters and Receivers which deals with how GSM and RF based transmission helps in sensing activities, the GSM Network Service section deals with the available services, the Database section deals with the centralized data storage method. Monitoring of objects is explained in the USE case section. The Working section details the layered stack structure and overall interaction of all the components of the design. The Simulation and Analysis section complements the Working section with careful analysis of the results. The Security Essentials section details the various methods that can improve the security of the design. Wide range of applications in emergency services, landmarks locating and warehouse management are explained in the Application section which is followed by the Conclusion section.

## BACKGROUND

The idea of interacting with real world entities has been explored and studied over the past decade and several approaches have been proposed such as the detection and tracking of moving vehicles in image sequences from traffic scenes recorded by a stationary camera In order to exploit the knowledge about the shape and the physical motion of vehicles in traffic scenes, a parameterized vehicle model is used for an intra frame matching process and a recursive estimator based on a motion model and is used for motion estimation (Koller & Daniilidis, 1992). The energy management issue in a sensor network application - Object Tracking Sensor Networks (OTSNs) is of great importance (Lin, Peng, & Tseng, 2006; Vincent- Tseng & Kawuu-Lin, 2005). Based on the fact that the movements of the tracked objects are sometimes predictable, system propose a Prediction-based Energy Saving scheme, called PES, to reduce the energy consumption for object tracking under acceptable conditions (Xu, Winter, & Lee, 2004; Carle & Simplot, 2004). Exploring Data mining alternatives for discovering the temporal movement of patterns of objects in sensor networks, location prediction strategies that utilize the discovered temporal movement patterns so as to reduce the prediction errors for energy savings (Padmanabhan & Tuzhilin, 1996) As energy forms an important criteria for measuring a models efficiency various strategies are discussed and proposed that provide an insight in this regard (Xu, Winter, & Lee, 2004).

Other efficient methods have been proposed and supported with results for the design and implementation of a single-chip GSM (Global System for Mobile communications) transceiver RF integrated circuit. The chip includes the RF-to-Baseband and Baseband-to-RF functions (Stetzler, Havens, & Koyama, 1995) and the use of direct conversion receiver topologies has been marked by many technical problems. The design of a GSM multi-band transceiver that also features the development of the offset PLL transmitter

*Figure 1. System structure*

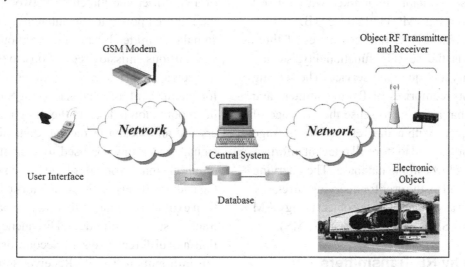

together with a fast locking fractional-N synthesizer (Strange & Atkinson, 2000) have all served an efficient guide to understand the concepts of object tracking.

## SYSTEM'S STRUCTURE AND FUNCTIONALITY

### System Architecture

The system comprises of RF transmitters incorporated into the objects, RF receivers located in their respective region, a database to store all the relevant information regarding the objects, a central system to process all the USE cases and act as interface for the system, and a GSM modem used for sending and receiving SMS regarding querying of the objects and its location (Figure 1).

Generally RF transmitters operate in a radio frequency range of 30 KHz to 300 GHz. They are programmed at different frequencies and integrated with objects. Each RF transmitter will be associated with a unique ID called Object ID; indicating the object being tagged. RF transmitters can be used to operate in low frequency range as well as high frequency range depending upon its

application. A balanced system would consist of both long range as well as short range frequencies increasing the flexibility of the system via effective frequency utilization. Incorporating a frequency range switch allows the system to select the desired frequency range for each transmitter thereby maneuvering the frequency as desired. RF transmitter consists of a microprocessor that transmits the Object ID across the selected frequency; this ID is detected by the corresponding area RF receiver. RF receivers are electronic devices that separate radio signals from one another and use demodulator to convert specific signals into audio, video, or data formats (Andreas, Hans-Florian, & Ulrich, 2005). In the system demodulation technique is restricted to only data format in order obtain the transmitted Object ID. Radio technique used by RF transmitters and receivers limit localized interference and noise. Each RF receiver is associated with a unique Location ID. This ID is used to determine the location of the object; this ID is different from the Object ID as the latter is used to represent the Object being detected. RF receivers display the unique Object ID being detected; all the relevant data of this object are stored in the database along with its location via the central system. RF receivers are also available in market with various specifications e.g., MC33591, U4311B.

The system architecture comprises of three parts that is the sensing functionality, storing functionality and querying service. The sensing functionality comprises of RF transmitters and receivers that are used to sense the presence of tagged objects within the region. The storage functionality is used to store all relevant information of the object in the database. The querying service is used by the user for querying necessary information associated with the object using GSM network and Short Message Service (SMS).

## Sensing by RF Transmitters and Receivers

In this section, efficient RF transmitters that are widely used in radio frequency communication systems are presented. Such a wireless communication device includes a built-in radio transceiver or is coupled to an associated radio transceiver. An RF switch is placed between the RF power amplifier and the antenna, so that the received communication signals from the antenna may be switched to a receive circuit. RF transmitter performs modulation, up conversion and power amplification. The two commonly used modulation technique formats are constant and variable-envelope modulation. The modulated signal is expressed as

$$x(t) = A \cos[\ Wct + \emptyset(t)\ ]$$

it has a constant envelope if $A$ doesn't vary with time. If $A$ is a function of time then it is variable-envelope modulation, such signals can be expressed as

$$x(t) = A(t) \cos[\ Wct + \emptyset(t)\ ]$$

The Baseband RF interface in the transmitter is used to perform the desired modulation and up conversion. Finally the PA/Antenna consists of a amplifier and matching network to deliver the required power to the antenna. However the signal transmitted by the antenna should comply with various emission regulations so that it does not corrupt other user's communication. GSM, for example enforces the mask, constraining both the modulation index and the design. RF receivers use an antenna to receive transmitted radio signals and a tuner is used to separate specific required signal from all of the other signals that the antenna receives. Demodulators or Detectors then extract information that was encoded before transmission. Amplitude and Frequency Modulation are different ways to decode or modulate this information. The RF Receiver amplification performs two functions, the low noise amplification of the signal coming from the antenna and the input termination. RF Voltage controlled Oscillator phase noise required for RF applications use the LC-type VCOs instead of the more classical and compact solutions such as ring or relaxation oscillators. In LC-type VCOs the performance depends strongly on the quality of the passive inductors and varactors used (Bietti, Svelto, & Castello, 2000). The most popular solution for integrated oscillators is the cross coupled differential pair which realises a differential negative resistance to compensate the losses in the LC tank.

## GSM Network Service

Basic services are of two types: Tele services and Bearer services. Tele services are application specific and thus make use of all seven layers of the ISO/OSI reference model. Table 1 provides an overview of the available Tele services.

Bearer service comprises of all services that enable transparent transmission of data between the interfaces to the network. Bearer services make use of only lower three layers of the ISO/OSI reference model. Table 2 provides an overview of the available Bearer services. The two bearer service groups are sub-divided into a variety of bearer services with different characteristics.

*Table 1. Overview of available tele-services*

| Tele service | Description |
|---|---|
| 11 | Telephony |
| 12 | Emergency calls |
| 21 | Short message MT |
| 22 | Short message MO |
| 23 | Cell broadcast |
| 61 | Alternate speech and fax group 3 |
| 62 | Automatic fax group 3 |
| 91 | Voice group call |
| 92 | Voice broadcast |

*Table 2. Overview of available bearer services*

| Bearer service | Description |
|---|---|
| 20 | Asynchronous data bearer service |
| 30 | Synchronous data bearer service |

An additional set of services called the supplementary services are provided to enhance the basic services. Supplementary services may be provisioned for an individual basic service or for a group of basic services, e.g., a subscriber may be barred from all outgoing calls for all tele-services and all bearer services, except Short Message Service. Some supplementary services may be activated or deactivated by the user. Further an operator can bar certain subscribers or subscriber groups from modifying their supplementary services.

A GSM modem is a wireless modem that works with a GSM wireless network (Jochen, 2003). The GSM modem is used for sending text messages in PDU mode using standard AT commands. In addition to the standard AT commands, GSM modems support an extended set of AT commands. Most of the GSM handsets have built in support for AT commands. Although GSM handset are not recommended for performance oriented tasks, for the following reasons, (*i*) SMS application has to run for longer durations, GSM phones normally run on battery power & it is not recommended to connect phone with the charger all the time, and (*ii*) Phone gets heated up while application has to send large number of messages.

A GSM terminal is a special type of modem used for sending and receiving data, an efficient route for the transmission of message can be determined novel methods (Lee & Wang, 2003), using SMS text messages via the GSM network. The GSM terminal requires a SIM card which can be obtained from a mobile operator or from the market. The cost of sending a text message is same as you send SMS from one cell-phone to another. For small companies and departments with relatively few messages passing in and out from them, the GSM terminal is feasible. However for large message throughput where several hundred messages per minute could be anticipated then the HTTP connection is preferred over GSM terminal. Advantage of Short Message Service is timely delivery of information and low cost. The General command for sending the SMS: *AT+CMGS="... recipient phone number..."* The GSM modem responds to the AT+CMGS command with a single '>' character. Then enter a text message and end the text with CTRL-Z. Thereafter, the GSM modem sends the SMS to the recipient. Similarly, the command to indicate a new message received is AT+CMTI, to read a new message is AT+CMGR and to delete a message is AT+CMGD (Jochen, 2003).

## Database

The mobile network operator provides a user with a Central Database service in which the user can store application data such as reports and other information regarding the objects. This service can be used, for example, to record a log of objects reported by the RF receivers in the given region. Similarly it can be used to keep track of out-of-range objects. Database is essential for querying and fetching information about required objects using GSM modem. The size of the Central Database is proportional to the number of objects being detected at that point of time. Database comprises of secondary memory devices such as hard disk. Size of the database is directly proportional to the area of application. RF receivers are used to store the description and location of the objects in the database via central system,. and the GSM modem is used to query the objects from the database. The database can be created using software toolkit like MySQL. The database is integrated to RF receivers and GSM modem via central system using programming languages like.NET, J2ME, Java running on Windows or Linux Operating System (OS). Database should be frequently updated to provide valid appropriate data to the users of the system. Updating process can be done via the Update Use case.

## MONITORING OBJECTS USING USE CASES

Monitoring and location of objects is an essential and calibrated part of the system, accuracy with which it is done determines the overall efficiency and the ease at which the system can be handled. Therefore due care is taken in defining the usage of USE case involved with the system along with its application. Interface between the mobile phone, GSM, and the RF transmitters and Receivers with the central system is achieved via these USE cases. Use cases are specified for various functions such as detecting of objects within the sensing range, notify objects that have left the sensing range, querying about objects, identifying new objects entering and leaving the sensing range. The USE cases are defined as follows:

• **OUTRANGE**: This is a typical USE case that is used to notify the central system regarding the objects leaving the sensing range. The user will be notified with the objects that leave the sensing region

as per the query, therefore simplifying the task of tracking the outflow of objects from the sensing region. The basic function of this USE case includes tracing of object location before and after the loss event. Most recent location of the object will be stored and notification will be sent when an object leaves the sensing region using OUTRANGE (Object) function.

• **INRANGE**: This is a typical USE case that is used to notify the central system regarding the objects entering the sensing range. The user will be notified with the objects that enter the sensing region as per the query, therefore simplifying the task of tracking the inflow of objects into the sensing region. The basic function of this USE case includes tracing of object location and storing it in the database. Current location of the object will be stored and notification will be sent to indicate the presence of object in the sensing region using INRANGE (Object) function.

• **LOCATE**: This USE case allows the user to locate an object from the list of objects that are already being detected and stored in the database. The system will respond to this query via GSM modem to the potential user, if the corresponding object is located in the database. Various object search strategies can be employed such as string matching, keyword followed by the object name e.g., Locate Mobile Canteen. The GSM modem will notify the user with a short message containing object's location and its description using OBJINFO() and OBJLOC() functions.

• **OBJINFO**: This USE case is used to fetch the data of the object from the central system and send it to the user via GSM modem.

• **OBJLOC**: This USE case is used to fetch location of the object from the database and send it to the user via GSM modem.

*Table 3. Use cases and their parameters*

| Function Description | Time | In | Out | Range | Location | Information |
|---|---|---|---|---|---|---|
| INRANGE | A1 | B1 | ∞ | C1 | OBJLOC() | OBJINFO() |
| OUTRANGE | A2 | ∞ | D2 | ∞ | ∞ | ∞ |
| LOCATE | A3 | B3 | ∞ | C3 | OBJLOC() | OBJINFO() |
| FIND | A4 | B4 | D4 | - | - | - |
| GUARD | A5 | B5 | D5 | ∞ | ∞ | ∞ |

*The symbols ∞, - indicate infinity and Not Applicable respectively.

- **GUARD**: This USE case acts as a watch guard keeping track of new objects entering the sensing region and old objects leaving the sensing region by sending corresponding notifications. The data is correspondingly updated in the database. This Use case provides useful information that helps in location query.
- **FIND**: This USE case handles all the query requests sent from the potential user.
- **REPORT**: This USE case prepares the report of the object found, corresponding data is sent to the user by the central system.

Query services and functions implementing the above USE cases are provided in Table 3. The parameter A(1,2,3,4,5); B(1,3,4,5); C(1,3) and D(2,4,5) in the table represent the values used in actual system. 'Time" service is used to indicate time taken to respond to the query. "In" service indicates the object within the range and "Out" service indicate the objects outside the range. "Range" service provides the range of the object from the central system. "Location" and "Information" service provide the object location and information respectively. The FIND USE case should produce a result which in turn will allow the LOCATE USE case to fetch the information of the object. If the queried object is in the sensing region then query produces a result otherwise it sends a default message "Object Not Found/ Object Out of Range" depending on the status. REPORT USE

case specifies the information the central system should generate if the object is found. The report consists of location and complete description of the object e.g., a Mobile Canteen object consists of a report specifying the location, type, theme and facilities. Different system limits can be specified using the following parameters,

1. **TimeMax**: maximum time taken by a query or the maximum query time.
2. **NumberMax**: maximum number of objects detected by the receiver.
3. **QueryMax**: maximum no of reports that can be generated by the system.

For OUTRANGE USE case TimeMax, NumberMax, QueryMax all are irrelevant since the object is outside the sensing region of the system. In INRANGE USE case NumberMax is relevant as it specifies the maximum number of objects the receiver can detect and if the system exceeds this limit then unused objects or out of region objects should be replaced by new ones. In FIND USE case TimeMax and QueryMax are relevant. TimeMax is dependent upon the network traffic and distance between the user and the GSM modem. TimeMax is also limited by the effective search strategy employed in the system. GUARD USE case is concerned with NumberMax parameter, when a object is identified it verifies with the NumberMax before its information is stored in the database. If this object exceed the total number of objects currently being detected then the unused entries will be replaced by this entry.

## WORKING

The layered stack comprises of RF transmitters, that operate in a radio frequency range of 30 KHz to 300 GHz. RF transmitters (Behzad, 1999) are programmed at different frequencies and integrated with the objects to be located. The wide range of radio frequency enables the system to cover greater area, further by coupling the system with RF re-transmitters allows the system to expand its area beyond the given normal range without loss of data or attenuation of the signal. RF re-transmitters act as relay junctions forwarding the RF data transmitted from one RF transmitter to another. Mere relay function among the RF transmitters can be accomplished by adjusting the RF transmitters in the same frequency and coupling the Micro controllers with the unique Object ID. Henceforth RF transmitters can be used to operate in low frequency range as well as high frequency range depending upon its application. Frequency range switch is used to select the desired frequency range for each transmitter. Therefore grouping among RF re-transmitters can be achieved via this Frequency switch. Further flexibility can be incorporated into the system by making the RF transmitters function in dual mode that is as re-transmitters as well as transmitters transmitting its tagged object ID. RF receivers on the other hand detect the incoming Object IDs and transmit this Object ID along with its Location ID to the central system. RF transmitters consists of electronic devices that separate incoming radio signals from one another and use demodulator to convert specific signals into audio, video, or data formats. RF receivers display the unique object ID being detected; all the relevant data of this object are stored in the database along with its location. RF receivers are also available in market with various frequency ranges.

An object could be a stationary or a mobile object. It could represent any real world entity. The objects are incorporated with transmitters that are programmed with different frequencies and identified by a unique ID. The RF receivers are used to identify these objects within its localized region. The range of the region depends upon the receiver used. Database is synchronized with the GSM modem via central system. Link manager effectively manages the interface between the RF transmitters and RF receivers with the central system ensuring smooth functioning of the overall system. All the queries regarding the objects are sent to GSM SIM via the GSM network in Short Message Service (SMS) format there by facilitating the automatic functioning of the central system. A potential user at the querying end interacts with the central system via the GSM at the other end. If valid the queried operation is carried out by the central system, on completion it sends a SMS to the potential user containing the information relating to the queried object. The layered stack structure representing various parts of the system is shown in Figure 2. This stack structure can be divided into two parts - the hardware specification, which describes the protocols for RF transmitters, RF receivers and database, and the software specification, which provides protocols and function needed to query information from the GSM modem via the central system. The stack structure comprises of RF transmitters are integrated to the objects. They are programmed to use a particular frequency in the radio frequency range. An RF transmitter generally includes a modulator that modulates an input signal and a radio frequency power amplifier that is coupled to the modulator to amplify the modulated input signal. The radio frequency power amplifier is coupled to an antenna that transmits the amplified modulated input signal. Each RF transmitter transmits a unique ID that represents the object associated with it to the corresponding receiver. This data signal is transmitted using wireless network with TCP/IP and UDP protocol.

*Figure 2. Layered stack structure*

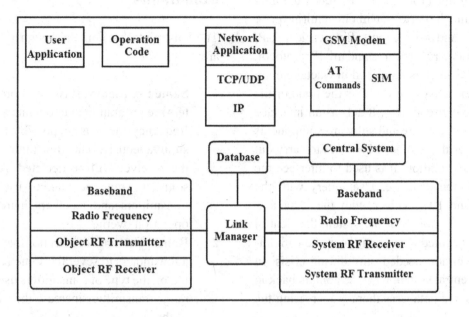

RF receivers are located in each region with a unique location ID. This Location ID is used to locate the region in which RF receiver is placed. RF receivers use wireless network to receive the signals transmitted by the RF transmitters within its sensing region. RF receiver displays the unique ID of the object detected along with its location ID. These IDs are sent to the front end. Visual studio along with programming language like Java,.NET can be used to create a front end forms to enter the description of the object and send it to the database. RF receivers detect all signals transmitted in the radio frequency range. Database comprises of secondary memory devices such as hard disk. Size of the database is directly proportional to the area of application. RF receivers are used to store the description and location of the objects in the database via central system and the GSM modem is used to query the objects from the database. The database can be created using software toolkit like MySQL. The database is integrated to RF receivers and GSM modem via central system using programming languages like. NET, J2ME, Java running on Windows or Linux Operating System (OS). Database should be frequently updated to provide valid appropriate data to the users of the system. Updating process can be done via the Update Use case.

Radio frequency specifies the air interface, which also includes the frequencies used, modulation technique employed and power transmitted. Radio frequency operates over a range of 3 KHz to 300 GHz. Baseband provides a description of basic connection establishment, packet formats, timing and basic Quality of Service (QoS) parameters used by the system. Link manager is used for link set-up and management of connections between devices like database, GSM modem, transmitters and receivers. Link manager performs functions such as synchronization, capability management, quality of service, power control, transmission mode and authentication. Therefore Link manager represents a software tool or software kit that is very important in providing trouble free operation of processing the queries and transmitting data across the system components. The network applications make use of standard TCP and UDP protocol. Transmission Control Protocol (TCP) sends the data directly between two components, and stay connected for the duration of the transfer.

User Datagram Protocol (UDP) sends the data packets it into the network without requiring prior connections and provides best effort service. TCP/UDP protocol runs over standard IP protocol. TCP/UDP over IP is described in Gaetano et al. (2004). Central system presents the brain or the core of the entire system. It acts as an interface between the hardware and software components and executes the various USE cases to carry out the desired operation. It is used to interface the GSM with the database, User query with the database and RF receiver with the Database. To increase flexibility of the system the Central system is operated in automation mode with human intervention needed only during crisis. In addition Central system will also contain backup processors and databases thereby increasing the reliability of the system.

GSM modem is integrated to the database via Central system. For example, .NET platform is used to interface GSM with the database. It makes use of standard AT modem commands for sending and receiving Short Message Service (SMS). SMS is an essential part of the GSM facilitating the automated operations of the central system. Each GSM modem is associated with a Subscriber Identity Module (SIM) that provides a number used by the user to query information about an object with the central system. All SMS are sent to the SIM of the GSM modem. Wireless GSM network is used for transmission of messages from the mobile station to the GSM modem. Multiple GSM modems can be incorporated to ease the work load on the system. A User interface consists of agent (potential user) who sends short message queries regarding the object using mobile devices such as cell phone and PDA. The queries are sent to the GSM modem using the Find and Query Use case. The GSM modem interacts with the database through the central system and sends back the corresponding information of the object to the user via the GSM network indicating that requested operation has been completed.

## Constraints

The constraints that limit the design of the system are:

1. **Same Frequency**: If two or more transmitters are programmed to operate at the same frequency, then it is not possible to run them simultaneously. The transmitter closer to the receiver will be detected. Alternative solution would be making use of tuned transmitters and receivers increasing the operation overhead.
2. **Range**: The range of detection may vary from a few meters to several kilometres depending on the type of transmitters used. Higher range transmitters increase the overall cost of the system.
3. **Reliability of Circuits**: Since electronic circuits are used, reliability is a factor of concern. The operation of equipments depends on the lifetime of hardware. Additional back up may be required increasing the overall cost.
4. **Power Consumption**: The power consumption is linearly proportional to the capacity of transmitters and receivers used. Higher the capacity greater is the consumption and vice versa.

## Simulation Model

A Simulation model is needed to develop and test designs before a physical prototype is constructed and to check the feasibility of the project. The simulator used is Proteus VSM Software (Brewer, Dellarocas, Colbrook, & Weihl, 1991). Proteus provides a full virtual debugging interface which is available in software, obviating the need for expensive hardware. The Proteus Design Suite offers unique offering to co-simulate high and low-level micro-controller code in the context of a mixed-mode SPICE circuit simulation. The simulation takes place in real time. This is a hardware simulation in which the inbuilt IC's of Proteus simulator are used to check the communication between the transmitter and the receiver. The data mapped to a particular transmitter is displayed on a LCD by means of a microcontroller. The functioning of the microcontroller can be found in Jochen (2003). The components used in the simulation model are as follows:

1.  **Encoder HT640**: The data given to the encoder is the ID for a transmitter. Two encoders are used in the simulation.
2.  **Decoder HT648**: The decoder is the receiver.
3.  **Microcontroller AT89C51**: The AT89C51 is a low-power, high-performance CMOS 8-bit microcomputer with 4K bytes of Flash Programmable and Erasable Read Only Memory (PEROM). Port 0 is used as input port to receive the transmitted data and Port 1 as output port to select the address lines for a particular data.
4.  **LCD LM016L**: The LCD to display which object is located.

The unique ID of the transmitters is given as input data to the encoder. The maximum length of the ID depends upon the Microcontroller chosen. Multiple micro processors and Micro controllers can be used for larger systems. The encoder of RF transmitter transmits the corresponding data to the decoder of the RF receiver. After decoding the received ID, the decoder forwards the ID to the microcontroller. The identified object along with its ID is displayed on the LCD connected to the RF receiver. Simulation resulted in high success rates and hence it is possible to implement it as a real world experiment. When none of the inputs to the encoder are connected and also buttons connecting the encoder and decoder are open, it suggests no transmission is taking place. Hence the LCD displays the 'NO OBJECT' message. When all of the inputs to the encoder are connected to VCC (high) and also buttons connecting encoder and decoder are closed, it suggests transmission is

taking place. The reception is detected and the LCD displays the 'VALID OBJECT 1' message. The results of the simulation are depicted in Figure 3.

## Analysis

In this section, the analysis of the performance under various test cases is carried out. These Test cases are used to test the efficiency, accuracy, speed and throughput of the system. Consider a system consisting of four RF transmitters and two RF receivers. RF transmitters have a unique ID's ID1, ID2, ID3, and ID4 respectively. RF receiver1 has a location ID R1 representing location R1 and RF receiver2 has a location ID R2 representing location R2. RF receivers are tuned into the same sensing range to detect the objects within the range. Consider the following cases (Table 4).

Summary of performance analysis of the system is given below.

*   **Efficiency**: The system is 80% efficient in detecting all objects within the sensing region. The drop in 20% is due to the network congestion and interference of signals during transmission leading to loss of data signal.
*   **Throughput**: In small scale application the system has a throughput of almost 100%. But in case of large scale system the throughput achieved is only 90% due to the network traffic.
*   **Accuracy**: Accuracy of the system is 100% in locating fixed objects. However in locating movable objects accuracy is only 85%, since it cannot provide pin point details of the location.
*   **Speed**: In small scale application the system has a high speed performance. But in large scale system the speed performance may drop due to network traffic and overall load of the system.

*Figure 3. Results of simulation*

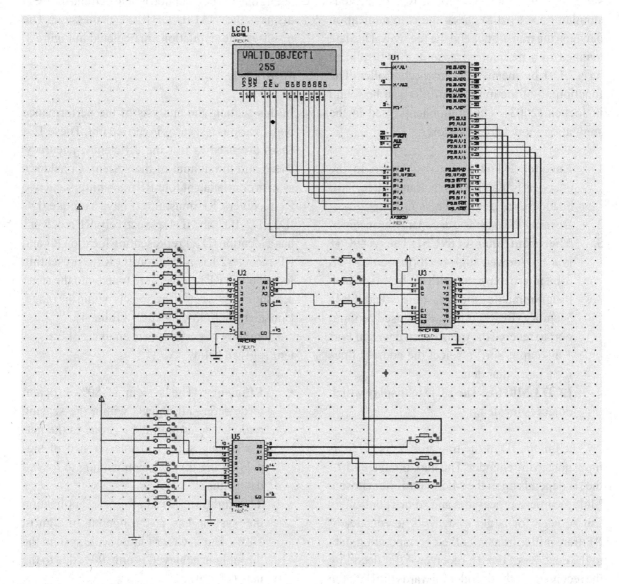

## SECURITY ESSENTIALS

An essential that surrounds any network based application is indeed the security aspect of it. Strengthening the security aspect of the system strengthens its application, thereby indicating a linear relation among the two. It is important to realize and ensure a reliable secured network based application, without which the entire system and its application tend to be meaningless. Therefore security presents sacrosanct part of the system, hence needs to be addressed with due diligence. Keeping this in view, little security related operations of the system will increase the overall credibility and reliability of the task being carried out. Three authenticated security process that is Iris scan, Finger print secured access and Encrypted password access is explained in the following sections:

*Table 4. Test cases and results*

| Test Cases | Result |
|---|---|
| All Objects within the sensing region R1. | All objects are detected in location R1 by Receiver R1. |
| All Objects within the sensing region R2. | All objects are detected in location R2 by Receiver R2. |
| Only Object ID3 and ID4 are within the sensing region R1. | Object ID3 and ID4 are detected in location R2 by receiver R2. |
| Object ID1 and ID3 are within the sensing region R1 and ID2 and ID4 are within the sensing region R1 | Object ID1 and ID3 are detected in location R1, and object ID2 and ID4 are detected in location R2. |
| Object ID4 is within the sensing region R1 and Object ID2 and ID3 are on the boundary of sensing region R2. | Object ID4 is detected in location R1, and object ID2 and ID3 are detected in location R2. |
| Object ID3 is within the sensing region R2 and Object ID1 and ID2 are on boundary of sensing region R1. | Object ID3 is detected in location R2, and object ID1 and ID2 are detected in location R1. |
| All Objects outside the sensing region R1 and R2. | No object is detected. |
| Initially object ID1 was in sensing Region R1, but later went to sensing region R2. | Object ID1 is currently detected at location R2. |

*Figure 4. Iris representation*

 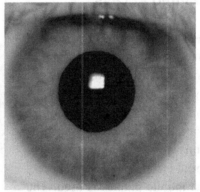

## Iris Scan

In today's world it is difficult to find a mobile device without being accompanied by a camera; an intelligent thought followed by a smart device could indeed revolutionize the camera technology in mobile devices. An Iris scanner cum Camera recorder could provide an authenticated and secured access to network applications. It is indeed a technological innovative thought that could extend the use of existing camera way beyond its intended purpose, providing a flexible system. The iris scan consists of three sections: Iris scanning, Iris bit converter and Iris matching. Details are represented below:

**Iris scanning**: The user iris along with the pupil is captured and sent to a central system for processing. For the purpose of capturing the Iris a camera cum Iris scanner should be integrated to the mobile device. An Iris is represented by two parts a inner pupil and outer ciliary body. Pupil is generally black in color and the remaining part of the iris is green, blue or brown depending on the individual's genetic makeup. Radial ridges extend from inner pupil to the periphery to carry blood to the Iris. The image of the Iris is captured in the infra red version during scanning producing a black and white image is shown in Figure 4.

**Iris bit converter**: On receiving the scanned image the central computer performs the bit analysis using Daugman's algorithm. The image captured consists of two concentric circle one of the pupil and other of the Iris. This algorithm converts the Iris image into 2048 bit representation consist of complex sign bits of the Gabor-domain representation and stores the corresponding data in the central computer. The bit image consists of complex numbers based on local amplitude and phase information of the Iris.

**Iris matching**: During the process of authentication the incoming scanned Iris image is complied into 2048 bit data and matched with the template of the stored data of the authenticate user. If only they match, user is provided access to the application. Otherwise access is denied.

## Fingerprint Secured Access

Finger prints, a well known forensic application used to distinguish every individual from another could be a very handy tool in the growing network world. As security being the core aspect of any network based application, secured finger print access would ensure a tamper proof system as desired. Keeping this in view, a unique technique using the finger print technology to gain access to the network via mobile phones is used. The fingerprint analysis consists of three sections: Fingerprint sensing, Fingerprint representation and analysis, and Fingerprint matching.

**Fingerprint sensing**: The most evident structural characteristic of a fingerprint is a pattern of interleaved ridges and valleys. In a fingerprint image, ridges are presented as dark lines and valleys by white lines. Ridges vary in width from 100 micro meters to 300 micro meters. When analyzed at the global level, the fingerprint pattern exhibits one or more regions, where the ridge lines assume distinct shapes that may be classified as one of the three topologies: loop, delta and whorl as given in (Gaetano et al., 2004). The characteristics of these topologies indicate that the fingerprint images depict various patterns about the center of the image. The aforementioned concepts are used

to put forward the idea of fingerprint recognition, by extracting features from bitmap images of the finger impressions. In result, a black and white image representing the fingerprint is produced as the end product of this phase.

**Fingerprint representation and analysis**: Black and white images when processed can be represented as an array of bits, i.e. matrix of 0's and 1's. The bit representation of image aids an effective analysis. The image is magnified according to a set template size in order to provide a detailed analysis. The obtained grayscale image is converted to black and white after processing the darker grayscale regions, which indicate the ridges on the fingerprint, as black and the lighter grayscale regions, which indicate the valleys on the fingerprint, as white. This converts a 8 bit per pixel image into a 1 bit per pixel image. Extended version of the scan-line rendering algorithm is used (Philippe & Marc, 1994) to scan through the image and obtain a graphical bit matrix:

Create an empty matrix $(X, Y)$ which is comparable to the size of the image, having elements with default values set to Zero. Scan the image for pixels row by row from top to bottom. For each scan line ($i$) find the intersections of the scan line with all curves of the fingerprint image as shown in Figure 5(a). ($ii$) Set the positions of the matrix, those correspond to these intersection points, to 1. The center of the fingerprint contour structure is obtained by finding the value y which represents the position of the scan-line consisting of maximum intersecting points obtained in the image, hence indicating the presence of maximum number of curves. Using the matrix data, the angular inclination of contours and the interleaved distance between the ridges and valleys of the corresponding fingerprint are calculated and stored along with the matrix values in the database. The center portion of the bit matrix is cropped out with a square dimension of distance 'a' units which can be represented by $(x/2 \pm a, y \pm a)$ covering only the central region of the fingerprint as shown in Figure 5(b).

*Figure 5. (a) Scan line intersections; (b) Rectangular cropping of major portions*

a                                  b

In order to ensure high security and avoid erroneous matching of data, the database is filled with fingerprints of the user in various orientations and positions. Many such square region inputs are taken from the valid user and the train process using a neural network is used to obtain a threshold value which identifies the fingerprint's matrix values.

**Fingerprint matching**: The new input fingerprint image is analyzed and represented in the form of a matrix using the extended scan-line algorithm. Since the central region would only provide the basic design of the fingerprint, its features such as interleaved distance and angular details are matched with that of the data stored by the train process in the database. The resulting value should at least match 90% of the threshold value in order to authenticate the user as a valid user. Once the user is authenticated as valid, the user is allowed to operate the device. If the requested user's fingerprint image is not matched, the user is denied access. In this way, security is enabled to provide an authenticated access to operate the electronic devices

**Encrypted Password access**: Simple and well established technique of accessing the network via user chosen password coupled with encryption provides a robust secured access to the network. Encryption at the user end converting user password into a cipher followed by decryption at central system ensures the authenticity of the user accessing the network. Standard encryption and

decryption algorithm is available in two types that are Symmetric and Asymmetric. In case of Symmetric encryption algorithm same key is used at both encryption and decryption process but in case of Asymmetric algorithm different key is used at encryption and decryption process. Asymmetric algorithm key is further classified into public and private key depending on whether the key has been made public or not. It's convenient to apply both Asymmetric and Symmetric key algorithm in the system presented above. In order to use the encryption and decryption algorithm the potential user has to abide the following steps:

1. **Registration**: The potential user is required to send a Symmetric encryption key to the central system by encrypting it with its publicly available Asymmetric key via the network. On receiving this encrypted message the central system applies its private Asymmetric decryption key to obtain the users Symmetric encryption key. Now the central system stores the Symmetric encryption key of this user for further authenticated communication between them.

2. **Password exchange**: The potential user creates a suitable password and encrypts in using the common exchanged symmetric key. This encrypted message is transmitted to the central system via the network. On receiving this message, the central system decrypts it and stores the password of the

user for further authenticated communication between them.

3. **Authentication**: Henceforth, for each access to the network user is required to make use of the password, thereby ensuring an authenticated communication. The password chosen by the user should be used before initiating a communication with the central system on each occasion, thereby increasing the robustness of the system. On authentication the communication between the user and central system will be active for a fixed time period.

It is important to note that the above mentioned security essentials can be incorporated into the system based on the importance of the application. Access to important data can be done with more stringent security access and for simpler systems a less stringent security access can be utilized. Combination of the above security essentials will increase the security aspect of the authenticated access thereby making the system more robust and reliable.

Simplicity of the system with an omnipresent activity indicates the true potential and capability of the system. The user's authenticated password, Iris data and finger print data are all stored in the central data base via the central system. The essence of security essentials is that it enables the network and the system to be accessed only by the authenticated individual.

## APPLICATIONS

The system being developed has a wide range of applications. Based on the capacity of transmitters and receiver, the usability can range from a small house or building to a large area such as a city. Some of the applications are as follows:

1. **Locating landmarks**: A transmitter can be attached to each landmark and its location and related description can be stored in a database. A visitor to the place can query from any mobile phone and get information about the location.

2. **Warehouse management**: Finding an item in a huge warehouse is a task that can be accomplished using the design and hence help in organizing the overall management activity.

3. **Helpline for emergency services**: A helpline number attached to the GSM can be used by people to locate hospitals, fire station and police station in case of emergencies. It also ensures a 24/7 service via the automated central system.

4. **Personal object management**: Objects used by people on a daily basis like spectacles, wallet, and so on may get displaced very often. Using transmitter and receiver, these objects can be tracked

5. **Enterprise support**: Within an enterprise it can be used to locate libraries, canteens, departments and offices. Thereby facilitates the movement of an individual by coordinating his directions with appropriate activity.

The system comprises of wide range of applicability and flexibility, thereby making it possible to be implemented in a small possible dimension like a room to a large dimension like a city. The beauty of the system lies in its evolving nature which allows it to change its basic organisation and structure based on application requirements.

## CONCLUSION

In present competitive world innovation with simplicity and reliability encompasses success. Thereby a comprehensive inexpensive system for operating essential and everyday electronic objects relying on mobile phones as omnipresent object-sensing devices is presented. Significant effort was spent on testing the hardware circumstances of detecting the objects using Proteus simulator. The results are encouraging but due to the cost constraints, implementation was not done on large scale. However, the simulation result envisages that this product could ease many activities done every day in many sectors. A highly flexible wireless controlled system would enable us to bridge the distance between the user and the electronic device and hence provide an easier access to real world entities, by just maneuvering the mobile phones, without the need to be physically present near the object. Also, it can be enhanced to ensure an authenticated access to these resources there by taking care of the security aspect. High degree of flexibility and convenience encourages more people to incorporate electronic devices in their day to day activities. The ease of tracking the device and its relevant information through inexpensive GSM technology provides vital strength to the system. In conclusion, an innovative multidimensional technological system is used to make life comfortable and simple. Since the advent of mobile technologies and the reliability of the network the above discussed concept is futuristic. As part of future enhancements the above mentioned applications can be implemented to explore the real time constraints, further applications can also be explored.

## REFERENCES

Behzad, R. (1999). *RF transmitter architectures and circuits.* In IEEE 1999, Custom Integrated Circuits Conference.

Bietti, I., Svelto, F., & Castello, R. (2000). Towards fully integrated CMOS RF receivers. In *Gallium Arsenide Applications Symposium*, GASS 2000, 2-6, Oct 2000.

Brewer, E. A., Dellarocas, C. N., Colbrook, A., & Weihl, W. E. (1991). *Proteus: A high-performance parallel-architecture simulator.* Cambridge, MA: Massachusetts Institute of Technology.

Carle, J., & Simplot, D. (2004). Energy-efficient area monitoring for sensor networks. *IEEE Computer Journal, 37*(2), 40–46. doi:10.1109/MC.2004.1266294

Christian, F., Philipp, B., Christof, R., & Wolfgang, K. (2008). Objects calling home: Locating objects using mobile phones. In *Proceedings of the 5th International Conference on Pervasive and Mobile Computing* (vol. 4, Issue 3, pp. 421-427). Toronto, Canada.

Eisenblatter, A., Geerdes, H.-F., & Turke, U. (2005). Public UMTS radio network evaluation and planning scenarios. *International Journal on Mobile Network Design and Innovation, 1*(1), 40-53. ISSN: 1744-2869

Gaetano, B., Waylon, B., Matthew, H., Carl, H., & Cameron, T. (2004). Reminding about tagged objects using passive RFIDs. In *Proceedings of the 6th International Conference on Ubiquitous Computing.* Nottingham, UK.

Jochen, S. (2003). *Mobile communications.* Pearson Education Limited.

Koller, D., & Daniilidis, K. (1992). Model-based object-tracking in traffic scenes in Computer Vision. In *Second European Conference on Computer Vision.* Santa Margherita Ligure, Italy.

Lee, J. T., & Wang, Y. T. (2003). Efficient data mining for calling path patterns in GSM networks. *Information Systems, 28*(8), 929–948. doi:10.1016/S0306-4379(02)00112-6

Lin, C. Y., Peng, W. C., & Tseng, Y. C. (2006). Efficient in-network moving object tracking in wireless sensor networks. *IEEE Transactions on Mobile Computing, 5*(8).

Padmanabhan, B., & Tuzhilin, A. (1996). Pattern discovery in temporal databases: A temporal logic approach. In *Proceedings of the Second International Conference on Knowledge Discovery and Data Mining.*

Philippe, L., & Marc, L. (1994). Stanford University, Fast volume rendering using a shear-warp factorization of the viewing transformation. In *SIGGRAPH '94 Proceedings of the 21st Annual Conference on Computer Graphics and Interactive Techniques.*

Stetzler, T. D., Havens, J. H., & Koyama, M. (1995). *A 2.7-4.5 V single chip GSM transceiver RF integrated circuit.* AT&T Bell Labs, Reading, PA. Dec 1995.

Strange, J., & Atkinson, S. (2000). *A direct conversion transceiver for multi-band GSM application in radio frequency integrated circuits.* In RFIC Symposium.

Tseng, V. S., & Lin, K. W. (2005). Energy efficient strategies for object tracking in sensor networks: A data mining approach. In *Proceedings of the International Workshop on Ubiquitous Data Management* (held with ICDE'05), April, 2005. Tokyo, Japan.

Xu, Y., Winter, J., & Lee, W. C. (2004). Prediction-based strategies for energy saving in object tracking sensor networks. In *Proceedings of the Fifth IEEE International Conference on Mobile Data Management (MDM'04)* (pp. 346–357).

## ADDITIONAL READING

CAMEL. (2006). *Intelligent Networks for the GSM, GPRS and UMTS Network Rogier Noldus.* John Wiley & Sons Ltd.

Christian, F., Christof, R., Philiip, B., Chie, N., & Wolfgang, K. (2007). A service architecture for monitoring physical objects using mobile phones. In *Proceedings of the 7th International Workshop.* Santander, Spain, May 2007.

Frank, S., & Christian, F. (2003). Interaction in Pervasive Computing Settings using Bluetooth-enabled Active Tags and Passive RFID Technology together with Mobile Phones. In *Proceedings of the First IEEE International Conference on Pervasive Computing and Communications* (pp. 378).

Han, J., & Kamber, M. (2000). *Data mining: Concepts and Techniques.* Morgan Kaufman Publishers.

Hara, T., Murakami, N., & Nishio, S. (2004). Replica allocation for correlated data items in Ad hoc Sensor Networks. *SIGMOD Record, 33*(1), 38–43. doi:10.1145/974121.974128

Heinzelman, W. R., Chandrakasan, A., & Balakrishnan, H. (2000). Energy- efficient communication protocol for Wireless Microsensor Networks. In *Proceedings of the 33rd Hawaii International Conference on System Sciences.*

Phillip, B. Gibbons, Brad Karp, Yan Ke, Suman Nath, & Srinivasan Seshan.(2007) IrisNet: An Architecture for a worldwide sensor web. In *IEEE Pervasive Computing, IEEE Multimedia* (Vol. 14, No. 4, pp. 8-13).

Tseng, S. M., & Tsui, C. F. (2004). Mining multi-level and location-aware associated service patterns in mobile environments. In *IEEE Transactions on Systems, Man and Cybernetics: Part B 34 (6).*

Tseng, Y. C., Kuo, S. P., Lee, H. W., & Huang, C. F. (2004). Location tracking in a wireless sensor network by mobile agents and its data fusion strategies. *The Computer Journal 47 (4). WINS Project.* Rockwell Science Center, http://wins.rsc.rockwell.com.

Woo, A., & Culler, D. (2001). A transmission control scheme for media access in sensor networks. In: *Proceedings of Seventh ACM Annual International Conference on Mobile Computing and Networking (Mobicom'01)* (pp. 221–235).

Yang, Q., Li, T., & Wang, K. (2004). Building association rule based sequential classifiers for web document prediction. *Journal of Data mining and Knowledge Discovery, 8 (3)*, 253–273.

Ye, W., Heidemann, J., & Estrin, D. (2002). An energy-efficient mac protocol for wireless sensor networks. In *Proc. of the 21st IEEE Infocom*, pp. 1567–1576.

## KEY TERMS AND DEFINITIONS

**GSM (Global System for Mobile Communications):** Most widely used digital mobile telephony system that uses a variation of Time Division Multiple Access (TDMA).

**GSM MODEM:** (GSM Modulator and Demodulator): A specialized type of modem which accepts a SIM card, and operates over a subscription to a mobile operator, just like a mobile phone. From the mobile operator perspective, a GSM modem looks just like a mobile phone.

**GSM SIM:** GSM Subscriber Information Module is a programmable card that stores all of a cell phone subscriber's personal information and phone settings. It is otherwise known as a smart card.

**LC:** L(inductor) C(capacitor): An electric circuit where capacitor is charged to a potential difference by connecting it across a battery and then is allowed to discharge through an inductor.

**PDU (Protocol Description Unit):** Secure method of sending and receiving SMS messages.

**SPICE Circuit Simulation:** General-purpose open source analog electronic circuit simulator. It is a powerful program that is used in integrated circuit and board-level design to check the integrity of circuit designs and to predict circuit behavior.

**TCP/UDP (Transmission Control Protocol/ User Datagram Protocol):** Protocols developed for the internet to get data from one network device to another.

**Use Cases:** Description of Potential Series of Interactions between two entities.

**VCO (Voltage Controlled Oscillators-Voice Carry Over):** A service that relays voice communication to those illiterate and deaf.

# Compilation of References

Abadi, M., & Fournet, C. (2001). Mobile values, new values, and secure communications. In *Proceedings of 28th ACM SIGPLAN-SIGACT Symposium on Principles of Programming Languages (POPL'01)* (pp. 104-115).

Abadi, D. J., Marcus, A., Madden, S. R., & Hollenbach, K. (2009). Sw-store: A vertically partitioned DBMS for Semantic Web data management. *The VLDB Journal*, *18*(2), 385–406. doi:10.1007/s00778-008-0125-y

Abadi, M., Blanchet, B., & Fournet, C. (2007). Just fast keying in the Pi calculus. [TISSEC]. *ACM Transactions on Information and System Security*, *10*(3), 1–59. Retrieved December 24, 2010 doi:10.1145/1266977.1266978

Abadi, M., & Gordon, A. D. (1999). A calculus for cryptographic protocols: The Spi-Calculus. *Journal of Information and Computation*, *148*(1), 1–70. doi:10.1006/inco.1998.2740

Abowd, G. D., Dey, A. K., Brown, P. J., Davies, N., Smith, M., & Steggles, P. (1999). Towards a better understanding of context and context-awareness. In *Huc '99: Proceedings of the 1st International Symposium on Handheld and Ubiquitous Computing* (pp. 304-307). London, UK: Springer-Verlag.

Abowd, A., Atkeson, C., Hong, J., Long, S., & Pinkerton, M. (1997). Cyberguide: A mobile context-aware tour guide. *Wireless Networks*, *3*(5), 421–433. doi:10.1023/A:1019194325861

Adomavicius, G., Sankaranarayanan, R., Sen, S., & Tuzhilin, A. (2005). Incorporating contextual information in recommender systems using a multidimensional approach. *ACM Transactions on Information Systems*, *23*(1), 103–145. doi:10.1145/1055709.1055714

Adomavicius, G., & Tuzhilin, A. (2005). Towards the next generation of recommender systems: A survey of the state-of-the-art and possible extensions. *IEEE Transactions on Knowledge and Data Engineering*, *17*(6). doi:10.1109/TKDE.2005.99

Adomavicius, G., & Tuzhilin, A. (2010). Context-aware recommender systems. In Kantor, P., Ricci, F., Rokach, L., & Shapira, B. (Eds.), *Recommender systems handbook: A complete guide for research scientists and practitioners*. Berlin, Germany: Springer.

Adrian, P. (2001). Wireless integrated microsystems will enhance information gathering and transmission. *Sensor Business Digest* (September issue).

Aglets, I. B. M. (2002). *Aglets documentation home page*. Retrieved January 2, 2011, from http://www.trl.ibm.com/aglets/documentation_e.htm

Agrawal, D. P., Deng, H., Poosarla, R., & Sanyal, S. (2003). Secure mobile computing. *Invited Talk, Proceedings of the 4th International Workshop on Distributed Computing (IWDC 2003)*, December 28-31, 2003, Kolkata, India.

Agrawal, D. P., & Zeng, Q.-A. (2002). *Introduction to wireless and mobile systems*. Brooks/Cole Publisher.

Agrawal, M., Rao, H. R., & Sanders, G. L. (2003). Impact of mobile computing terminals in police work. *Journal of Organizational Computing and Electronic Commerce*, *13*(2), 73–89. doi:10.1207/S15327744JOCE1302_1

Akter, S., D'Ambra, J., & Ray, P. (2011). Trustworthiness in mHealth information services: An assessment of a hierarchical model with mediating and moderating effects using Partial Least Squares (PLS). *Journal of the American Society for Information Science and Technology*, *62*(1). doi:10.1002/asi.21442

Al- Jaljouli. R. (2006). *A proposed security protocol for data gathering mobile agents*. Masters Thesis Dissertation. University of New South Wales, School of Computer Science and Engineering, Australia.

Algesheimer, J., Cachin, C., Camenisesh, J., & Karjoth, G. (2000). *Cryptographic security for mobile code*. IBM Research Report, Zurich, Switzerland.

Al-Jaljouli, R. (2005). Boosting m-business using a truly secured protocol for information gathering mobile agents. In *Proceedings of 4th International Conference on Mobile Business*. IEEE Computer Society Press.

Al-Jaljouli, R. (2005). *Formal methods in the enhancement of the data security protocols of mobile agents* (Technical Report TR 520). University of New South Wales, School of Computer Science and Engineering, Australia. Retrieved on December 24, 2010, from http://cgi.cse.unsw.edu.au/~reports

Al-Jaljouli, R. (2006). *A proposed security protocol: security protocol protocol* for data gathering mobile agents. Master's Dissertation, University of New South Wales, School of Computer Science and Engineering, Australia.

Al-Jaljouli, R., & Abawajy, J. (2010). Negotiation strategy for mobile agent-based e-negotiation. In *Proceedings of the 13th International Conference on Principle and Practice of Information Mobile Agents (PRIMA 2010)*.

Al-Jaljouli, R., & Abawajy, J. (2009). *Agents based e-commerce and securing exchanged information. Pervasive Computing: Innovations in Intelligent Multimedia and Applications, Computer Communications and Networks Series* (pp. 383–404). Berlin, Germany: Springer-Verlag.

Alzoubi, K. M., Wan, P-J., & Frieder, O. (2002). Distributed heuristics for connected dominating set in wireless ad hoc networks. *IEEE / KICS Journal on Communication Networks, 4*(1), 22-29.

Anagnostopoulos, C. B., Ntarladimas, Y., & Hadjiefthymiades, S. (2007). Situational computing: An innovative architecture with imprecise reasoning. *Journal of Systems and Software, 80*(12), 1993–2014. doi:10.1016/j.jss.2007.03.003

Anagnostopoulos, C. B., Tsounis, A., & Hadjiefthymiades, S. (2007). Context awareness in mobile computing environments. *Wireless Personal Communications: An International Journal, 42*(3), 445–464. doi:10.1007/s11277-006-9187-6

Anand, S. S., & Mobasher, B. (2003). *Intelligent techniques for Web personalization*. Revised Selected Papers of the 2nd Workshop on Intelligent Techniques in Web Personalization (ITWP 2003), Heidelberg, Germany: Springer.

Anderson & Perin. (2009). *Case studies from the Vital Wave mHealth report*. Retrieved November 10, 2010, from http://www.cs.washington.edu/homes/anderson/docs/2009/ mHealthAnalysis_v1.pdf

*Android developer's guide*. (2010). Retrieved December 13, 2010, from http://developer.android.com/guide/index.html

Anegg, H., Kunczier, H., Michlmayr, E., Pospischil, G., & Umlauft, M. (2002). LoL@: Designing a location based UMTS application. *Elektrotechnik und Informationstechnik, 119*(2), 48–51.

Arkko, J., Deverapalli, V., & Dupont F. (2004). *Using IPsec to protect mobile IPv6 signaling between mobile nodes and home agents*. IETF RFC 3776.

Arkko, J., Kempf, J., Zill, B., & Nikander, P. (2005). *Secure network discovery*. IETF RFC 3971.

Auer, S., Bizer, C., Kobilarov, G., Lehmann, J., & Zachary, I. (2007). DBpedia: A nucleus for a Web of open data. In *Proceedings of the 6th International Semantic Web Conference ISWC* (pp. 11-15).

Austin, G. L., & Bellon, A. (1982). Very-short-range forecasting of precipitation by the objective extrapolation of radar and satellite data. In Browning, K. A. (Ed.), *Nowcasting* (pp. 177–190). Academic Press.

Aziz, B., Gray, D., Hamilton, G., Oehl, F., Power, J., & Sinclair, D. (2001). Implementing protocol verification for e-commerce. In *Proceedings of International Conference on Advances in Infrastructure for Electronic Business, Science, and Education on the Internet (SSGRR 2001)*.

Baader, F., Calvanese, D., McGuinness, D. L., Nardi, D., & Patel-Schneider, P. F. (Eds.). (2003). *The description logic handbook: Theory, implementation, and applications*. Cambridge University Press.

Bader, R., Neufeld, E., Woerndl, W., & Prinz, V. (2011). Context-aware POI recommendations in an automotive scenario using multi-criteria decision making methods. *Workshop on Context-aware Retrieval and Recommendation (CaRR 2011), Conference on Intelligent User Interfaces (IUI 2011).* Palo Alto, CA, USA.

Bader, R., Woerndl, W., & Prinz, V. (2010). *Situation awareness for proactive in-car recommendations of points-of-interest (POI).* Workshop Context Aware Intelligent Assistance (CAIA 2010), 33rd Annual German Conference on Artificial Intelligence (KI 2010). Karlsruhe, Germany.

Baldauf, M., Dustdar, S., & Rosenberg, F. (2007). A survey on context-aware systems. *International Journal of Ad Hoc and Ubiquitous Computing, 2*(4), 263–277. doi:10.1504/IJAHUC.2007.014070

Baldessari, R., Ernst, T., Festag, A., & Lenardi, M. (2009). *IETF draft.* draft-ietf-mext-nemo-ro-automotive-req-02.

Baldessari, R., Festag, A., & Abeille, J. (2007a). *NEMO meets VANET: A deployability analysis of network mobility in vehicular communications.* International Conference on ITS Telecommunications (ITST). Sophia. France.

Baldessari, R., Festag, A., & Lenardi, M. (2007b). *C2C-C consortium requirements for usage of NEMO in VANETs.* IETF draft, draft-baldessari-c2ccc-nemo-req-00.

Banks, K. (2007). Then came the Nigerian elections: The story of frontline SMS. *SAUTI: The Stanford Journal of African Studies, (Spring/Fall)*, 1–4.

Banno, A., & Teraoka, F. (2006). *LIN6: An efficient network mobility protocol in IPv6. Information Networking: Advances in Data Communication and Wireless Networks, LNCS 3961* (pp. 3–10). Berlin, Germany: Springer-Verlag.

Becker, C., & Bizer, C. (2008). *DBpedia mobile: A location-enabled linked data browser.* In Workshop on linked data on the Web (LDOW 2008).

Becker, C., & Bizer, C. (2009b). *Marbles.* Retrieved from http://marbles.sourceforge.net/

Becker, C., & Bizer, C. (2009a). Exploring the geospatial semantic web with dbpedia mobile. *Journal of Web Semantics, 7*(4), 278–286. doi:10.1016/j.websem.2009.09.004

Becket, D. (2004). *RDF/XML syntax specication* (W3C recommendation 10 February 2004). Retrieved November 19, 2010, from http://www.w3.org/TR/2004/REC-rdf-syntax-grammar-20040210

Beckett, D., & Berners-Lee, T. (2008). *Turtle - Terse RDF triple language* (W3C Team Submission 14 January 2008). Retrieved November 25, 2010, from http://www.w3.org/TeamSubmission/turtle/

Behzad, R. (1999). *RF transmitter architectures and circuits.* In IEEE 1999, Custom Integrated Circuits Conference.

Beigl, M., Gray, P., & Salber, D. (2001). Location modeling for ubiquitous computing. *Workshop Proceedings, Ubicomp 2001.* Atlanta.

Bellifemine, F., Poggi, A., & Rimassa, G. (1999). Jade - FIPA - complaint agent framework. In *Proceedings of 4th International Conference on Practical Applications and Intelligent Agents and Multi-Agents (PAAM)*, (pp. 97-108).

Bellon, A., & Austin, G. L. (1978). The evaluation of two years of real-time operation of a short-term precipitation Forecasting Procedure (Sharp). *Journal of Applied Meteorology, 17*, 1778–1787. doi:10.1175/1520-0450(1978)017<1778:TEOTYO>2.0.CO;2

Benjamin, S. G., Brown, J. M., Brundage, K. J., Devenyi, D., Schwartz, B. E., & Smirnova, T. G. … Dimego, G. L. (1998). The operational RUC-2. In *Sixteenth Conference on Weather Analysis and Forecasting*, (pp. 249-252).

Berenson, H., Bernstein, P., Gray, J., Melton, J., O'Neil, E., & O'Neil, P. (1995). A critique of ANSI SQL isolation levels. In *Proceedings of ACM SIGMOD Conference* (pp. 1-10).

Berg, M., Wariero, J., & Modi, V. (2009). *Every child counts - The use of SMS in Kenya to support community based management of acute malnutrition and malaria in children under five.* Retrieved December 25, 2010, from http://www.childcount.org/reports/ChildCount_Kenya_InitialReport.pdf

Bergman, M. K. (2005). *Untapped assets: The $3 trillion value of U.S. enterprise documents*. BrightPlanet Corporation White Paper, July 2005, 42 pp.

Berkovsky, S., Kuflik, T., & Ricci, F. (2007). *Distributed collaborative filtering with domain specialization*. ACM Conference on Recommender Systems. Minneapolis, MN, USA.

Bernardos, C. J., Soto, I., Calderón, M., Boavida, F., & Azcorra, A. (2007). VARON: Vehicular ad hoc route optimization for NEMO. *Computer Communications, 30*, 1765–1784. doi:10.1016/j.comcom.2007.02.011

Berners-Lee, T., Fielding, R., & Masinter, L. (2005). *Uniform resource identier (URI): Generic syntax (RFC 3986)*. Retrieved 11th December, 2010, from http://www.faqs.org/rfcs/rfc3986.html

Berners-Lee, T., Hendler, J., & Lassila, O. (2001). The Semantic Web. *Scientific American, 284*(5), 34–43. doi:10.1038/scientificamerican0501-34

Bernstein, P., Hsu, M., & Mann, B. (1990). Implementing recoverable requests using queues. *SIGMOD Record, •••*, 112–122. doi:10.1145/93605.98721

Bhagwat, P., Perkins, C., & Tripathi, S. (1996). Network layer mobility: An architecture and survey. *IEEE Personal Communications, 3*(3), 54–64. doi:10.1109/98.511765

Biegel, G., & Cahill, V. (2004). A framework for developing mobile, context-aware applications. In *Proceedings of the Second IEEE International Conference on Pervasive Computing and Communications (PerCom '04)* (pp. 361-365). Washington, DC, USA.

Bieszczad, A., White, T., & Pagurek, B. (1998). Mobile agents for network management. *Journal of IEEE Communications Surveys, 1*(1).

Bietti, I., Svelto, F., & Castello, R. (2000). Towards fully integrated CMOS RF receivers. In *Gallium Arsenide Applications Symposium*, GASS 2000, 2-6, Oct 2000.

Bijsmans, J. (2000). *Future mobile computing. Research Report 7.4.2000, Hewlett Packard Laboratories Bristol*. HP Invent, Hewlett-Packard Company.

Birnkammerer, S., Woerndl, W., & Groh, G. (2009). *Recommending for groups in decentralized collaborative filtering*. Tech. Report TUM-I0927, Institut fuer Informatik, Technische Universitaet Muenchen, Germany.

Bizer, C. (2009). The emerging Web of linked data. *IEEE Intelligent Systems, 24*, 87–92. doi:10.1109/MIS.2009.102

Bizer, C., Heath, T., & Berners-Lee, T. (2009). Linked data - The story so far. *International Journal on Semantic Web and Information Systems, 5*(3), 1–22. doi:10.4018/jswis.2009081901

Bjerknes, G., Ehn, P., Kyng, M., & Nygaard, K. (1987). *Computers and democracy: A Scandinavian challenge*. Gower Pub Co.

Blanchet, B., & Aziz, B. (2003). A calculus for secure mobility. In *Proceedings of 18th Asian Computing Science Conference (ASIAN '03), Lecture Notes in Computer Science, 2896* (pp.188-204). Springer-Verlag.

Boardman, R., & Sasse, M. A. (2004). Stuff goes into the computer and doesn't come out: A cross-tool study of personal information management. In *Proceedings of Sigchi Conference on Human Factors in Computing Systems* (pp. 583-590). New York, NY: ACM.

Boehm, S., Koolwaaij, J., & Luther, M. (2008). Share whatever you like. *Electronic Communication of the EASST (ECEASST)*, 11.

Boehm, S., Koolwaaij, J., Luther, M., Souville, B., Wagner, M., & Wibbels, M. (2008). Introducing IYOUIT. In *The Semantic Web* (pp. 804–817). ISWC.

Bolchini, C., Curino, C. A., Quintarelli, E., Schreiber, F. A., & Tanca, L. (2007). A data-oriented survey of context models. *SIGMOD Record, 36*(4), 19–26. doi:10.1145/1361348.1361353

Boley, H., Hallmark, G., Kifer, M., Paschke, A., Polleres, A., & Reynolds, D. (2010). *RIF core dialect (W3C recommendation 22 June 2010)* [Computer software manual]. Retrieved from http:// www.w3.org/TR/rif-core/

Borchers, A., Herlocker, J., Konstan, J., & Riedl, J. (1998). Ganging up on information overload. *Computer, 31*(4), 106–108. doi:10.1109/2.666847

Boreale, M. (2001). Symbolic trace analysis of cryptographic protocols. In *Proceedings of 28ᵗʰ International Colloquium on Automata, Languages and Programming (ICALP), Lecture Notes in Computer Science, 2076* (pp.667-681). Springer-Verlag.

Boreale, M., & Buscemi, M. (2001). *STA: A tool for trace analysis of cryptographic protocols - ML object code and examples*. Retrieved December 24, 2010, from http://www.dsi.unifi.it/~boreale/tool.html

Boreale, M., & Buscemi, M. (2002a). A framework for the analysis of security protocols. In *Proceedings of the 13ᵗʰ International Conference on Concurrency Theory (CONCUR)* (pp.483-498). *Lecture Notes in Computer Science, 2076* (pp. 667-681). Heidelberg, Germany: Springer-Verlag.

Boreale, M., & Buscemi, M. (2002b). Experimenting with STA, a tool for automatic analysis of security protocols. *In Proceedings of ACM Symposium on Applied Computing (SAC),* (pp. 281-285). ACM Press.

Boreale, M., & Gorla, D. (2002). Process calculi and the verification of security protocols. *Journal of Telecommunications and Information Technology – Special Issue on Cryptographic Protocol Verification (JTIT), 4*, 28-40.

Bournelle, J., Valadon, G., Binet, D., Zrelli, S., Laurent-Maknavicius, M., & Combes, J. M. (2006). *AAA considerations within several NEMO deployment scenarios*. International Workshop on Network Mobility (WONEMO). Sendai, Japan.

Boyer, J. (2007) *XForms 1.0* (3ʳᵈ ed.). W3C. Retrieved December 25, 2010, from http://www.w3.org/TR/xforms/

Braa, J., & Humberto, M. (2007). *Building collaborative networks in Africa on Health Information Systems and open source software development–Experiences from the HISP/BEANISH Network*. BEANISH Network.

Braa, K., & Purkayastha, S. (2010). Sustainable mobile information infrastructures in low resource settings. *Studies in Health Technology and Informatics, 157*, 127.

Bradley, N. A., & Dunlop, M. D. (2005). Toward a multidisciplinary model of context to support context-aware computing. *Human-Computer Interaction, 20*(4), 403–446. doi:10.1207/s15327051hci2004_2

Bradshaw, J. (1997). An introduction to software agents. In Bradshaw, J. M. (Ed.), *Software agents* (pp. 3–46). AAAI Press.

Breibart, Y., Garcia-Molina, H., & Silberschatz, A. (1992). Overview of multidatabase transaction management. *The VLDB Journal, 2*, 181–239. doi:10.1007/BF01231700

Brewer, E. A., Dellarocas, C. N., Colbrook, A., & Weihl, W. E. (1991). *Proteus: A high-performance parallel-architecture simulator*. Cambridge, MA: Massachusetts Institute of Technology.

Brewington, B., Gray, R., Moizumi, K., Kotz, D., Cybenko, G., & Rus, D. (1999). Mobile agents for distributed information retrieval. In Klusch, M. (Ed.), *Intelligent agents* (pp. 355–395). Springer-Verlag. doi:10.1007/978-3-642-60018-0_19

Brickley, D., & Guha, R. (2004). RDF vocabulary description language 1.0: RDF schema (W3C Recommendation February 10, 2004) [Computer software manual].

Brickley, D., & Miller, L. (2007). *FOAF vocabulary specication 0.91*.

Broekstra, J., Kampman, A., & van Harmelen, F. (2002). Sesame: A generic architecture for storing and querying RDF and RDF schema. In I. Horrocks & J. Hendler (Eds.), *Proceedings of the First International Semantic Web Conference* (pp. 54-68). Springer Verlag.

Brusilovsky, P., Kobsa, A., & Nejdl, W. (2007). *The adaptive Web*. Berlin, Germany: Springer. doi:10.1007/978-3-540-72079-9

Burke, R. (2007). Hybrid Web recommender systems. In Brusilovsky, P., Kobsa, A., & Nejdl, W. (Eds.), *The adaptive Web* (pp. 377–408). Berlin, Germany: Springer. doi:10.1007/978-3-540-72079-9_12

Burrows, M., Abadi, M., & Needham, R. (1990). Logic of authentication. *ACM Transactions on Computer Systems, 8*(1), 18–36. doi:10.1145/77648.77649

Butenko, S., Cheng, X., Du, D.-Z., & Pardalos, P. M. (2003a). On the construction of virtual backbone for ad hoc wireless networks. In Butenko, S., Murphey, R., & Pardalos, P. M. (Eds.), *Cooperative control: Models, applications and algorithms* (pp. 43–54). Berlin, Germany: Springer Publishers.

Butenko, S., Cheng, X., Oliviera, C., & Pardalos, P. M. (2003b). A new heuristic for the minimum connected dominating set problem on ad hoc wireless networks. In Butenko, S., Murphey, R., & Pardalos, P. M. (Eds.), *Recent developments in cooperative control and optimization* (pp. 61–73). Berlin, Germany: Springer Publishers. doi:10.1007/978-1-4613-0219-3_4

Buyukkokten, O., Garcia-Molina, H., & Paepcke. A., (2000). Focused Web searching with PDAs. *Computer Networks – The International Journal of Computer Telecommunications Networking, 33*(1-6), 213-230.

Cai, G., & Xue, Y. (2006). Activity-oriented context-aware adaptation assisting mobile geo-spatial activities. In *IUI '06: Proceedings of the 11th International Conference on Intelligent User Interfaces* (pp. 354-356). New York, NY: ACM.

Calderón, M., Bernardos, C. J., & Bagnulo, M. (2005). Securing route optimization in NEMO. *International Symposium on Modeling and Optimization in Mobile, Ad Hoc and Wireless Networks, WIOPT2005* (pp.248).

Caldwell, D., & Koch, J. L. (1998). *Mobile computing and its impact on the changing nature of work and organization*. Leavey School of Business and Administration. Working Paper. Santa Clara University, Santa Clara, CA.

Capra, L., Emmerich, W., & Mascolo, C. (2002), Middleware for mobile computing. In *Tutorial Proceedings of International Conference of Networking 2002*, Springer (pp 1-39).

Carle, J., & Simplot, D. (2004). Energy-efficient area monitoring for sensor networks. *IEEE Computer Journal, 37*(2), 40–46. doi:10.1109/MC.2004.1266294

Carnegie Melon University. (2000). *Symbolic model prover*. Retrieved December 24, 2010, from http://www-2.cs.cmu.edu/~modelcheck/symp.html

Carroll, J. J., Bizer, C., Hayes, P., & Stickler, P. (2005). Named graphs. *Journal of Web Semantics, 3*(4), 247–267. doi:10.1016/j.websem.2005.09.001

Carton, A., Clarke, S., Senart, A., & Cahill, V. (2007). Aspect-oriented model-driven development for mobile context-aware computing. In *SEPCASE '07: Proceedings of the 1st International Workshop on Software Engineering for Pervasive Computing Applications, Systems, and Environments*. Washington, DC: IEEE Computer Society.

Chalmers, D., & Sloman, M. (1999). A survey of quality of service in mobile computing environments. *IEEE Communications Surveys. Second Quarter, 1999*, 1–10.

Chan, V., Ray, P., & Parameswaran, N. (2008). Mobile e-health monitoring: An agent based approach. *Telemedicine and e-Health Communication Systems, 2*(2), 223–230.

Chen, H. (2004). *An intelligent broker architecture for pervasive context-aware systems*. Unpublished doctoral dissertation, University of Maryland, Baltimore County.

Chen, W., Lin, C., & Lien, Y. (1999). A mobile agent infrastructure with mobility and management support. In *Proceedings of Workshop on Parallel Processing*, (pp. 508–513).

Chen, B., Cheng, H., & Palen, J. (2009). Integrating mobile agent technology with multi-agent systems for distributed traffic detection and management systems. *Journal of Transportation Research Part C: Emerging Technologies, 17*(1), 1–10. doi:10.1016/j.trc.2008.04.003

Cheng, G., & Qu, Y. (2009). Searching linked objects with falcons: Approach, implementation and evaluation. *International Journal on Semantic Web and Information Systems, 5*(3), 49–70. doi:10.4018/jswis.2009081903

Cheng-Zhong, X., & Song, F. (2003). Privilege delegation and agent-oriented access control in Naplet. In *Proceedings of IEEE ICDCS Workshop on Mobile Distributed Computing (MDC)*, (pp. 493-497).

Chen, H., & Joshi, A. (2004). An ontology for context-aware pervasive computing environments. *Special Issue on Ontologies for Distributed Systems. The Knowledge Engineering Review, 18*(3), 197–207. doi:10.1017/S0269888904000025

Chen, L., & Pu, P. (2009). Interaction design guidelines on critiquing-based recommender systems. [UMUAI]. *User Modeling and User-Adapted Interaction, 19*(3), 167–206. doi:10.1007/s11257-008-9057-x

Chess, D., Harrison, C., & Kershenbaum, A. (1997). Mobile agents: Are they a good idea? *Mobile Object Systems: Towards the Programmable Internet, LNCS, 1222,* 25–47. Springer-Verlag.

Chien, B.-C., He, S.-Y., Tsai, H.-C., & Hsueh, Y.-K. (2010). An extendible context-aware service system for mobile computing. *Journal of Mobile Multimedia, 6*(1), 49–62.

Choy, M. C., Srinivasan, D., & Cheu, R. L. (2003). Cooperative, hybrid agent architecture for real-time traffic signal control. *Journal of IEEE Transactions on Systems. Man and Cybernetics Part A: System and Humans, 33*(5), 597–607. doi:10.1109/TSMCA.2003.817394

Christian, F., Philipp, B., Christof, R., & Wolfgang, K. (2008). Objects calling home: Locating objects using mobile phones. In *Proceedings of the 5ᵗʰ International Conference on Pervasive and Mobile Computing* (vol. 4, Issue 3, pp. 421-427). Toronto, Canada.

Consulting, V. W. (2009). *mHealth for development: The opportunity of mobile technology for healthcare in the developing world.* Washington, DC: UN Foundation-Vodafone Foundation Partnership.

Cooperative Vehicle-to-Infrastructure System. (2006). *CVIS project.* Retrieved from http://www.cvisproject.org

Cormen, T. H., Leiserson, C. E., Rivest, R. L., & Stein, C. (2001). *Introduction to algorithms.* Cambridge, MA: MIT Press.

Corradi, A., Montanari, R., & Stefanelli, C. (1999a). Mobile agents integrity in e-commerce applications. In *Proceedings of the 19ᵗʰ IEEE International Conference on Distributed Computing Systems Workshop (ICDCS'99),* (pp. 59-64). IEEE Computer Society Press.

Corradi, A., Montanari, R., & Stefanelli, C. (1999b). Mobile agents protection in the Internet environment. In *Proceedings of the 23ʳᵈ Annual International Computer Software and Applications Conference (COMPSAC '99)* (pp. 80 - 85).

Costa, R., Correia, A., & Moital, M. (2008). *The determinants of intention to adopt mobile electronic tourist guides.* First International Meeting for Tourism Management: The Public and Private Sector (EIGTUR 2008). Minas Gerais, Ouro Preto, Brazil.

Coutaz, J., Crowley, J. L., Dobson, S., & Garlan, D. (2005). Context is key. *Communications of the ACM - Special Issue. The Disappearing Computer, 48*(3), 49–53.

Crounse, B. (2006). *Mobile devices usher in new era in healthcare delivery: House calls for health professionals.* Microsoft Corporation.

Crumbley, J. (2003). *Trust us.* Access Control and Security systems, Government Security. Retrieved March 23, 2011 from http://securitysolutions.com/mag/security_trust_us/index.html

Cugola, G., Ghezzi, C., Picco, G., & Vigna, G. (1997). Analyzing mobile code languages. *Mobile Object Systems: Towards the Programmable Internet,* LNCS 1222, Springer-Verlag, 93-110.

Dahlbom, B., & Ljungberg, F. (1998). Mobile informatics. *Scandinavian Journal of Information Systems, 10*(1&2), 227–234.

Davies, N., Blair, G. S., Cheverst, K., & Friday, A. (1996). Supporting collaborative applications in a heterogeneous mobile environment. [Elsevier.]. *Computer Communications Special Issue on Mobile Computing, 19,* 346–358.

Davies, N., Friday, A., Blair, G. S., & Cheverst, K. (1996). Distributed systems support for adaptive mobile applications. [ACM Press.]. *Mobile Networks and Applications, 1*(4), 399–408.

Deshpande, M., & Karypis, G. (2004). Item-based top-n recommendation algorithms. *ACM Transactions on Information Systems, 22*(1), 143–177. doi:10.1145/963770.963776

Devarapalli, V., Wakikawa, R., Petrescu, A., & Thubert, P. (2005). *Network mobility (NEMO) basic support protocol.* IETF RFC 3963.

Dey, A. K. (2000). *Providing architectural support for building context-aware applications.* Unpublished doctoral dissertation, Georgia Institute of Technology.

Dey, A. K., Abowd, G. D., Pinkerton, M., & Wood, A. (1997). Cyberdesk: A framework for providing self-integrating ubiquitous software services. In *ACM Symposium on User Interface Software and Technology* (pp. 75-76).

Dey, A. K. (2001). Understanding and using context. *Personal and Ubiquitous Computing, 5*(1), 4–7. doi:10.1007/s007790170019

Dey, A. K., Abowd, G. D., & Salber, D. (2001). A conceptual framework and a toolkit for supporting the rapid prototyping of context-aware-applications. *Human-Computer Interaction, 16*(2), 97–166. doi:10.1207/S15327051HCI16234_02

Diaper, D. (1990). *Task analysis for human-computer interaction.* Upper Saddle River, NJ: Prentice Hall PTR.

Documentation, S. T. A. (2001). *Symbolic trace analyzer.* Retrieved December 24, 2010, from http://www.dsi.unifi.it/~boreale/documentation.html

Dömel, P. (1997). Interaction of Java and Telescript agents. *Mobile Object Systems: Towards the Programmable Internet, LNCS, 1222,* 295–314. Springer-Verlag.

Donner, J. (2008). Research approaches to mobile use in the developing world: A review of the literature. *The Information Society, 24*(3), 140–159. doi:10.1080/01972240802019970

Dourish, P. (2004). What we talk about when we talk about context. *Personal and Ubiquitous Computing, 8*(1), 19–30. doi:10.1007/s00779-003-0253-8

Ducheneaut, N., Partridge, K., Huang, Q., Price, R., Roberts, M., Chi, E. H., et al. (2009). *Collaborative filtering is not enough? Experiments with a mixed-model recommender for leisure activities.* 17th International Conference on User Modeling, Adaptation, and Personalization (UMAP). Trento, Italy.

Dunham, M. H., & Helal, A. (1995). Mobile computing and databases: Anything new? *SIGMOD Record, 24*(4), 5–9. doi:10.1145/219713.219727

Durante, L., Sisto, R., & Valenzano, A. (2000). A state-exploration technique for Spi-calculus testing equivalence verification. *In Proceedings of the IFIP International Joint Conference on Formal Description Techniques for Distributed Systems and Communication Protocols (FORTE XIII) and Protocol Specification, Testing and Verification (PSTV XX),* (pp. 155-170). Dordrecht, The Netherlands: Kluwer Academic Publishers.

Eddy, W., Ivanvic, W., & Davis, T. (2009). *Network mobility route optimization requirements for operational use in aeronautics and space exploration mobile networks.* IETF RFC 5522.

Egenhofer, M. J. (2002). Toward the semantic geospatial Web. In Voisard, A., & Chen, S.-C. (Eds.), *ACM-GIS* (pp. 1–4). ACM.

Eisenblatter, A., Geerdes, H.-F., & Turke, U. (2005). Public UMTS radio network evaluation and planning scenarios. *International Journal on Mobile Network Design and Innovation, 1*(1), 40-53. ISSN: 1744-2869

Electric, M. (1997). Concordia: An infrastructure for collaborating mobile agents. In *Proceedings of 1st International Workshop on Mobile Agents (MA 97).*

Erling, O., & Mikhailov, I. (2007). RDF support in the Virtuoso DBMS. In S. Auer, C. Bizer, C. Müller, & A. V. Zhdanova (Eds.), *1st Conference on Social Semantic Web,* (vol. 113, pp. 59-68). Leipzig, Germany.

Ernst, T. (2001). *Network mobility support in IPv6.* Doctoral Dissertation, University Joseph Fourier.

Ernst, T., & Charbon, J. (2004). *Multi-homing with NEMO basic support.* In International Conference on Mobile Computing and Ubiquitous Computing.

Ernst, T., & Lach. H.-Y. (2007). *Network mobility support terminology.* IETF RFC 4885.

ETSI EN 302 665. (2010). *Intelligent transport systems (ITS): Communications architecture.*

Euzenat, J. (2005). *Alignment infrastructure for ontology mediation and other applications* (*Vol. 168,* pp. 81–95). Mediate.

Euzenat, J., Pierson, J., & Ramparany, F. (2008). Dynamic context management for pervasive applications. *The Knowledge Engineering Review, 23*(1), 21–49. doi:10.1017/S0269888907001269

Fall, K., & Varadhan, K. (2001). *NS notes and documentation.* Unpublished Manual – The VINT Project at LBL, Xerox PARC, UCB, and USC/ISI.

Farmer, W., Guttman, J., & Swarup, V. (1996). Security for mobile agents: Issues and requirements. In *Proceedings of 19th National Information Systems Security Conference*, (pp. 591-597).

Fathi, H., Shin, S.-H., Kobara, K., & Imai, H. (2007). Secure AAA and mobility for nested mobile networks. In *International Conference on Intelligent Transportation Systems Telecommunications, ITST'07* (pp. 1-6). IEEE Communications Society.

Fathi, H., Shin, S.-H., Kobara, K., & Chakraborty, S.-S. (2006). LR-AKE-based AAA for network mobility (NEMO) over wireless links. *IEEE Journal on Selected Areas in Communications, 24*(9), 1725–1737. doi:10.1109/JSAC.2006.875111

Federal Communications Commission (FCC). (2010). *Report and order on open Internet rules* Federal Communications Commission, December 2010, Washington, DC. Retrieved from http://www.fcc.gov/Daily_Releases/Daily_Business/2010/db1223/FCC-10-201A1.pdf

Feller, J., & Fitzgerald, B. (2002). *Understanding open source software development*. Boston, MA: Addison-Wesley Longman Publishing Co., Inc.

Fielding, R., Gettys, J., Mogul, J., Frystyk, H., Masinter, L., Leach, P., et al. (1999). *Hypertext transfer protocol - HTTP/1.1 (RFC 2616)* [Computer software manual].

Fiore, M., & Abadi, M. (2001). Computing symbolic models for verifying cryptographic protocols. In *Proceedings of the 14th IEEE Computer Security Foundations Workshop (CSFW 2001)*, (pp. 160-173). IEEE Computer Society Press.

Fischer, L. (2003). *Protecting integrity and secrecy of mobile agents on trusted and non-trusted agent places*. Thesis Dissertation, University of Bremen, Germany.

Fontelo, P. A., & Chismas, W. G. (2005). PDAs, handheld devices and wireless healthcare environments: Minitrack introduction. *Proceedings of the 38th Hawaii International Conference on System Sciences*.

for UMTS. (2002). In Paterno, F. (Ed.), *Mobile Human-Computer Interaction* (pp. 140–154). Berlin, Germany: Springer.

Formal Systems (Europe) Ltd. (2000). *Failures divergence refinement*. FDR2 user manual. Retrieved from http://www.formal.demon.co.uk/fdr2manual/index.html

Forman, G. H., & Zahorjan, J. (1994). The challenges of mobile computing. *Computer, 27*(4), 38–47. doi:10.1109/2.274999

Fournier, D., Mokhtar, S. B., Georgantas, N., & Issarny, V. (2006). Towards ad hoc contextual services for pervasive computing. In *MW4SOC '06: Proceedings of 1st Workshop on Middleware for Service Oriented Computing (MW4SOC 2006)* (pp. 36-41). New York, NY: ACM.

Frank, L. (2008). Architecture for mobile ERP systems. In *Proceedings of 7th International Conference on Applications and Principles of Information Science (APIS2008)* (pp.412-415).

Frank, L. (2010b). Architecture for integrated mobile logistics management and control. In *Proceedings of 2nd International Conference on Information Technology Convergence and Services (ITCS 2010)*. IEEE Computer Society.

Frank, L. (2011b). Architecture for integrating heterogeneous distributed databases using supplier integrated e-commerce systems as an example. In *Proceedings of the International Conference on Computer and Management (CAMAN 2011)*. Wuhan, China.

Frank, L. (2011c). Countermeasures against consistency anomalies in distributed integrated databases with relaxed ACID properties. In *Proceedings of Innovations in Information Technology (Innovations 2011)*. Abu Dhabi, UAE.

Frank, L., & Andersen, S. K. (2010). Evaluation of different database designs for integration of heterogeneous distributed electronic health records. In *Proceedings of the International Conference on Complex Medical Engineering (CME2010)*. IEEE Computer Society.

Frank, L. (1985). *Databaser*. Copenhagen: Samfundslitteratur.

Frank, L. (1988). *Database theory and practice*. Addison-Wesley.

Frank, L. (2003). *Patent application*. Copenhagen Business School.

Frank, L. (2005). Replication methods and their properties. In Rivero, L. C., Doorn, J. H., & Ferraggine, V. E. (Eds.), *Encyclopedia of database technologies and applications*. Hershey, PA: Idea Group Inc.doi:10.4018/978-1-59140-560-3.ch092

Frank, L. (2010a). *Design of distributed integrated heterogeneous or mobile databases* (pp. 1–157). Germany: LAP LAMBERT Academic Publishing AG & Co. KG.

Frank, L. (2011a). Architecture for ERP system integration with heterogeneous e-government modules. In Chhabra, S., & Kumar, M. (Eds.), *Strategic enterprise resource planning models for e-government: Applications & methodologies*. Hershey, PA: IGI Global. doi:10.4018/978-1-60960-863-7.ch007

Frank, L., & Pape-Haugaard, L. (2011). *Integration of health records by using relaxed ACID properties between hospitals, physicians and mobile units like ambulances and doctors. International Journal of Handheld Computing Research (IJHCR), 2(4)*. IGI Global.

Frank, L., & Zahle, T. (1998). Semantic ACID properties in multidatabases using remote procedure calls and update propagations. *Software, Practice & Experience, 28*, 77–98. doi:10.1002/(SICI)1097-024X(199801)28:1<77::AID-SPE148>3.0.CO;2-R

Franklin, S., & Graesser, A. (2006). Is it an agent, or just a program? A taxonomy for autonomous agents. *Intelligent Agents III Agent Theories, Architectures, and Languages. LNCS, 1193*, 21–35. Springer-Verlag.

Frank, R. (2000). *Understanding smart sensors* (2nd ed.). Norwood, MA: Artech House.

Franz, T., Ansgar, S., & Staab, S. (2009). Are semantic desktops better? Summative evaluation comparing a semantic against a conventional desktop. *In Proceedings of Fifth International Conference on Knowledge Capture* (pp. 1-8). New York, NY: ACM

Fuchs, M., Rasinger, J., & Höpken, W. (2007). Exploring information services for mobile tourist guides – Results from an expert survey. In Dimanche, F. (Ed.), *Tourism mobility & technology* (pp. 4-14). Travel and Tourism Research Association (TTRA), Nice, France.

Gaddah, A., & Thomas Kunz, T. (2003). *A survey of middleware paradigms for mobile computing.* Carleton University Systems and Computing Engineering Technical Report SCE-03-16, July 2003

Gaetano, B., Waylon, B., Matthew, H., Carl, H., & Cameron, T. (2004). Reminding about tagged objects using passive RFIDs. In *Proceedings of the 6th International Conference on Ubiquitous Computing*. Nottingham, UK.

Ganapathy, K., & Ravindra, A. (2008). *mHealth: A potential tool for health care delivery in India*. Rockefeller Foundation.

Garcia-Molina, H., & Salem, K. (1987). Sagas. In *ACM SIGMOD Conference* (pp. 249-259).

Gellersen, H. W., Schmidt, A., & Beigl, M. (2002). Multi-sensor context-awareness in mobile devices and smart artifacts. *Mobile Networks and Applications, 7*(5). doi:10.1023/A:1016587515822

Giansiracusa, M. (2003). *Mobile agent protection mechanisms and the trusted proxy server architecture (TAPS). Technical Report*. Information Security Research Centre, Queensland University of Technology.

Glass, G. (1999). *Object space voyager core package technical overview*, (pp. 611-627). Mobility, ACM Press/Addison-Wesley Publishing Co.

Goerss, J. S., Velden, C. S., & Hawkins, J. D. (1998). The impact of multispectral GOES-8 wind information on Atlantic tropical cyclone track forecasts in 1995. *Monthly Weather Review, part II. Nogaps Forecasts, 126*, 1219–1227.

Gold, J. (2005). *Managing mobility in the enterprise*. A J. Gold Associates White Paper, July 2005. Retrieved from www.jgoldassociates.com

Grace-Martin, M., & Gay, G. (2001). Web browsing, mobile computing and academic performance. *Journal of Educational Technology & Society, 4*(3), 95–107.

Gray, R. (1996). Agent TCL: A flexible and secure mobile-agent system. In *Proceedings of 4th Annual TCL/TK Workshop (TCL)*.

Gray, R. S. (2000). Soldiers, agents and wireless networks: A report on a military application. In *Proceedings of the 5th International Conference and Exhibition on the Practical Application of Intelligent Agents and Multi-Agents (PAAM 2000)*.

Gray, R. S., Cybenko, G., Kotz, D., Peterson, R. A., & Rus, D. (2002). D'Agents: Applications and performance of a mobile-agent system. *Journal of Software - Practice and Experience, 32*(6), 543-573.

Gray, J., & Reuter, A. (1992). *Transaction processing: Concepts and techniques* (1st ed.). San Francisco, CA: Morgan Kaufmann Publishers Inc.

Greenbaum, J. M., & Kyng, M. (1991). *Design at work: Cooperative design of computer systems*. CRC.

Grosz, B. J., & Kraus, S. (1996). Collaborative plans for complex group action. *Artificial Intelligence, 86*(2), 269–357. doi:10.1016/0004-3702(95)00103-4

Grove, T. (2003). *Summary analysis: The final HIPAA security rule*. HIPAA Advisory, White Paper, Phoenix Health Systems, February 2003.

Grun, C., Werthner, H., Proll, B., Retschitzegger, W., & Schwinger, W. (2008). *Assisting tourists on the move – An evaluation of mobile tourist guides*. In 7th International Conference on Mobile Business (ICMB '08). Barcelona, Spain.

Gundavelli, S., Devarapalli, V., Chowdhury, K., & Patial, B. (2008). *Proxy mobile IPv6*. IETF RFC 5213.

Gupta, A. K. (2008). Challenges of mobile computing. *Proceedings of 2nd National Conference on Challenges & Opportunities in Information Technology (COIT-2008), RIMT-IET* (pp. 86-90). Mandi Gobindgarh, March 29, 2008.

Gupta, R. (2011). A SURVEY of comparative study of mobile agent platforms. *Journal of Engineering Science Systems, 3*(3), 1943–1948.

Hameed, K. (2003). The application of mobile computing and technology to health care services. *Telematics and Informatics, 20*, 99–106. doi:10.1016/S0736-5853(02)00018-7

Hannotin, X., Maggi, P., & Sisto, R. (2001). Lecture Notes in Computer Science: *Vol. 2240. Formal specification and verification of mobile agent data integrity properties: A case study* (pp. 42–53). Springer-Verlag.

Hanseth, O., & Monteiro, E. (2001). *Understanding information infrastructure*. Retrieved from http://www.ifi.uio.no/*oleha/Publications/bok.html

Hanseth, O., & Ciborra, C. (2007). *Risk, complexity and ICT*. Edward Elgar Publishing.

Harris, S., & Gibbins, N. (2003). 3store: Efficient bulk RDF storage. In *Practical and Scalable Semantic Systems, Proceedings of the First International Workshop on Practical and Scalable Semantic Systems*.

Hashimi, S. Y., Komatineni, S., & MacLean, D. (2010). *Pro Android 2*. Apress. doi:10.1007/978-1-4302-2660-4

Havenstein, H. (2005). *Health care: Doctors and PDAs proved a good match, helping give the industry an early lead with wireless*. Cerner Bridge Medical, Computerworld News.

Heath, T., Dzbor, M., & Motta, E. (2005). Supporting user tasks and context: Challenges for Semantic

Heath, T., Motta, E., & Dzbor, M. (2005). Context as a foundation for a semantic desktop. In S. Decker, J. Park, D. Quan, & L. Sauermann (Eds.), *Proceedings of the Semantic Desktop Workshop at the ISWC*, Galway, Ireland, (vol. 175).

Heath, T., & Motta, E. (2008). Revyu: Linking reviews and ratings into the Web of data. *Journal of Web Semantics: Science. Services and Agents on the World Wide Web, 6*, 266–273. doi:10.1016/j.websem.2008.09.003

Henricksen, K., Indulska, J., & Rakotonirainy, A. (2002). Modeling context information in pervasive computing systems. In *Pervasive '02: Proceedings of First International Conference on Pervasive Computing* (pp. 167-180). London, UK: Springer-Verlag.

Henricksen, K., Indulska, J., McFadden, T., & Balasubramaniam, S. (2005). *Middleware for distributed context-aware systems*. In OTM Conference.

Herlocker, J., Konstan, J. A., Borchers, A., & Riedl, J. (1999). *An algorithmic framework for performing collaborative filtering*. In 22nd Annual International ACM SIGIR Conference on Research and Development in Information Retrieval. Berkeley, CA.

Herlocker, J., Konstan, J. A., Terveen, L. G., & Riedl, J. T. (2004). Evaluating collaborative filtering recommender systems. *ACM Transactions on Information Systems, 22,* 5–53. doi:10.1145/963770.963772

Herrera, T. (2006). *Solutions for Health Insurance Portability and Accountability Act (HIPAA) compliance.* White Paper, Juniper Networks, Inc.

Hertel, A., Broekstra, J., & Stuckenschmidt, H. (2008). *RDF storage and retrieval systems.* Online. Retrieved from http://ki.informatik.uni-mannheim.de/fileadmin/publication/Hertel08RDFStorage.pdf

Hinze, A., & Buchanan, G. (2006). The challenge of creating cooperating mobile services: Experiences and lessons learned. In *ACSC '06: Proceedings of the 29th Australasian Computer Science Conference* (pp. 207-215). Darlinghurst, Australia.

Hofer, T., Schwinger, W., Pichler, M., Leonhartsberger, G., Altmann, J., & Retschitzegger, W. (2003). Context-awareness on mobile devices - The Hydrogen approach. In *HICSS '03: Proceedings of 36th Annual Hawaii International Conference on System Sciences (HICSS'03).* Washington, DC, USA: IEEE Computer Society.

Hofmann-Wellenhof, B., Lichtenegger, H., & Collins, J. (2004). *Global positioning system: Theory and practice.* Berlin, Germany: Springer Publishers.

Howard, A., Mataric, M. J., & Sukhatme, G. J. K. (2002). An incremental self deployment algorithm for mobile sensor networks. *Autonomous Robots, 13*(2), 113–126. doi:10.1023/A:1019625207705

Hu, D. H., Dong, F., & Wang, C.-L. (2009). A semantic context management framework on mobile device. In *IC-ESS 2009: Proceedings of the 2009 International Conference on Embedded Software and Systems* (pp. 331-338). Washington, DC: IEEE Computer Society.

Huang, C. M., Lee, C. H., & Tseng, P. H. (2009a). Multihomed SIP-based network mobility using IEEE 802.21 media independent handover. In *IEEE International Conference on Communications ICC2009,* (pp. 1-5).

Huang, C. M., Lee, C. H., & Tseng, P. H. (2009b). Multiple router management for SIP-based network mobility. In *International Symposium on Computers and Communications, ISCC2009* (pp. 863-868). IEEE Communications Society.

Huang, C., Tseng, Y., & Lo, L. (2004). *The coverage problem in three-dimensional wireless sensor networks.* In Global Telecommunications Conference, GLOBECOM '04.

Huang, C. M., Lee, C. H., & Zheng, J. R. (2006). A novel SIP-based route optimization for network mobility. *IEEE Journal on Selected Areas in Communications, 24*(9), 1682–1691. doi:10.1109/JSAC.2006.875113

Huebscher, C., & McCann, A. (2005). An adaptive middleware framework for context-aware applications. *Personal and Ubiquitous Computing, 10*(1), 12–20. doi:10.1007/s00779-005-0035-6

IEEE Std. 802.21. (2008). *IEEE standard for local and metropolitan area networks-Part 21: Media independent handover.* May 2008. ISO-21217-CALM-Architecture. (2010). *Intelligent transport systems - Communications access for land mobiles (CALM) - Architecture.*

IEEE. (1999). *LAN standards of the IEEE Computer Society.* Wireless LAN Medium Access Control (MAC) and PHysical Layer (PHY) specification. IEEE Standard 802.11, 1999 Edition.

Indulska, J., & Sutton, P. (2003). Location management in pervasive systems. In *ACSW Frontiers '03: Proceedings of Australasian Information Security Workshop Conference on ACSW Frontiers 2003* (pp. 143-151). Darlinghurst, Australia: Australian Computer Society, Inc.

Irizarry, J. (2008). Potential applications of emerging portable computing platform for information sharing in construction projects. *Proceedings International Conference on Construction and Real Estate Management, (CD-ROM), ICCREM,* Toronto, Canada.

ITU (International Telecommunication Union). (2009). *World telecommunication/ICT indicators database 2009.* Geneva, Switzerland: International Telecommunication Union (ITU). Retrieved from www.itu.int/ITU-D/ict

ITU (International Telecommunication Union). (2010). *World telecommunication/ICT indicators database 2010.* Geneva, Switzerland: International Telecommunication Union (ITU). Retrieved from www.itu.int/ITU-D/ict

Jacobsson, M., Rost, M., & Holmquist, J. E. (2006). *When media gets wise: Collaborative filtering with mobile media agents.* In 11th International Conference on Intelligent User Interfaces (IUI '06). Sydney, Australia.

James, G., Cohen, D., Dodier, R., Platt, G., & Palmer, D. (2006). A deployed multi-agent framework for distributed energy applications. In *the Proceedings of the 5th International Joint Conference on Autonomous Agents and Multi-agents Systems (AAMAS 2006).*

Jannach, D., Zanker, M., Felfernig, A., & Friedrich, G. (2010). *Recommender systems: An introduction.* Cambridge University Press.

Jansen, W., & Karygiannis, T. (2000). Mobile agent security. *Special publication 800-19 of National Institute of Standards & Technology (NIST),* 1-43.

Jansen, W. (2000). *Countermeasures for mobile agent security.* Computer Communications, Special Issue on Advances in Research and Applications of Network Security.

Javan, N. T., & Dehghan, M. (2007). Reducing end-to-end delay in multi-path routing algorithms for mobile ad hoc networks. In Zhang, H., Olariu, S., Cao, J., & Johnson, D. B. (Eds.), *Proceedings of the International Conference on Mobile Ad hoc and Sensor Networks, Lecture Notes in Computer Science: Vol. 4864* (pp. 703-712). Berlin, Germany: Springer Publishers.

Jemma, I. B., Tsukada, M., Menouar, H., & Ernst, T. (2010). *Validation and evaluation of NEMO in VANET using geographic routing.* In International Conference on Intelligent Transportation Systems Telecommunications, ITST. Kyoto, Japan.

Jiang, Y., Shi, M., & Shen, X. (2006). Multiple key sharing and distribution scheme with (n,t) threshold for NEMO group communications. *IEEE Journal on Selected Areas in Communications, 24*(9), 1738–1747. doi:10.1109/JSAC.2006.875114

Jo, M., & Inamura, H. (2008). Secure route optimization for mobile network node using secure address proxying. *IEEE Network Operations and Management Symposium NOMS2008, 7,* 137-143. IEEE Communications Society.

Jochen, S. (2003). *Mobile communications.* Pearson Education Limited.

Johansen, D., Renesse, R., & Schneider, F. (1995). Operating system support for mobile agents. In *Proceedings of 5th IEEE Workshop on Hot Topics in Operating Systems (HotOS-V),* (pp. 42-45).

Johnson, D., Perkins, C., & Arkko, J. (2004). *Mobility support for IPv6.* IETF RFC 3775.

Johnson, D. B., Maltz, D. A., & Broch, J. (2001). *DSR: The dynamic source routing protocol for multi-hop wireless ad hoc networks. Ad Hoc Networking* (pp. 139–172). Boston, MA: Addison Wesley.

Jones, M., & Marsden, G. (2005). *Mobile interaction design.* John Wiley & Sons.

Jung, S., Zhao, F., Felix Wu, S., & Kim, H. (2004). *Threat analysis of network mobility (NEMO). Information and Communication Security, LNCS (Vol. 3264,* pp. 333–337). Berlin, Germany: Springer-Verlag.

Kaenampornpan, M., & Ay, B. B. (2004). *An intergrated context model: Bringing activity to context.* In Workshop on Advanced Context Modelling, Reasoning and Management - UBICOMP.

Kakihara, M., & Sorensen, C. (2006). Practicing mobile professional work: Tales of locational, operational, and interactional mobility. *Emerald, 6*(3), 180–187.

Kamvar, M., & Baluja, S. (2006). A large scale study of wireless search behavior: Google mobile search. In *CHI '06: Proceedings of Sigchi Conference on Human Factors in Computing Systems* (pp. 701-709).

Kantor, P., Ricci, F., Rokach, L., & Shapira, B. (2010). *Recommender systems handbook: A complete guide for research scientists and practitioners*. Berlin, Germany: Springer.

Kar, K., & Banerjee, S. (2003). Node placement for connected coverage in sensor networks. In *Proceedings of the Workshop on Modeling and optimization in Mobile, Ad hoc and Wireless Networks (WIPOT'03)*. Sophia Antipolis, France, 2003.

Karagiannis, G., Wakikawa, R., Kenny, J., Bernardos, C. J., & Kargl, F. (2010). *Traffic safety applications requirements*. IETF Draft. Draft-karagiannis-traffic-safety-requirements-02.

Karjoth, G. (2000). Secure mobile-based merchant brokering in distributed marketplaces. *Agent Systems, Mobile Agents, and Applications. LNCS, 1882*, 427–441. Springer-Verlag.

Karjoth, G., Asokan, N., & Gülcü, C. (1998). Protecting the computation results of free-roaming agents. *Journal of Personal and Ubiquitous Computing, 2*(2), 92–99.

Karjoth, G., Lange, D., & Oshima, M. (1997). A security model for Aglets. *IEEE Internet Computing, 1*(4), 68–77. doi:10.1109/4236.612220

Karnik, N. (1998). *Security in mobile agent systems*. PhD dissertation, Department of Computer Science and Engineering, University of Minnesota.

Karnik, N., & Tripathi, A. (1999). *Security in the Ajanta mobile agent system* (Technical Report TR-5-99). University of Minnesota, Minneapolis.

Karygiannis, T., & Owens, L. (2002). *Wireless network security 802.11: Bluetooth and handheld devices*. NIST Special Publication 800-848.

Katz, R. H. (1994). Adaptation and mobility in wireless Information Systems. *IEEE Personal Communication, 1*(1), 6-17.

Kearney, A. T. (2001). *Network publishing: Creating value through digital content*. ATKearney White Paper, April 2001, 32 pp.

Kent, S. (2005). *IP encapsulating security payload (ESP)*. IETF RFC 4303.

Kenteris, M., Gavalas, D., & Economou, D. (2008). Evaluation of mobile tourist guides. In *First World Summit on the Knowledge Society (WSKS 2008)* (pp. 603-610). Athens, Greece.

Kenteris, M., Gavalas, D., & Economou, D. (2010). Mytilene e-guide: A multiplatform mobile application tourist guide exemplar. *Multimedia tools and applications*. ISSN 1380-7501 (in press).

Kenteris, M., Gavalas, D., & Economou, D. (2011). Electronic mobile guides: A survey. *Personal and Ubiquitous Computing, 15*(1), 97–111. doi:10.1007/s00779-010-0295-7

Kim, H., & Moreau, T. (2001). Mobile code for key propagation. In *Proceedings of 1st International Workshop on Security of Mobile Multi-Agent Systems (SEMAS'2001)*.

Kim, S. W., Park, S. H., Lee, J., Jin, Y. K., Park, H.-M., & Chung, A. (2004). Sensible appliances: Applying context-awareness to appliance design. *Personal and Ubiquitous Computing, 8*(3-4), 184–191. doi:10.1007/s00779-004-0276-9

Kin, T. K., & Samsudin, A. (2007). *Efficient NEMO security management via CA-PKI*. In International Conference on Telecommunications and Malaysia International Conference on Communications, ICT-MICC2007.

Klungsøyr, J., Wakholi, P., Macleod, B., Escudero-Pascual, A., & Lesh, N. (2008). *Open-ROSA, JavaROSA, GloballyMobile - Collaborations around open standards for mobile applications*. International Conference on M4D Mobile Communication Technology for Development, Karlstad University, Sweden.

Klyne, G., & Carroll, J. J. (2004). *Resource description framework (RDF): Concepts and abstract syntax* (W3C recommendation 10 February 2004) [Computer software manual].

Kobsa, A. (2007). Privacy-enhanced personalization. *Communications of the ACM, 50*(8), 24–33. doi:10.1145/1278201.1278202

Kok, J. K., Warmer, C. J., & Kamphuis, I. G. (2005). Multiagent control in the electricity infrastructure. In *the Proceedings of 4th International Joint Conference on Autonomous Agents and Multiagent Systems* (pp. 75-82).

Koller, D., & Daniilidis, K. (1992). Model-based object-tracking in traffic scenes in Computer Vision. In *Second European Conference on Computer Vision.* Santa Margherita Ligure, Italy.

Koodli, R. (2009). *Mobile IPv6 fast handovers.* IETF RFC 5568.

Koo, J.-D., Oh, S. H., & Lee, C. (2010). Authenticated route optimization scheme for network mobility (NEMO) support in heterogeneous networks. *International Journal of Communication Systems, 23,* 1252–1267. doi:10.1002/dac.1112

Korpipää, P., Malm, E., Salminen, I., Rantakokko, T., Kyllönen, V., & Känsälä, I. (2005). Context management for end user development of context-aware applications. In *MDM '05: 6th International Conference on Mobile Data Management* (pp. 304-308). ACM.

Korpipää, P., Mantyjarvi, J., Kela, J., Keranen, H., & Malm, E. (2003). Managing context information in mobile devices. [IEEE.]. *Pervasive Computing, 2*(3), 42–51. doi:10.1109/MPRV.2003.1228526

Koudounas, V., & Iqbal, O. (1996). *Mobile computing: Past, present and future.* Retrieved April 16, 2011, from http://www.doc.ic.ac.uk/~nd/surprise_96/journal/vol4/vk5/report.html

Kou, L., Markowsky, G., & Berman, L. (1981). A fast algorithm for Steiner trees. *Acta Informatica, 15*(3), 141–145. doi:10.1007/BF00288961

Kray, C., & Baus, J. (2003). A survey of mobile guides. In *Proceedings of HCI in Mobile Guides, Fifth International Symposium on Human Computer Interaction with Mobile Devices and Services,* Udine, Italy.

Kray, C., Laakso, K., Elting, C., & Coors, V. (2003). Presenting route instructions on mobile devices. In *Proceedings of the 2003 International Conference on Intelligent User Interfaces (IUI'03).* Miami, FL, USA.

Kristoffersent, S., & Ljungberg, F. (1998). Representing modalities in mobile computing: A model of IT use in mobile settings. In Urban, B., Kirste, T., & Ide, R. (Eds.), *Proceedings of Interactive Applications in Mobile Computing.* Germany: Fraunhofer Institute for Computer Graphics.

Krogstie, J., Lyytinen, K., Opdahl, A., Pernici, B., Siau, K., & Smolander, K. (2004). Research areas and challenges for mobile Information Systems. *International Journal of Mobile Communications, 2*(3), 220–234.

Krüger, A., Baus, J., Heckmann, D., Kruppa, M., & Wasinger, R. (2007). Adaptive mobile guides. In Brusilovsky, P., Kobsa, A., & Nejdl, W. (Eds.), *The adaptive Web.* Berlin, Germany: Springer. doi:10.1007/978-3-540-72079-9_17

Kuhn, F., Moscibroda, T., & Wattenhofer, R. (2004). Unit disk graph approximation. In Basagni, S., & Phillips, C. (Eds.), *Proceedings of the Joint Workshop on Foundations of Mobile Computing* (pp. 3201-3205). Philadelphia, PA: ACM.

Kukec, A., Bagnulo, M., & Oliva, A. (2010). CRYPTRON: CRYptographic prefixes for route optimization in NEMO. In *International Conference on Communications, ICC2010* (pp. 1-5).

Kumar, V. (n.d.). *Mobile computing: A brief history of personal communication system.* University of Missouri-Kansas City, USA. Retrieved April 27, 2011 from http://k.web.umkc.edu/kumarv/cs572/PCS-history.pdf

Kumar, K., & van Hillegersberg, J. (2000). ERP experiences and evolution. *Communications of the ACM, 43*(4), 22–26. doi:10.1145/332051.332063

Kuutti, K. (1995). *Activity theory as a potential framework for human-computer interaction research.* Cambridge, MA: Massachusetts Institute of Technology.

Laganier, J., & Eggert, L. (2008). Host identity protocol (HIP) rendezvous extension. *IETF RFC 5204.*

Lange, D. (2002). Introduction to special issue on mobile agents. *Autonomous Agents and Multi-Agent Systems, 5*(1), 5–6. doi:10.1023/A:1013400230894

Lange, D., & Oshima, M. (1998). *Programming and developing Java mobile agents with Aglets.* Addison-Wesley.

Lange, D., & Oshima, M. (1999). Seven good reasons for mobile agents. *Journal of Communications of ACM, 42*(3), 88–89. doi:10.1145/295685.298136

Lee, C. W., Chen, M. C., & Sun, Y. S. (2008). *A network mobility management scheme for fast QoS handover.* In International Wireless Internet Conference. Hawaii. USA.

Lee, S., & Gerla, M. (2001). Split multi-path routing with maximally disjoint paths in ad hoc networks. In *Proceedings of the IEEE International Conference on Communications, Vol. 10* (pp. 3201-3205). Helsinki, Finland: IEEE.

Lee, J. T., & Wang, Y. T. (2003). Efficient data mining for calling path patterns in GSM networks. *Information Systems, 28*(8), 929–948. doi:10.1016/S0306-4379(02)00112-6

Lehmann, J., Bizer, C., Kobilarov, G., Auer, S., Becker, C., & Cyganiak, R. (2009). DBpedia - A crystallization point for the Web of data. *Journal of Web Semantics, 7*(3), 154–165. doi:10.1016/j.websem.2009.07.002

Lemire, D., & Maclachlan, A. (2005). *Slope one predictors for online rating-based collaborative filtering*. In SIAM Conference on Data Mining (SDM 2005). Newport Beach, USA.

Leonhardt, U. (1998). *Supporting location-awareness in open distributed systems*. Ph.D. Thesis, Department of Computing, Imperial College, London, 1998.

Leu, F. Y. (2009). A novel network mobility handoff scheme using SIP and SCTP for multimedia applications. [Elsevier.]. *Journal of Network and Computer Applications, 32*, 1073–1091. doi:10.1016/j.jnca.2009.02.007

Lewis, R. (2007). *Dereferencing HTTP URIs*. Retrieved January 9, 2010, from http://www.w3.org/2001/tag/doc/httpRange-14/2007-05-31/HttpRange-14

Li, F., Yang, Y., & Wu, J. (2009). *Mobility management in MANETs: Exploit the positive impacts of mobility. Guide to Wireless Ad Hoc Networks, Computer Communications and Networks Series* (pp. 211–235). London, UK: Springer-Verlag.

Lin, C. Y., Peng, W. C., & Tseng, Y. C. (2006). Efficient in-network moving object tracking in wireless sensor networks. *IEEE Transactions on Mobile Computing, 5*(8).

Linden, G., Smith, B., & York, J. (2003). Amazon.com recommendations: Item-to-item collaborative filtering. *IEEE Internet Computing, 7*(1), 76–80. doi:10.1109/MIC.2003.1167344

Lin, Y.-B., & Chlamtac, I. (2001). *Wireless and mobile network architectures*. John Wiley & Sons.

Loureiro, S., Molva, R., & Roudier, Y. (2000). Mobile code security. In *Proceedings of ISYPAR (4ème Ecole d'Informatique des Systems Parallèles et Répartis)*. Retrieved January 2, 2011, from http://agentlab.swps.edu.pl/agent_papers/NATO_2006.pdf

Lowe, G. (1997). Casper: A compiler for the analysis of security protocols. In *Proceedings of the 10th Computer Security Foundation Workshop (PCSFW)*. IEEE Computer Society Press.

Lowe, G. (1996). Breaking and fixing the Needham-Schroeder public key protocol using FDR. In *Proceedings of Tools and Algorithms for the Construction and Analysis of Systems (TACAs)* (*Vol. 1055*, pp. 147–166). Lecture Notes in Computer Science Springer-Verlag.

Luther, M., Mrohs, B., Wagner, S., Steglich, M., & Kellerer, W. (2005). Situational reasoning - A practical OWL use case. In *Proceedings of 7th International Symposium on Autonomous Decentralized Systems (ISADS2005)* (pp. 96-103). Chengdu, China.

Luther, M., Fukazawa, Y., Wagner, M., & Kurakake, S. (2008). Situational reasoning for task-oriented mobile service recommendation. *The Knowledge Engineering Review, 23*(1), 7–19. doi:10.1017/S0269888907001300

Ma, L., & Tsai, J. J. P. (2000). Formal verification techniques for computer communication security protocols. In S. K. Chang (Ed.), *Handbook of software engineering and knowledge engineering*. Retrieved December 24, 2010, from ftp://cs.pitt.edu/chang/handbook/12.pdf

Maggi, P., & Sisto, R. (2002a). Experiments on formal verification of mobile agent data integrity properties. In *Proceedings of Workshop from Data to Agents (WOA)* (pp. 131-136).

Maggi, P., & Sisto, R. (2002b). Using SPIN to verify security properties of cryptographic protocols. In *Proceedings of 9th International Spin Workshop on Model Checking of Software (SPIN 2002), Lecture Notes in Computer Science, 2318* (pp.187-204). Springer-Verlag.

Maggi, P., & Sisto, R. (2003). A configurable mobile agent data protection protocol. In *Proceedings of Autonomous Agents and Multiagent Systems* (pp. 851–858). New York, NY: ACM Press. doi:10.1145/860710.860712

McCarthy, B., Edwards, C., & Dunmore, M. (2009). *Using NEMO to support the global reachability of MANET nodes*. In IEEE International Conference on Computer Communications (INFOCOM). Rio de Janeiro, Brazil.

McGinty, L., & Reilly, J. (2010). On the evolution of critiquing recommenders. In Kantor, P., Ricci, F., Rokach, L., & Shapira, B. (Eds.), *Recommender systems handbook: A complete guide for research scientists and practitioners*. Berlin, Germany: Springer.

McGinty, L., & Smyth, B. (2006). Adaptive selection: An analysis of critiquing and preference-based feedback in conversational recommender systems. *International Journal of Electronic Commerce*, *11*(2), 35–57. doi:10.2753/JEC1086-4415110202

McGuinness, D. L., & van Harmelen, F. (Eds.). *OWL Web ontology language overview* (W3C recommendation). World Wide Web Consortium. Retrieved January 10, 2010, from http://www.w3.org/TR/2004/REC-owl-features-20040210/

Meadows, C. (1994). Formal verification of cryptographic protocols: A survey. In *Proceedings of 4th International Conference on the Theory and Applications of Cryptology-Advances in Cryptography (ASIACRYPT)* (pp. 135-150).

Meadows, C. (1996). Language generation and verification in the NRL protocol analyzer. In *Proceedings of 9th IEEE Computer Society Foundations Workshop (CSFW)* (pp. 48-61).-

Meghanathan, N. (2009). A beaconless node velocity-based stable path routing protocol for mobile ad hoc networks. In *Proceedings of the Sarnoff Symposium Conference*. Princeton, NJ: IEEE.

Meghanathan, N. (2008). Exploring the stability-energy consumption-delay-network lifetime tradeoff of mobile ad hoc network routing protocols. *Journal of Networks*, *3*(2), 17–28. doi:10.4304/jnw.3.2.17-28

Meghanathan, N. (2010). Benchmarks and tradeoffs for minimum hop, minimum edge and maximum lifetime per multicast tree in mobile ad hoc networks. *International Journal of Advancements in Technology*, *1*(2), 234–251.

Mehrotra, S., Rastogi, R., Korth, H., & Silberschatz, A. (1992). A transaction model for multi-database systems. In *Proceedings of International Conference on Distributed Computing Systems* (pp 56-63).

Meier, R. (2010). *Professional Android 2 application development*. Wiley Publishing.

Melen, J., Yitalo, J., & Samela, P. (2008). *Security parameter index multiplexed network address translation (SPINAT)*. IETF Draft, draft-melenspinat-01.

Melen, J., Yitalo, J., Samela, P., & Henderson, T. (2009). *Host identity protocol-based mobile router (HIPMR)*. IETF draft, draft-melen-hip-mr-02.

Mell, P. (1999). Understanding the world of your enemy with I-CAT. In *Proceedings of 22nd National Information System Security Conference*, (pp.432-443).

Michelini, M., Hijazi, S., Nassar, C. R., & Zhiqiang, W. (2003). Spectral sharing across 2G-3G systems. Signals, Systems and Computers, 2003. In *Conference Record of the Thirty-Seventh ASILOMAR Conference*.

Mihalic, K., & Tscheligi, M. (2007). 'Divert: Mother-in-law': Representing and evaluating social context on mobile devices. In Mobilehci '07: *9th International Conference on Human Computer Interaction with Mobile Devices & Services* (pp. 257-264). ACM.

Millen, J. K., Clark, S. C., & Freedman, S. B. (1987). The interrogator: protocol security analysis. *Journal of IEEE Transactions on Software Engineering*, *13*(2), 274–288. doi:10.1109/TSE.1987.233151

Miller, B., Konstan, J., & Riedl, J. (2004). PocketLens: Toward a personal recommender system. *ACM Transactions on Information Systems*, *22*(3), 437–476. doi:10.1145/1010614.1010618

Miller, E., Shen, D., Liu, J., & Nicholas, C. (2000). Performance and scalability of a large-scale n-gram based information retrieval system. *Journal of Digital Information*, •••, 1.

Milner, R., Tofte, M., Harper, R., & MacQueen, D. (1997). *The definition of standard ML* (rev. ed.). MIT Press.

Milojicic, D. (1999). Trend wars - Mobile agent applications. *Concurrency - IEEE, 7*(3), 80 – 90.

Mitchell, J. C., Mitchell, M., & Stern, U. (1997). Automated analysis of cryptographic protocols using Mur. In *Proceedings of Symposiums on Security and Privacy* (pp. 141–153). IEEE Computer Society Press. doi:10.1109/SECPRI.1997.601329

Mitra, A., Sardar, B., & Saha, D. (2011). Efficient management of fast handoff in wireless network mobility (NEMO). Indian Institute of Management Calcutta. *Working Paper Series WPS, N° 671.*

Mobach, D. (2007). *Agent-based mediated service negotiation.* PhD Thesis, Computer Science Department, Vrije University Amsterdam.

Mobility Workbench. (2004). *A tool for manipulating and analyzing mobile concurrent systems described in the Pi.* Retrieved December 24, 2010, from http://www.it.uu.se/research/group/mobility/mwb

Monteiro, E., & Hanseth, O. (1995). Social shaping of information infrastructure: On being specific about the technology. *Information Technology and Changes in Organizational work. Proceedings of the IFIP WG8. 2 Working Conference on Information Technology and Changes in Organizational work, December 1995* (pp. 325–343).

*Moscow, ML.* (2004). Retrieved December 24, 2010, from http://www.dina.dk/~sestoft/mosml.html

Motik, B., Patel-Schneider, P. F., & Parsia, B. (2009, October). *OWL 2 Web ontology language structural specication and functional-style syntax* (W3C recommendation 27 October 2009) [Computer software manual]. Retrieved from http://www.w3.org/TR/owl2-syntax/

Mukherjee, A., & Purkayastha, S. (2010). Exploring the potential and challenges of using mobile based technology in strengthening health information systems: Experiences from a pilot study. *AMCIS 2010 Proceedings*, (p. 263).

Mussabir, Q. B., Yao, W., Niu, Z., & Fu, X. (2007). Optimized FMIPv6 using IEEE 802.21 MIH services in vehicular networks. [IEEE Communications Society.]. *IEEE Transactions on Vehicular Technology, 56*, 3397. doi:10.1109/TVT.2007.906987

Myles, A., & Skellern, D. (1993). Comparing four IP based mobile host protocols. In *Joint European Networking Conference* (pp. 191–196). Macquarie University, Sydney, Australia.

Na, J., Cho, S., Kim, C., Kee, S., Kang, H., & Koo, C. (2004). *Route optimization scheme based on path control header.* IETF Draft, draft-na-nemo-path-control-header-00.

Nardi, B. A. (Ed.). (1995). *Context and consciousness: Activity theory and human-computer interaction.* Cambridge, MA: Massachusetts Institute of Technology.

Nehrkorn, T. H., Thomas, C., & Lawrence, W. (1993). *Nowcasting methods for satellite imagery.* In Atmospheric and Environmental Research Inc. Cambridge.

Ng, C., Thubert, P., Watari, M., & Zhao, F. (2007). *Network mobility route optimization problem statement.* IETF RFC 4888.

Ng, C., Zhao, F., Watari, M., & Thubert, P. (2007). *Network mobility route optimization solution space analysis.* IETF RFC 4889.

Ng, S. (1999). Protecting mobile agents against malicious hosts by intension of spreading. In *Proceedings of International Conference on Parallel and Distributed Processing Techniques and Applications (PDPTA)*, (pp. 725-729).

Nikander, P., & Arkko, J. (2004). Delegation of signaling rights. *Security Protocols, LNCS, 2845*, 203–214. doi:10.1007/978-3-540-39871-4_17

Ning, W., & Naiqian, Z. (2006). Wireless sensors in agriculture and food industry - Recent development and future perspective. *Computers and Electronics in Agriculture, 50*(1).

Nitschke, L., Paprzyeki, M., & Ren, M. (2006). Mobile agent security. In Thomas, J., & Essaidi, M. (Eds.), *Information assurance and computer security* (pp. 102–123).

Nokia. (2006). *The route to true competitive advantage: Today's evolution of workforce mobility.* White Paper. Nokia.

Nováczki, S., Bokor, L., Jeney, G., & Imre, S. (2008). Design and evaluation of novel HIP-based network mobility protocol. *Journal of Networks, 3*(1), 10–24. doi:10.4304/jnw.3.1.10-24

Oiwa, T., Kunishi, M., Ishiyama, M., Kohno, M., & Teraoka, F. (2003). A network mobility protocol based on LIN6. In *IEEE Vehicular Technology Conference, vol. 3,* (pp. 1984-1988). IEEE Communications Society.

Oren, E., Delbru, R., Catasta, M., Cyganiak, R., Stenzhorn, H., & Tummarello, G. (2008). Sindice.com: A document-oriented lookup index for Open Linked data. *International Journal of Metadata, Semantics and Ontologies, 3*(1).

Outtagarts, A. (2009). Mobile agent-based applications. *Journal of Computer Science and Network Security, 9*(11).

Ozaki, T., Kim, J.-B., & Suda, T. (2001). Bandwidth-efficient multicast routing for multi-hop, ad hoc wireless networks. In *Proceedings of the International Conference on Computers and Communication Networks,* vol. 2 (pp. 1182-1192). Anchorage, AK, USA: IEEE.

Pacitti, E., & Simon, E. (2000). Update propagation strategies to improve freshness in lazy master replication databases. *The VLDB Journal, 8,* 305–318. doi:10.1007/s007780050010

Pack, S., Kwon, T., Choi, Y., & Paik, E. K. (2009). An adaptive network mobility support protocol in hierarchical mobile IPv6 networks. In *IEEE Transcations on Vehicular Technology Conference, 58,* (p. 3627).

Padmanabhan, B., & Tuzhilin, A. (1996). Pattern discovery in temporal databases: A temporal logic approach. In *Proceedings of the Second International Conference on Knowledge Discovery and Data Mining.*

Papastavrou, S., Samaras, G., & Pitoura, E. (2000). Mobile agents for World Wide Web distributed database access. *IEEE Transactions on Knowledge and Data Engineering, 12*(5), 802–820. doi:10.1109/69.877509

Paul, M. W., Frances, C. H., Timothy, J., Schmit, R., & Aune, M. (1998). Application of GOES-8/9 soundings to weather forecasting and nowcasting. *Bulletin of the American Meteorological Society.*

Paulson, L. C. (1997). Proving properties of security protocols by induction. In *Proceedings of the 10th Computer Society Foundations Workshop (CSFW)* (pp. 70-83).

Paurobally, S., & Jennings, N. R. (2005). Developing agent web service agreements. In *Proceedings of the IEEE/WIC/ACM International Conference on Intelligent Agent Technology* (pp. 464-470).

Pawar, P., van Halteren, A. T., & Sheikh, K. (2007, March). Enabling context-aware computing for the nomadic mobile user: A service oriented and quality driven approach. In *IEEE Wireless Communications and Networking Conference WCNC 2007* (pp. 2531-2536). IEEE Communication Society.

Perkins, C. (1997). Mobile IP. *IEEE Communication Magazine.*

Perkins, C. E., & Royer, E. (1999). Ad hoc on-demand distance vector routing. In *Proceedings of the 2nd Annual International Workshop on Mobile Computing Systems and Applications* (pp. 90-100). New Orleans, LA: IEEE.

Petrescu, A., Olivereau, A., Janneteau, C., & Lach, H.-Y. (2004). *Threats for basic network mobility support (NEMO threats).* draft-petrescu-nemo-threats-xx.

Pham, V., & Karmouch, A. (1998). Mobile software agents: An overview. *Journal of IEEE Communications, 36*(7), 26–37. doi:10.1109/35.689628

Phang, S. Y., Lee, H. J., & Lim, H. (2007). *A secure deployment framework of NEMO (network mobility) with firewall traversal and AAA server.* In International Conference on Coverage Information Technology. IEEE Computer Society.

Philippe, L., & Marc, L. (1994). Stanford University, Fast volume rendering using a shear-warp factorization of the viewing transformation. In *SIGGRAPH '94 Proceedings of the 21st Annual Conference on Computer Graphics and Interactive Techniques.*

Pietriga, E., Bizer, C., Karger, D., & Lee, R. (2006). Fresnel: A browser-independent presentation vocabulary for RDF. In *Proceedings of the 5th International Semantic Web Conference ISWC 2006* (Vol. 4273, pp. 158-171). Springer-Verlag.

Portale, O. (2002). *Healthcare: the mobile opportunity.* Mobile Enterprise, Sun Microsystems, Feature story, November 08, 2002.

Poslad, S., Laamanen, H., Malaka, R., Nick, A., Buckle, P., & Zipl, A. (2001). *CRUMPET: Creation of user-friendly mobile services personalised for tourism.* In Second International Conference on 3G Mobile Communication Technologies. London, UK.

Pospischil, G., Umlauft, M., & Michlmayr, E. (2002). Designing LoL@, a mobile tourist guide

Prasad, R., & Muñoz, L. (2003). *WLANs and WPANs towards 4G wireless*. Norwood, MA: Artech House.

Preuveneers, D., den Bergh, J. V., Wagelaar, D., Georges, A., Rigole, P., Clerckx, T., et al. (2004). Towards an extensible context ontology for ambient intelligence. In P. Markopoulos, B. Eggen, E. Aarts, & J. L. Crowley (Eds.), *Second European Symposium on Ambient Intelligence* (vol. 3295, pp. 148-159). Eindhoven, The Netherlands: Springer.

Process Algebra Compiler. (2000). Verification tool. Retrieved December 24, 2010, from http://www.reactive-systems.com/pac

Prud'hommeaux, E., & Seaborne, A. (2008*). SPARQL query language for RDF* (W3C recommendation January 15, 2008) [Computer software manual]. Retrieved from http://www.w3.org/TR/rdf-sparql-query/

Purdom, J. F. W. (1976). Some uses of high resolution goes imagery in the Mesoscale forecasting of convection and its behavior. *Monthly Weather Review*, *104*, 1474–1483. doi:10.1175/1520-0493(1976)104<1474:SUOHRG>2.0.CO;2

Pyramid Research. (2010). *Health check: Key players in mobile healthcare*. Pyramid Research. Retrieved from http://www.pyramidresearch.com/store/RPMHEALTH.htm

Raento, M., Oulasvirta, A., & Eagle, N. (2009). Smartphones. *Sociological Methods & Research*, *37*(3), 426. doi:10.1177/0049124108330005

Rahimi, S., Angryk, R., Bjursell, J., Paprzycki, M., Ali, D., Cobb, M., et al. (2001). Comparison of mobile agent frameworks for distributed geospatial data integration. In *Proceedings of 4th AGILE Conference on Geographic Information Science,* (pp. 643-655).

Ramparany, F., Euzenat, J., Broens, T. H. F., Bottaro, A., & Poortinga, R. (2006, April). *Context management and semantic modelling for ambient intelligence* (Technical Report No. TR-CTIT-06-52). Enschede.

Raptis, D., Tselios, N., & Avouris, N. (2005). *Context-based design of mobile applications for museums: A survey of existing practices.* In MobileHCI '05: 7th International Conference on Human Computing Interaction with Mobile Devices & Services. ACM.

Rasinger, J., Fuchs, M., & Höpken, W. (2007). Information search with mobile tourist guides: A survey of usage intention. *Journal of Information Technology & Tourism*, *9*(34).

Raubal, M., & Panov, I. (2009). A formal model for mobile map adaptation. *Location Based Services and Tele Cartography II - From Sensor Fusion to Context Models. Selected Papers from the 5th International Symposium on LBS & TeleCartography.* Salzburg, Austria.

Raymond, E. S. (2001). *The cathedral and the Bazaar: Musings on Linux and open source by an accidental revolutionary*. Sebastopol, CA: O'Reilly & Associates, Inc.

Rebolj, D., Magdič, A., & Čuš Babič, N. (2001). Mobile computing in construction. In *Advances in Concurrent Engineering: Proceedings of the 8th ISPE International Conference on Concurrent Engineering: Research and Applications.* West Coast Anaheim Hotel, California, USA, July 28 - August 1, 2001.

Rebolj, D., Magdič, A., & Čuš Babič, N. (2004). Mobile computing – The missing link to effective construction IT. *Proceedings of the International Conference on Construction Information Technology, Langkawi,* February 17-21, 2004, (pp.327-334).

Redish, J., & Wixon, D. (2003). In Jacko, J. A., & Sears, A. (Eds.), *The human-computer interaction handbook* (pp. 922–940). Hillsdale, NJ: L. Erlbaum Associates Inc.

Rescorla, E. (1999). *Diffie-Hellman key agreement method*. IETF RFC 2631.

Rhazi, A., Pierre, S., & Boucheneb, H. (2007). Secure protocol based on sedentary agent for mobile agent environments. *Journal of Computer Science*, *3*(1), 35–427. doi:10.3844/jcssp.2007.35.42

Ricci, F. (2010). (in press). Mobile recommender systems. *International Journal of Information Technology and Tourism*.

Ricci, F., & Nguyen, Q. N. (2006). MobyRek: A conversational recommender system for on-the-move travelers. In Fesenmaier, D. R., Werthner, H., & Wober, K. W. (Eds.), *Destination recommendation systems: Behavioural foundations and applications* (pp. 281–294). CABI Publishing. doi:10.1079/9780851990231.0281

Ricci, F., & Nguyen, Q. N. (2007). Acquiring and revising preferences in a critique-based mobile recommender system. *IEEE Intelligent Systems, 22*(3), 22–29. doi:10.1109/MIS.2007.43

Rogers, R., Lombardo, J., Mednieks, Z., & Meike, B. (2009). *Android application development: Programming with the Google SDK*. Beijing, China: O'Reilly.

Rosenberg, J., Camarillo, G., Johnston, A., Peterson, J., Sparks, R., Handley, M., & Schooler, E. (2002). *SIP: Session initiation protocol*. IETF RFC 3261.

Roth, V. (2001). Programming Satan's agents. *Journal of Electronic Notes in Theoretical Computer Science (ENTCS), 63*.

Roth, V. (2002). Empowering mobile software agents. In *Proceedings of 6th IEEE Mobile Agents Conference, Lecture Notes in Computer Science, 2535* (pp. 47-63). Springer-Verlag.

Roth, V. (1999). Mutual protection of co-operating agents. In Vitek, J., & Jensen, C. D. (Eds.), *Secure Internet programming: Security issues for mobile and distributed objects* (pp. 275–285).

Rötting, O., Röpke, W., Becker, H., & Gärtner, C. (2001). Polymer microfabrication technologies. *Springer Link, 8*, 32–36.

Royer, E., & Perkins, C. E. (1999). Multicast operation of the ad-hoc on-demand distance vector routing protocol. In *Proceedings of the 5th Annual Conference on Mobile Computing and Networking* (pp. 207-218). Seattle, WA: ACM.

Rushby, J. (2006). Tutorial: Automated formal methods with PVS, SAL, and Yices. In *Proceedings of the 4th IEEE International Conference on Software Engineering and Formal Methods (SEFM'06)* (p. 262).

Ryan, N. S., Pascoe, J., & Morse, D. R. (1998). Enhanced reality fieldwork: The context-aware archaeological assistant. In V. Ganey, M. van Leusen, & S. Exxon (Eds.), *Computer applications in archaeology 1997*. Oxford, UK: Tempus Reparatum. Retrieved from http://www.cs.kent.ac.uk/pubs/ 1998/616

Sandberg, A. (1985). *Socio-technical design, trade union strategies and action research. Research Methods in Information Systems* (pp. 79–92). Amsterdam, The Netherlands: North-Holland.

Sarwar, B., Karypis, G., Konstan, J., & Riedl, J. (2001). *Item-based collaborative filtering recommendation algorithms*. In 10th International Conference on World Wide Web (WWW10), Hong Kong, China.

Satoh, I. (2002). Physical mobility and logical mobility in ubiquitous computing environments. *Proceedings of International Conference on Mobile Agents (MA'2002), LNCS*. Springer, October, 2002.

Satoh, I. (2003). Spatial agents: Integrating user mobility and program mobility in ubiquitous computing environments. *Wireless Communications and Mobile Computing, 3*, 411–423. doi:10.1002/wcm.126

Satyanarayanan, M. (1993). *Mobile computing*. In Hot Topics, IEEE Computer, September 1993.

Satyanarayanan, M. (1996). Fundamental challenges in mobile computing. In *Proceedings of the Fifteenth Annual ACM Symposium on Principles of Distributed Computing* (Philadelphia, Pennsylvania, United States, May 23-26, 1996) *PODC '96* (pp. 1-7). New York, NY: ACM Press.

Satyanarayanan, M. (2002). The evolution of CODA. *ACM Transactions on Computer Systems - TOCS, 20*(2), 85-124.

Sauermann, L., & Cyganiak, R. (2008). *Cool URIs for the Semantic Web*. W3C Interest Group Note. Retrieved from http://www.w3.org/TR/cooluris/

Sauermann, L., Bernardi, A., & Dengel, A. (2005). Overview and outlook on the semantic desktop. In S. Decker, J. Park, D. Quan, & L. Sauermann (Eds.), *Proceedings of the 1st Workshop on the Semantic Desktop at the ISWC 2005 Conference* (vol. 175, pp. 1-18). CEUR-WS.

Schafer, J., Frankowski, D., Herlocker, J., & Sen, S. (2007). Collaborative filtering recommender systems. In Brusilovsky, P., Kobsa, A., & Nejdl, W. (Eds.), *The adaptive Web* (pp. 377–408). Berlin, Germany: Springer. doi:10.1007/978-3-540-72079-9_9

Schandl, B. (2010). Replication and versioning of partial RDF graphs. In *Proceedings of 7th European Semantic Web Conference (ESWC 2010)*.

Schilit, B., & Theimer, M. (1994, Sep/Oct). Disseminating active map information to mobile hosts. [IEEE.]. *Network, 8*(5), 22–32.

Schmidt, A., Beigl, M., & Gellersen, H.-W. (1998). There is more to context than location. *Computers & Graphics, 23*, 893–901. doi:10.1016/S0097-8493(99)00120-X

Schneiderman, B. (1994). Dynamic queries for visual information seeking. *IEEE Software, 11*(6), 70–77. doi:10.1109/52.329404

Schraefel, M. C., Smith, D. A., Owens, A., Russell, A., Harris, C., & Wilson, M. (2005). The evolving mspace platform: Leveraging the Semantic Web on the trail of the memex. In *Hypertext 2005: Proceedings of Sixteenth ACM Conference on Hypertext and Hypermedia* (pp. 174-183). New York, NY: ACM.

Seaborne, A. (2004). *RDQL - A query language for RDF*. W3C member submission. Retrieved from http://www.w3.org/Submission/2004/SUBMRDQL-20040109/

Sen, A. (1999). *Development as freedom*. Oxford University Press.

Serrano-Alvarado, P., Roncancio, C., & Adiba, M. (2004). A survey of mobile transactions. [DAPD]. *Distributed and Parallel Databases, 16*, 1–38. doi:10.1023/B:DAPD.0000028552.69032.f9

Sewell, P. (1998). Global/ local subtyping and capability inference for a distributed calculus. Automata, languages and programming. In *Proceedings of 25th International Colloquium (ICALP), Lecture Notes in Computer Science, 1443* (pp.695-706). Springer-Verlag.

Shahriar, A. Z. M., Atiquzzaman, M., & Ivanvic, W. (2010). Route optimization in network mobility: Solutions, classification, comparison and future research directions. *IEEE Communications Survey and Tutorials, 12*(1), 24–38. doi:10.1109/SURV.2010.020110.00087

Shekar, S., & Lin, D. (1994). Genesis and advanced traveller information systems (ATIS): Killer applications for mobile computing. *MOBIDATA: An Interactive Journal of Mobile Computing, 1*(1).

Shin, S., Kobata, K., & Imai, H. (2005). A simple leakage-resilient authenticated key establishment protocol, its extensions and applications. *IECE Transactions in Fundamentals. E (Norwalk, Conn.), 88-A*(3), 248–254.

Shukla, B. P., Pal, P. K., & Joshi, P. C. (2010). Extrapolation of sequence of geostationary satellite images for Weather nowcasting. *Geoscience and Remote Sensing Letters, 8*.

Silva, A. R., Romão, A., Deugo, D., & Silva, M. M. (2001). Towards a reference model for surveying mobile agent systems. *Journal of Autonomous Agents and Multi-Agent Systems, 4*(3), 187–231. doi:10.1023/A:1011443827037

Sohn, T., Li, K. A., Griswold, W. G., & Hollan, J. D. (2008). A diary study of mobile information needs. In *CHI '08: Proceedings of the Twenty-sixth Annual Sigchi Conference on Human Factors in Computing*

Soliman, H., Castellucia, C., El Malki, K., & Bellier, L. (2008). *Hierarchical mobile IPv6 (HMIPv6) mobility management*. IETF RFC 5380.

Springer, T., Wustmann, P., Braun, I., Dargie, W., & Berger, M. (2008). A comprehensive approach for situation-awareness based on sensing and reasoning about context. In *UIC '08: Proceedings of 5th International Conference on Ubiquitous Intelligence and Computing* (pp. 143-157). Berlin, Germany: Springer-Verlag.

Stetzler, T. D., Havens, J. H., & Koyama, M. (1995). *A 2.7-4.5 V single chip GSM transceiver RF integrated circuit*. AT&T Bell Labs, Reading, PA. Dec 1995.

Stojmenovic, I. (Ed.). (2002). *Handbook of wireless networks and mobile computing*. John Wiley & Sons, Inc.doi:10.1002/0471224561

Strang, T., & Linnhoff-Popien, C. (2004). A context modeling survey. *Workshop on Advanced Context Modelling, Reasoning and Management*. UbiComp Conference, Nottingham/England.

Strang, T., & Popien, C. L. (2004, September). A context modeling survey. In *Ubicomp 1st International Workshop on Advanced Context modelling, Reasoning and Management* (pp. 31-41). Nottingham.

Strange, J., & Atkinson, S. (2000). *A direct conversion transceiver for multi-band GSM application in radio frequency integrated circuits*. In RFIC Symposium.

Stroobants, R. (2006). *Mobile tourist guides. Technical Report*. Belgium: Katholieke Universiteit Leuven.

Subramanya, S. R., & Yi, B. K. (2007). Enhancing the user experience in mobile phones. *Computer, 40*(12), 114–117. doi:10.1109/MC.2007.420

Sun, J. Z., & Sauvola, J. (2002). On fundamental concept of mobility for mobile communications. *Proceedings of the 13th IEEE International Symposium on Personal, Indoor and Mobile Radio Communications, 2,* (pp. 799– 803). Lisbon, Portugal, 2002.

Suri, N., Bradshaw, J. M., Breedy, M. R., Groth, P. T., Hill, G., & Jeffers, R. (2000). Strong mobility and fine-grained resource control in NOMADS. In *Proceedings of 2nd International Symposium on Agent Systems and Applications and 4th International Symposium on Mobile Agents (ASA/MA)*, (pp. 2-15).

Susman, G. I., & Evered, R. D. (1978). An assessment of the scientific merits of action research. *Administrative Science Quarterly, 23*(4), 582–603. doi:10.2307/2392581

Su, W., Lee, S.-J., & Gerla, M. (2001). Mobility prediction and routing in ad hoc wireless networks. *International Journal of Network Management, 11*(1), 3–30. doi:10.1002/nem.386

Syverson, P., Goldschlag, M., & Reed, M. (1997). Anonymous connections and onion routing. *IEEE Symposium on Security and Privacy*, (pp. 44-54).

Teevan, J., Jones, W., & Bederson, B. B. (2006). Personal information management. *Communications of the ACM, 49*(1), 40–43. doi:10.1145/1107458.1107488

Teo, H.-S. (2008). An activity-driven model for context-awareness in mobile computing. In *MobileHCI '08: 10th International Conference on Human Computer Interaction with Mobile Devices & Services* (pp. 545-546). New York, NY: ACM.

Thaysen, J., Boisen, A., Hansen, O., & Bouwstra, S. (1999). Digest of technical papers, transducers 99. In *10th International Conference on Solid-State Sensors and Actuators* (pp. 1852-1855). Sendai, Japan.

Thomas, G. D. A. (2007). *A framework for secure mobile computing in healthcare*. Master's thesis. Nelson Mandela Metropolitan University, South Africa.

Tintarev, N., & Masthoff, J. (2010). Designing and evaluating explanations for recommender systems. In Kantor, P., Ricci, F., Rokach, L., & Shapira, B. (Eds.), *Recommender systems handbook: A complete guide for research scientists and practitioners*. Berlin, Germany: Springer. doi:10.1007/978-0-387-85820-3_15

TRAI (Telecom Regulatory Authority of India). (2010). *Annual telecom report*. India: Ministry of Telecom.

Tripathi, A., Ahmed, T., Pathak, S., Carney, M., & Dokas, P. (2001). *Paradigms for mobile agent-based active monitoring of network systems*. Technical Report, Department of Computer Science, University of Minnesota. Retrieved January 2, 2011, from http://www.cs.umn.edu/Ajanta

Tripathi, A. R., Karnik, N. M., Ahmed, T., Singh, R. D., Prakash, A., & Kakani, V. (2002). Design of the Ajanta system for mobile agent programming. *Journal of Systems and Software, 62*(2), 123–140. doi:10.1016/S0164-1212(01)00129-7

Tsai, C. S. (2009). Designing a novel mobility management scheme for enhancing the binding update of HMIPv6 with NEMO environment. In *International Conference on Future Networks* (pp. 87-91). IEEE Communications Society.

Tseng, V. S., & Lin, K. W. (2005). Energy efficient strategies for object tracking in sensor networks: A data mining approach. In *Proceedings of the International Workshop on Ubiquitous Data Management* (held with ICDE'05), April, 2005. Tokyo, Japan.

Turban, E. (2008). *Mobile, wireless, and pervasive computing. Information Technology for management: Transforming organizations in the digital economy* (pp. 167–206). New York, NY: John Wiley & Sons.

Van Olst, R., & Dwolatzky, B. (2004). Electronic data collection using mobile computing technologies. *GIS Technical, a Proceeding of SATNAC Conference*. Stellenbosch, 2004.

Vigna, G. (1998). Lecture Notes in Computer Science: *Vol. 1419. Cryptographic traces for mobile agents. Journal of Mobile Agent and Security* (pp. 137–153). Heidelberg, Germany: Springer-Verlag.

Vitek, J., & Gastagna, G. (1999). Seal: A framework for secure mobile computations. In *Proceedings of Internet Programming Language Workshop (ICCL), Lecture Notes in Computer Science, 1686* (pp. 47-77). Springer-Verlag.

Vizard, M. (2000). The way things work is the fundamental problem with mobile computing. *InfoWorld, 22*(40), 83.

Volz, J., Bizer, C., Gaedke, M., & Kobilarov, G. (2009). Discovering and maintaining links on the Web of data. In *The Semantic Web - ISWC 2009: 8th International Semantic Web Conference* (pp. 650-656). Chantilly, VA, USA.

Wakikawa, R., & Watari, M. (2004). *Optimized route cache control protocol (ORC).* IETF Draft, draft-wakikawa-nemo-orc-01.

Wakikawa, R., Thurbet, P., Boot, T., Bound, J., & McCarthy, B. (2007). *Problem statement and requirements for MANEMO.* IETF Draft, draft-wakikawa-manemo-problem-statement.

Wales, J. (2003). *The pocket PC's prescription for health care.* Smart Phone & Pocket PC Magazine, May 2003.

Walsham, G. (1995). Interpretive case studies in IS research: Nature and method. *European Journal of Information Systems, 4*(2), 74–81. doi:10.1057/ejis.1995.9

Warneke, B. A., & Pister, K. S. J. (2002). *MEMS for distributed wireless sensor networks.* In 9th IEEE International Conference on Electronics, Circuits and Systems (ICECS).

Warschauer, M. (2004). *Technology and social inclusion: Rethinking the digital divide.* The MIT Press.

Wayne, A. (2002). Jansen: Intrusion detection with mobile agents. *Journal of Computer Communications, 25*(15), 1392–1401. doi:10.1016/S0140-3664(02)00040-3

Web research. *Proceedings of the ESWC2005: Workshop on End-User Aspects of the Semantic Web (UserSWeb).*

Weeler, T. (2005). *Voyager: High volume transaction processing in enterprise applications.* White Paper. Recursion Software, Inc. Retrieved January 2, 2011, from http://www.recursionsw.com/About_Us/inc/White_papers/Voyager_RSI/2005-02-16-Voyager_High-Volume-Transactions_RSI.pdf

Weikum, G., & Schek, H. (1992). Concepts and applications of multilevel transactions and open nested transactions. In Elmagarmid, A. (Ed.), *Database transaction models for advanced applications* (pp. 515–553). Morgan Kaufmann.

Weithöner, T., Liebig, T., Luther, M., Böhm, S., Henke, F., & Noppens, O. (2007). Real-world reasoning with OWL. In ESWC2007: *Proceedings of 4th European Conference on the Semantic Web* (pp. 296-310). Berlin, Germany: Springer-Verlag.

Widjaja, I., & Balbo, S. (2005). Spheres of role in Context-awareness. In *OZCHI '05: Proceedings of 17th Australia Conference on Computer-Human Interaction* (pp. 1-4). Narrabundah, Australia: Computer-Human Interaction Special Interest Group (CHISIG) of Australia.

Wilhelm, U. (1999). *A technical approach to privacy based on Mobile agents protected by tamper resistant hardware.* PhD Dissertation, École Polytechnique Fédérale De Lausanne (EPFL), Switzerland.

Wilson, M., Russell, A., Smith, D. A., Owens, A., & Schraefel, M. C. (2005). mSpace mobile: A mobile application for the Semantic Web. *End User Semantic Web Workshop, ISWC2005,* (p. 11).

Woerndl, W., & Woehrl, M. (2008). *SeMoDesk: Towards a mobile semantic desktop.* Personal Information Management (PIM) Workshop, CHI 2008 Conference, Florence, Italy.

Woerndl, W., Muehe, H., & Prinz, V. (2009). *Decentral item-based collaborative filtering for recommending images on mobile devices.* Workshop on Mobile Media Retrieval (MMR'09), MDM 2009 Conference. Taipeh, Taiwan.

Woerndl, W., Muehe, H., Rothlehner, S., & Moegele, K. (2009). *Context-aware recommendations in decentralized, item-based collaborative filtering on mobile devices.* Workshop on Innovative Mobile User Interactivity, MobiCASE Conference. San Diego, USA.

Woerndl, W., Schueller, C., & Wojtech, R. (2007). *A hybrid recommender system for context-aware recommendations of mobile applications.* In IEEE 3rd International Workshop on Web Personalisation, Recommender Systems and Intelligent User Interfaces (WPRSIUI'07). Istanbul, Turkey.

Woerndl, W., Brocco, M., & Eigner, R. (2009). Context-aware recommender systems in mobile scenarios. [IJITWE]. *International Journal of Information Technology and Web Engineering, 4*(1), 67–86. doi:10.4018/jitwe.2009010105

Woerndl, W., Groh, G., & Hristov, A. (2009). Individual and social recommendations for mobile semantic personal information management. *International Journal on Advances in Internet Technology, 2*(2&3).

Woerndl, W., & Schlichter, J. (2008). Contextualized recommender systems: Data model and recommendation process. In Pazos-Arias, J., Delgado Kloos, C., & Lopez Nores, M. (Eds.), *Personalization of interactive multimedia services: A research and development perspective.* Hauppauge, NY: Nova Publishers.

Woerndl, W., Schulze, F., & Yordanova, V. (2010). Modeling and learning relevant locations for a mobile semantic desktop application. [JMPT]. *Journal of Multimedia Processing and Technologies, 1*(1).

Wojciechowski, M., & Xiong, J. (2006). Towards an open context infrastructure. *Proceedings of Workshop on Context Awareness for Proactive Systems (CAPS'06)* (pp.125-136).

Woo, M. S., Lee, H. B., Han, Y. H., & Min, S. G. (2010). A tunnel compress scheme for PMIPv6-based nested NEMO. In *International Conference on Wireless Communications Networking and Mobile Computing, WiCOM2010* (p. 1).

World, J. Solutions for Java Developer. (2011). *Solve real problems with aglets, a type of mobile agent.* Retrieved January 2, 2011, from http://www.javaworld.com/javaworld/jw-05-1997/jw-05-hood.html?page=1

Xia, F., Yakovlev, A. V., Clark, I. G., & Shang, D. (2002). Data communication in systems with heterogeneous timing. *IEEE Micro, 22,* 58–69. doi:10.1109/MM.2002.1134344

Xu, Y., Winter, J., & Lee, W. C. (2004). Prediction-based strategies for energy saving in object tracking sensor networks. In *Proceedings of the Fifth IEEE International Conference on Mobile Data Management (MDM'04)* (pp. 346–357).

Yao, M., Foo, E., Peng, K., & Dawson, E. (2003). An improved forward integrity protocol for mobile agents. In *Proceedings of the 4th International Workshop on Information Security Applications (WISA), Lecture Notes in Computer Science, 2908* (pp. 272-285). Springer-Verlag.

Ye, Z., Krishnamurthy, S. V., & Tripathi, S. K. (2003). A framework for reliable routing in mobile ad hoc networks. In *Proceedings of the International Conference on Computer Communications, Vol. 1* (pp. 270-280). San Francisco, CA: IEEE.

Yitalo, J., Melen, J., Samela, P., & Petander, H. (2008). *An experimental evaluation of a HIP based network mobility scheme. WWIC 2008, LNCS 5031* (pp. 139–151). Berlin, Germany: Springer-Verlag.

Ylitalo, J. (2000). *Secure platforms for mobile agents.* Technical Report, Department of Computer Science, Helsinki University of Technology, Finland. Retrieved January 2, 2011 from http://www.tml.tkk.fi/Opinnot/Tik-110.501/1999/papers/mobileagents/mobileagents.html

Zander, S., & Schandl, B. (2010). A framework for context-driven RDF data replication on mobile devices. In *Proceedings of the 6th International Conference on Semantic Systems (I-Semantics).* Graz, Austria.

Zander, S., & Schandl, B. (2011). Context-driven RDF data replication on mobile devices. *Semantic Web Journal-Interoperability, Usability, Applicability, 1*(1).

Zeng, X., Bagrodia, R., & Gerla, M. (1998). GloMoSim: A library for parallel simulation of large-scale wireless networks. In *Proceedings of the 12th Workshop on Parallel and Distributed Simulations* (pp. 154). Banff, Canada: ACM.

Zhang, A., Nodine, M., Bhargava, B., & Bukhres, O. (1994). Ensuring relaxed atomicity for flexible transactions in multidatabase systems. In *Proceedings of the ACM SIGMOD Conference* (pp.67-78).

# About the Contributors

**A.V. Senthil Kumar** obtained his BSc Degree (Physics) in 1987, P.G.Diploma in Computer Applications in 1988, MCA in 1991 from Bharathiar University. He obtained his Master's of Philosophy in Computer Science from Bharathidasan University, Trichy during 2005 and his Ph.D in Computer Science from Vinayaka Missions University during 2009. To his credit he had industrial experience for five years as System Analyst in a Garment Export Company. Later he took up teaching and attached to CMS College of Science and Commerce, Coimbatore. He is currently working as Director in the Department of Post Graduate and Research in Computer Science, Hindusthan College of Arts and Science, Coimbatore, India. He has published more than 30 research articles in refereed international conferences and journals as well as book chapters. He has edited a book and serving as Editor-in-Chief for an international journal. He is an Editorial Board Member and Reviewer for various international journals. He is also a committee member for various international conferences.

**Hakikur Rahman**, PhD., is the Founder-Principal of Institute of Computer Management and Science (ICMS) and President of ICMS Foundation. He is currently serving as a Post-Doctoral researcher at the University of Minho, Portugal. He is an Adjunct Faculty of the Bangabandhu Sheikh Mujibur Rahman Agricultural University, Bangladesh (a public university) and a former faculty of the International University of Business Agriculture and Technology, Bangladesh (a private university). He served Sustainable Development Networking Foundation (SDNF) as its Executive Director (CEO) from January 2007 to December 2007, the transformed entity of the Sustainable Development Networking Programme (SDNP) in Bangladesh where he was working as the National Project Coordinator since December 1999, and the South Asia Foundation Bangladesh Chapter as the Coordinator-Secretary during 2001-2008. SDNP is a global initiative of UNDP and it completed its activity in Bangladesh on December 31, 2006. Before joining SDNP he worked as the Director, Computer Division, Bangladesh Open University. Graduating from the Bangladesh University of Engineering and Technology (Electrical & Electronics Engg.) in 1981, he has done his Master's of Engineering (Electrical Engg. with Computer specialization) from the American University of Beirut in 1986 and completed his PhD in Computer Engineering from the Ansted University, UK in 2001. He has contributed over 25 book chapters, authored/edited over 20 books, and published over 50 articles/papers on ICT for Development (knowledge management, e-governance, e-learning, data mining applications and Internet governance). He is the Chairman of SchoolNet Foundation Bangladesh; the Founder-Chairperson of Internet Society Bangladesh Chapter (2000-2010); Head Examiner (Computer), Bangladesh Technical Education Board (1996-2008); Editor, the Monthly Computer Bichitra; and Executive Director, BAERIN (Bangladesh Advanced Education Research and Information Network) Foundation.

\* \* \*

**Jemal H. Abawajy** is a faculty member at Deakin University, Australia. He is actively involved in funded research in robust, secure, and reliable resource management for pervasive computing (mobile, clusters, enterprise/data grids, Web services) and networks (wireless and sensors) and has published more than 150 research articles in refereed international conferences and journals as well as book chapters. He is currently the principal supervisor of 13 PhD and co-supervisor of 3 PhD students. Dr. Abawajy has guest-edited several international journals and served as an associate editor of international conference proceedings. In addition, he is on the editorial board of several international journals. Dr. Abawajy has been a member of the organizing committee for over 200 international conferences serving in various capacity including chair, general co-chair, vice-chair, best paper award chair, publication chair, session chair, and program committee.

**Raja Aljaljouli** is a PhD student in Science and Technology Faculty at Deakin University and a holder of Master's Degree in Software Engineering, University of New South Wales. She is currently in the stage of submitting PhD thesis. Her Master's Degree Thesis was on the security of data gathering mobile agents. She has published conference papers, technical reports, and a book chapter. She served as a reviewer of book chapters for publication. Her research interests are in multi-agent systems, security of mobile agents, security protocols, negotiation strategies, utility function in e-commerce application, e-negotiation, and formal verification methods.

**Arhatha B** graduated from Department of Computer Science and Engineering, M S Ramaiah Institute of Technology, Bangalore in the year 2010. Her special interests lie in embedded, Web, and mobile technologies.

**Harish Raddi C S** graduated from Department of Computer Science and Engineering, M S Ramaiah Institute of Technology, Bangalore in the year 2011. He is currently working as Associate Solution Architect at Akamai Technologies, Bangalore. His special interests lie in cloud, Web, and mobile technologies. He has contributed to MSRIT Linux Association as Chairman for the year 2010-2011 and initiated several co-curricular and extra-curricular activities.

**Lars Frank** has, for 20 years, been a Database Consultant for both private companies and organizations in the public sector. Since 1994 he has been Associate Professor at the Department of Informatics in Copenhagen Business School. His research areas are system integration, data warehousing, ERP architectures, Distributed health records, E-commerce, Mobile databases, Transaction models, Work flow management, multidatabases, and data modeling. In 2008 he received Dr. Merc. Degree from Copenhagen Business School for a dissertation about integration of heterogeneous IT-systems.

**Marivi Higuero** obtained her BSc and MSc degrees in Electrical Engineering in 1992 and her Ph.D. degree in 2005, all from UPV/EHU. She currently works as an Assistant Professor in the Department of Electronics and Telecommunications in the Faculty of Engineering of Bilbao in UPV/EHU, teaching telematics lab and fundamentals, as well as doctorate courses on Advanced Networks and Services and

Security in Wireless Networks. Her research interests include wireless networking, mobility management, and security applied to transportation systems.

**Srinivasa K G** received his B.E. degree in Information Science and Engineering from M S Ramaiah Institute of Technology, Bangalore University in the year 2000, and M.E. degree in Computer Science and Engineering from University Visvesvaraya College of Engineering, Bangalore University in the year 2003 with distinction. He obtained the PhD in Computer Science and Engineering at Department of Computer Science and Engineering, Bangalore University. He is now working as a Professor in the Department of Computer Science and Engineering, M S Ramaiah Institute of Technology, Bangalore. He is the recipient of All India Council for Technical Education - Career Award for Young Teachers, Indian Society of Technical Education – ISGITS National Award for Best Research Work Done by Young Teachers, Institution of Engineers (India) – IEI Young Engineer Award in Computer Science and Engineering, IEEE, USA – Best Paper Award, and IMS Singapore – Visiting Scientist Fellowship Award. He has published more than fifty research papers in international conferences and journals. He is the author of the book File Structures using C++, TMH, and Soft Computing for Data Mining Applications by Springer. He has been invited by many universities in India and abroad to deliver expert talks. He is also the recipient of BOYSCAST Fellowship from DST, Govt. of India for the year 2009-10.

**Anil Kumar M** graduated from Department of Computer Science and Engineering, M S Ramaiah Institute of Technology, Bangalore in the year 2011. He is currently working as Systems Engineer at Texas Instruments, Bangalore. His special interests lie in cloud, Web, and mobile technologies. He has contributed to MSRIT Linux Association as Treasurer for the year 2010-2011 and initiated several co-curricular and extra-curricular activities.

**Natarajan Meghanathan** is currently working as an Assistant Professor of Computer Science at Jackson State University, Jackson, MS. Dr. Meghanathan has published more than 100 peer-reviewed articles (more than half of them being journal publications). He has also received federal education and research grants from the U. S. National Science Foundation and the Army Research Lab. Meghanathan has been serving in the editorial board of several international journals and in the technical program and organization committees of several international conferences. More information about Natarajan Meghanathan can be found at http://www.jsums.edu/cms/nmeghanathan

**Korbinian Moegele** received a Bachelor's degree in Computer Science at Technische Universitaet Muenchen (TUM). His areas of interest include distributed applications, mobile computing, and recommender systems. While obtaining practical experience implementing the mobile city guide described in this chapter, he is continuing his studies with the Master's degree in Computer Science at TUM until 2011.

**Sunil Kumar N** graduated from Department of Computer Science and Engineering, M S Ramaiah Institute of Technology, Bangalore in the year 2010. Presently he is working as a Member of the Technical Staff at Netapp Pvt. Ltd, Bangalore. His special interests lie in embedded, Web, mobile, and cloud technologies. He has contributed to IEEE MSRIT Student branch as Chairman for the year 2009-2010 and initiated several co-curricular and extra-curricular activities.

**Saptarshi Purkayastha** is a Research Fellow at the Department of Computer & Information Science at the Norwegian University of Science & Technology, Norway. His research interests include health information systems, bioinformatics, mobile computing, virtual machines, GPGPU computing, artificial intelligence, & robotics. He has worked in different research positions in the software industry and teaches electronics and programming as visiting faculty at college-level. A key current research focus is design, development, and implementation of integrated health Information Systems in developing countries.

**Vivian Prinz** received her Diploma degrees in Media Informatics at Ludwig-Maximilians-University Munich (LMU). Now, she is Research Assistant at the Chair for Applied Informatics and Cooperative Systems at Technische Universitaet Muenchen (TUM). Her research interests include distributed systems, vehicle-to-vehicle communications, information management in vehicular networks and active safety, and deployment applications based on V2V communications.

**Vijayendra R** graduated from the Department of Computer Science and Engineering, M S Ramaiah Institute of Technology, Bangalore in the year 2009. His special interests lie in Web and mobile technologies. He has contributed to Association of Computer Engineers, MSRIT Chapter as Co-ordinator for the year 2008-2009, and initiated several co-curricular and extra-curricular activities.

**Harsha R** graduated from Department of Electronics and Engineering, M S Ramaiah Institute of Technology Bangalore, in the year 2010, and is currently working as Research Fellow at National Center of Biological Sciences, Bangalore. His special interests lie in robotics, Web, and mobile technologies.

**Abhishek S C** graduated from R V College of Engineering, Bangalore in the year 2010. His special interests lie in embedded, Web, and mobile technologies.

**Bernhard Schandl** is a Researcher and Entrepreneur in the field of semantic technologies. He received his Doctoral degree in Computer Science from University in Vienna in 2009, where he has carried out research on the Semantic Desktop and the Semantic Web and has been actively involved in several research projects. He is co-founder and CTO of Gnowsis.com, designing social productivity and collaboration software. Contact him at bernhard.schandl@gnowsis.com.

**Nerea Toledo** received her BSc and MSc degrees in Telecommunication Engineering in 2007 from the University of the Basque Country (UPV/EHU). She received another MSc in Information and Communication Systems in Wireless Networks in 2008 from the UPV/EHU. She is a PhD student in the Department of Electronics and Telecommunications in the Faculty of Engineering of Bilbao in UPV/EHU, holding a grant funded by UPV/EHU. She currently works as a Lecturer in the same department. Her research interests include wireless networking, mobility management, and security applied to the ITS scenario.

**Wolfgang Woerndl** obtained Diploma degrees in both Computer Science and Business Administration at Technische Universitaet Muenchen (TUM) and Fernuniversitaet Hagen. After working in the industry for several years, he returned to Technische Universitaet Muenchen as a Doctoral candidate and completed his dissertation on privacy in decentralized user profile management in 2003. Since then, he is a Senior Researcher and Lecturer at the Chair for Applied Informatics and Cooperative Systems at TUM. His current research interests include user modeling, personalization and recommender systems, and mobile and context-aware applications.

**Stefan Zander** is a Doctoral candidate and Research Associate at the Multimedia Information Systems' research group of the Faculty of Computer Sciences at the University of Vienna. His research interests cover the fields of context-aware and mobile computing, Semantic Web technologies for mobile platforms, and knowledge representation on the Semantic Web. He received a Master of Science degree from the University of Central Lancashire, UK, and a Master of Information Technology degree from the Department of Computer Science of the University of Applied Sciences Würzburg, Germany. Contact him at stefan.zander@univie.ac.at.

# Index